Distributed Energy Management of Electrical Power Systems

IEEE Press
445 Hoes Lane
Piscataway, NJ 08854

Distributed Energy Management of Electrical Power Systems

Yinliang Xu
Wei Zhang
Wenxin Liu
Wen Yu

Published by John Wiley & Sons, Inc., Hoboken, New Jersey.
Published simultaneously in Canada.

Library of Congress Cataloging-in-Publication Data

Names: Xu, Yinliang, author. | Zhang, Wei (Engineer), author. | Liu, Wenxin, 1978- author. | Yu, Wen, 1977- author.
Title: Distributed energy management of electrical power systems / Yinliang Xu, Wei Zhang, Wenxin Liu, Wen Yu.
Description: First edition. | Hoboken, NJ : John Wiley & Sons, Inc., [2021] | Series: IEEE Press series on power engineering ; 100 | Includes bibliographical references and index.
Identifiers: LCCN 2020024730 (print) | LCCN 2020024731 (ebook) | ISBN 9781119534884 (hardback) | ISBN 9781119534877 (adobe pdf) | ISBN 9781119534891 (epub)
Subjects: LCSH: Distributed generation of electric power. | Electric power systems–Management. | Distributed parameter systems.
Classification: LCC TK1006 .X83 2020 (print) | LCC TK1006 (ebook) | DDC 621.31/213–dc23
LC record available at https://lccn.loc.gov/2020024730
LC ebook record available at https://lccn.loc.gov/2020024731

Cover design by Wiley
Cover images: © Sam Robinson/Getty Images, © geniusksy/Shutterstock
Set in 9.5/12.5pt STIXTwoText by SPi Global, Chennai, India
10 9 8 7 6 5 4 3 2 1

To my wife Han Lin and my son NingyiXu

Yinliang Xu

For all of my loved ones who have gone on to a better life

Wei Zhang

Dedicated to my father Guilin Liu, mother Shujuan Wang, wife Lily Liu, and son Oscar Tianyi Liu

Wenxin Liu

To my children Huijia and Lisa

Wen Yu

Contents

About the Authors

Yinliang Xu received the BS and MS degrees in control science and engineering from Harbin Institute of Technology, China, in 2007 and 2009, respectively, and the PhD degree in Electrical and Computer Engineering from New Mexico State University, Las Cruces, NM, USA, in 2013. Dr. Xu is now an Associate Professor with Tsinghua-Berkeley Shenzhen Institute (TBSI), Tsinghua Shenzhen International Graduate School (Tsinghua SIGS), Tsinghua University, Shenzhen, Guangdong, China. He is also an adjunct faculty with the Department of Electrical and Computer Engineering at Carnegie Mellon University, Pittsburgh, PA, USA. Before joining TBSI, he was with the School of Electronics and Information Technology, Sun Yat-Sen University. From 2013 to 2014, he was a visiting scholar with the Department of Electrical and Computer Engineering, Carnegie Mellon University, Pittsburgh, PA, USA. His research interests include distributed control and optimization of power systems, renewable energy integration, and microgrid modeling and control. He is a senior member of IEEE. Dr. Xu serves as an associate editor for the CSEE Journal of Power and Energy Systems, IET Smart Grid, IET Generation, Transmission & Distribution, and IEEE Access.

Wei Zhang received the BS and MS degrees in power system engineering from the Harbin Institute of Technology, Harbin, China, in 2007 and 2009, respectively, and the PhD degree in Electrical and Computer Engineering from New Mexico State University, Las Cruces, NM, USA, in 2013. From 2014 to 2017, Dr. Zhang was Lecturer with School of Electrical Engineering and Automation, Harbin Institute of Technology and was an Associate Professor with the same department till 2019. Currently, He is a Postdoctoral Researcher with Department of Civil, Environmental, and Construction Engineering, University of Central Florida, Orlando, USA. His research interests include distributed control and optimization of power systems, Microgrids control and optimization, renewable energy and power system state estimation, and stability analysis.He is a senior member of

IEEE and he also serves as an associate editor for IET Smart Grids, IET Renewable Power Generation and Journal of Control, Automation and Electrical systems.

Wenxin Liu received the BS degree in industrial automation and MS degree in control theory and applications from Northeastern University, Shenyang, China, in 1996 and 2000, respectively, and the PhD degree in electrical engineering from the Missouri University of Science and Technology (formerly University of Missouri-Rolla), Rolla, MO, USA, in 2005. Dr. Liu worked as an assistant scholar scientist with the Center for Advanced Power Systems, Florida State University, Tallahassee, FL, USA, until 2009. From 2009 to 2014, he was an assistant professor with the Klipsch School of Electrical and Computer Engineering, New Mexico State University, Las Cruces, NM, USA. Currently, he is an associate professor with the Department of Electrical and Computer Engineering, Lehigh University, Bethlehem, PA, USA. His research interests include power systems, power electronics, and controls. Dr. Liu is an editor for IEEE Transactions on Smart Grid, IEEE Transactions on Industrial Informatics, and IEEE Transactions on Power Systems

Wen Yu received the BS degree from Tsinghua University, Beijing, China in 1990 and the MS and PhD degrees, both in electrical engineering, from Northeastern University, Shenyang, China, in 1992 and 1995, respectively. From 1995 to 1996, he served as a lecturer in the Department of Automatic Control at Northeastern University, Shenyang, China. Since 1996, he has been with the Centro de Investigaci ón y de Estudios Avanzados, Instituto Politécnico Nacional (CINVESTAV-IPN), Mexico City, Mexico, where he is currently a professor with the Departamento de Control Automatico. From 2002 to 2003, he held research positions with the Instituto Mexicano del Petroleo. He was a senior visiting research fellow with Queen's University Belfast, Belfast, UK, from 2006 to 2007, and a visiting associate professor with the University of California, Santa Cruz, from 2009 to 2010. He also holds a visiting professorship at Northeastern University in China from 2006. Dr. Wen Yu serves as an associate editor of IEEE Transactions on Cybernetics, Neurocomputing, and Journal of Intelligent and Fuzzy Systems. He is a member of the Mexican Academy of Sciences.

Preface

The ever-growing demand, the rising penetration level of renewable generation, and the increasing complexity of electric power systems pose new challenges to control, operation, management, and optimization of power grids. Conventional centralized control structure requires a complex communication network with two-way communication links and a powerful central controller to process a large amount of data, which reduces overall system reliability and increases its sensitivity to failures; thus, it may not be able to operate under the increased number of distributed renewable generation units. Distributed control strategy enables easier scalability, simpler communication network, and faster distributed data processing, which can facilitate highly efficient information sharing and decision making. The distributed approach is a promising candidate to address the features of modern power grids by providing fast, flexible, reliable, and cost-effective solutions.

Considering the foreseeable large deployment of distributed renewable generation in electric power systems, a comprehensive technical source book is needed for both academic and industrial fields. This book will be providing in-depth analysis and discussion of fully distributed control approaches and their applications for electric power systems. The book will cover a wide range of the topics, including both large- (hours) and short- (seconds) time-scale-level coordination control and optimization. The technical aspects in terms of control, operation management, and optimization of electric power systems will be elaborated with the fundamental knowledge and advanced techniques.

This book focuses on fully distributed approaches for control, operation management, and optimization of electrical power systems with high distributed renewable generation penetration level. With the integration of more and more controllable and dispatchable aggregator of distributed generation, energy storage systems, elastic and inelastic loads, not only the power supply-demand should be balanced in a timely manner but also the frequency and voltage must be properly regulated to ensure efficient, safe, and reliable operation of power systems. Conventionally, power systems are commonly managed via centralized control approaches because of the existing structure of the traditional bulk power system. However, with the increasing number of highly intermittent distributed renewable generation with a vast geographical span, distributed control approaches have been proposed and thereafter been deployed in some pilot projects. It is foreseeable that distributed control approaches for power systems will be a promising and effective solution to advance the existing control and energy management technologies for grid integration of renewable energy sources.

In Chapter 1, an introduction on centralized and distributed solutions to power management and several distributed algorithms are provided. Chapter 2 introduces the communication network topology configuration, the multi-agent framework-based hardware-in-the-loop controller and setup of the real-time digital simulation test-bed.

Chapter 3 discusses distributed active power control and optimization of power systems. For a microgrid with a high level of renewable energy penetration to operate autonomously, it must maintain the instantaneous supply-demand balance of active power. A distributed sub-gradient-based solution to coordinate the operations of different types of distributed renewable generators in a microgrid is developed for the primary and secondary frequency control. An effective distributed tertiary control strategy based on distributed dynamic programming algorithm is then proposed to optimally allocate the total power demand among different generation units using asynchronous communication, which can lead to simpler implementation and faster convergence speed. Finally, an improved distributed gradient algorithmic-based approach, which can address both equality and inequality constraints, is proposed to realize online power system control and optimization.

In Chapter 4, to accommodate the increasing penetration level of distributed generators in the power system, appropriate voltage and reactive power control and optimization approaches are investigated to improve voltage profiles and reduce the power loss and the opportunity cost of reactive power generation. Distributed model-free Q-learning-based algorithm, which is easy to carry out and is adaptable in various operating conditions, is developed, and a sub-gradient-based reactive power control algorithm is proposed to reduce the possibility of random control updates and increase the rate of convergence.

In Chapter 5, a distributed demand response solution to maximize the overall utility of users while meeting the requirement for load reduction requested by the system operator is proposed in emerging smart grids is presented. The proposed distributed dynamic programming algorithm is implemented via a distributed framework, which can reduce the computation and communication burden and protect users' private information. As an important load in future smart grids, the flexibility of PEV's charging behavior can be beneficial to the power systems by offering various ancillary services, such as load leveling, reserve provision, frequency regulation, etc. It is important to design proper charging strategies so as to satisfy the consumers in terms of the financial charging cost, charging time, and state of charge, without violating operational constraints of the power system.

In Chapter 6, a distributed energy management approach that considers both generation suppliers and load users is proposed to maximize the total social welfare of the whole power system while satisfying various system constraints. Furthermore, an online solution based on a multi-agent system framework is developed to achieve better consumer participation and efficient supply-demand balance with faster response than traditional centralized methods.

In Chapter 7, two approaches for distributed estimation approaches are presented. The consensus algorithm-based distributed estimation is a two-loop iteration architecture, which is developed for multi-area power systems. The inner loop is used to discover the information of gain matrix and gradient vector by applying the consensus confusion technique, and the outer loop is used to update the estimation based on the second-order Newton's method. The distributed sub-gradient algorithm-based state estimation is a fully distributed integrated solution for multi-area TI and SE solution, which can be implemented with MAS framework. Numerical studies for variable scales of systems demonstrate that it is adaptable, robust and is a promising option for the state estimation of large interconnected power systems.

In Chapter 8, the steps to evaluate the algorithm are discussed. We start with controller-hardware-in-the-loop simulation, then the power hardware-in-the-loop simulation, and finally the hardware experimentation. Hardware experimentation is necessary, which can tell us whether an algorithm will work in reality or not and how to make it work and work better. In addition, it can also tell us the implementation requirement.

In Chapter 9, more issues and constraints in power system such as coupling variables, non-convexity, and communication topology change are investigated. Towards to more mature implementation of distributed control solution in real world, many facets and technology details such as investment, reliability, optimality, feasibility, etc. need to be taken into consideration, and research efforts and contributions from all around the world are more than welcome.

This book is written for power system engineers who are moving into the area of renewable energy and distributed generation, smart grids, economic dispatch, demand response, electric vehicle charging management, virtual power plant, state estimation, etc., and researchers and practitioners working in these areas who are enthusiastic to see what benefits and advantages distributed control and optimization algorithms can bring. This book follows an evolutionary procedure. Particularly, the founding technologies and background knowledge are systematically introduced first for each topic, and then the technical advances and the specific and emerging technologies will be elaborated. Therefore, the readers can get a clear picture of all the knowledge following a relatively fast learning curve. Meanwhile, only the proven and feasible solutions will be introduced in this book so that the applicability of the technical solutions can be guaranteed. Most of the advanced control and optimization strategies presented in this book are accompanied by extensive simulation and experimental results. Therefore, this book is also very useful for practitioners in this area to see how distributed control and optimization strategies could improve system performance and work in practice. This book also provides a precious opportunity for graduate students and researchers who work in the area to become familiar with the up-to-date techniques. It can be adopted as a textbook for graduate programs on power system engineering, microgrids, renewable energy integration, and smart grid.

Tsinghua-Berkeley Shenzhen Institute (TBSI)
Tsinghua Shenzhen International Graduate School
(Tsinghua SIGS), Tsinghua University, Shenzhen, China
Yinliang Xu
University of Central Florida, Orlando, USA
Wei Zhang
Lehigh University, Bethlehem, PA, USA
Wenxin Liu
Instituto Politécnico Nacional, Mexico City, Mexico
Wen Yu

Acknowledgment

This work was supported by the U.S. National Science Foundation under Grant ECCS 1125776 and by the U.S. Office of Naval Research under award numbers N00014-18-1-2185, N00014-16-1-3121, N00014-16-1-2884, N00014-15-1-0047, N00014-13-1-0161, N00014-12-1-1010, and N00014-12-1-0761.

List of Figures

List of Tables

1

Background

1.1 Power Management

With the development of smart grids and the deep interconnection of multiple large-scale regional power grids, power systems are considered as the largest artificial networks that have ever been built. Power systems include coupled primary/secondary power equipment and are supported by advanced control technology and efficient communication networks to form an intelligent autonomous system. To meet the high-performance requirements from all aspects, power management is of great importance in the smart grid. On the one hand, with the increasing penetration level of distributed energy generators, the dispatchable power generation will be relatively less as the renewable energy generation has the characteristics of high uncertainty and intermittency. In order to realize the real-time supply-demand balance between power generation and consumption, effective power generation scheduling strategies should be designed to accommodate the integration of distributed energy resources. On the other hand, flexible and controllable loads can interact with the power grid through the smart meter, which is an important link to ensure the real-time balance between grid power generation and consumption. Thus, load-side power management plays a more important role.

The power management in the smart grid is to optimize the energy utilization efficiency by coordinating the various controllable units of the power grid through efficient communication network and advanced control techniques to ensure the safe, stable, reliable, and efficient operation of the entire power grid. Power management can be classified into multiple problems according to different objectives, time scales, and control targets. Proper and effective control strategies are the key to achieve safe, reliable, stable, efficient, and flexible operation of the power grid.

Distributed Energy Management of Electrical Power Systems, First Edition.
Yinliang Xu, Wei Zhang, Wenxin Liu, and Wen Yu.

Published 2021 by John Wiley & Sons, Inc.

In addition to the basic objectives of ensuring system stability, the power management in the smart grid also includes the following objectives:

(1) Enable the proportional distribution of active/reactive power output of distributed energy resources;
(2) Ensure that the voltage amplitude and frequency are kept within the allowable range and compensate for the fluctuations of the distributed energy generators output and the dynamic load power demand;
(3) Reduction of the circulating current between distributed energy resources and realize the desired power exchange with the external power grid;
(4) Be adaptive to the plug-and-play feature and the system topology changes;
(5) Participate in the regulation of the power market, realize the optimal scheduling and coordinated power allocation, and provide various auxiliary services when necessary; and
(6) Identify the topology of the smart grid promptly.

The above objectives require that the power grid, the distributed energy resources, and the controllable loads within the system coordinate their respective control decisions with each other to achieve this. To achieve the above objectives, many scholars propose a hierarchical control structure. The primary control layer aims to maintain the stability of the voltage, frequency, and output power according to the control command of the system. Distributed energy resources should select specific control strategies based on their characteristics and control mode. When no frequency and voltage support is required, the distributed energy resources operate in the constant power output mode. The existing coordination mechanisms of multiple distributed energy resources can be divided into master–slave control and peer-to-peer control mode. In the master–slave mode, the master-distributed energy resource with the largest generation capacity is controlled with constant voltage and frequency mode to provide voltage and frequency support to the power system. Other slave-distributed energy resources are controlled under the constant power mode to maintain the active/reactive power balance. In the peer-to-peer mode, the distributed energy resources are all controlled based on the droop control to maintain the stability of the voltage, frequency, and the active/reactive power balance in the power system by simulating the active frequency characteristics and reactive voltage characteristics of the conventional generator because droop control strategies can achieve power sharing by adjusting the droop coefficient and do not require any information exchange.

The secondary control layer aims to realize the regulation of voltage and frequency and focus to research on the first two aspects. Because the master distributed energy resource device in the master–slave control mode may not have enough capacity to compensate for the power fluctuations, the droop control in the peer-to-peer control mode essentially results in poor performance, the generation or load change will cause the system voltage and frequency deviations. The accumulated deviations of voltage and frequency may lead to the collapse of power systems, so it is necessary to adjust the distributed energy resources to eliminate the deviations and improve the overall dynamic performance of the power system. In addition, in the frequency droop control, the active power can be accurately shared among multiple distributed energy resources. However, in the voltage droop control, the reactive power sharing usually leads to unreasonable distribution because of the inconsistent output impedances of the inverters. Inaccurate reactive power sharing affects energy efficiency and the life span of power electronic equipment and causes current circulation, which will seriously impact the system reliability, stability, and economy. Generally, conventional power systems adopt centralized control strategies to achieve the secondary control target.

Recently, different control methods have been proposed for the implementation of secondary frequency and voltage control. Because the frequency is a global variable, it can either be controlled via a centralized or distributed method based on direct or indirect access to the global information or a decentralized control based on the local measurement. Although the voltage output of the inverter is a local variable, in order to achieve the accurate reactive power sharing and voltage recovery of the power system, the secondary voltage control is mainly centralized and distributed depending on the system information interaction.

The tertiary control layer aims to optimize the power grid economic operation and energy management. The total operating cost of the power systems is minimized, and the distributed energy resources utilization efficiency is maximized while ensuring a stable and reliable operation. There exist plenty of literature studies on tertiary energy management in terms of economic dispatch, demand response, and loss minimization, which can be mainly divided into three categories. The first category is the analytical methods, such as lambda iteration, linear, and nonlinear programming. The second category is the heuristic methods such as hybrid immune algorithm, particle swarm optimization, ant colony optimization, etc. The third category is the distributed optimization approaches, which only require information exchange through a sparse communication network and can achieve optimal or near-optimal solutions while satisfying various local and coupled constraints.

1.2 Traditional Centralized vs. Distributed Solutions to Power Management

Traditionally, the stable and economic operation of power systems has been achieved primarily through centralized or decentralized control with the little involvement of the distributed coordinated control. The existing centralized energy management requires each user and power generator to send the local information to the control center. After collecting all the information, the control center processes a huge amount of data and makes control decisions, and then decisions are transmitted to the local users and the power generation units. The centralized control structure has the advantages of simplicity, high convergence accuracy, and fast convergence.

These centralized approaches have been effective so far for conventional power systems. However, they may face severe challenges to manage future power systems with a high penetration level of distributed energy resources because of the following reasons. First, the centralized approaches require sophisticated communication infrastructure between the central controller and every single unit in the power network to collect information globally and a powerful central controller to process huge amount of data and make complicated control decisions. Thus, these solutions are computationally and communicationally expensive for implementation and highly rely on the capability and reliability of the control center, so they are less robust to single-point failures. Second, different vectors need to coordinate each other's respective control decisions to achieve the global system objective in the future power systems where participants may not be willing to reveal private information such as their generation and utility cost functions and power consumption patterns. Third, the operating conditions of the power systems may change rapidly and frequently because of the unexpected supply-demand imbalance and lowered inertia caused by the increasing penetration level of power electronics-based control devices, and centralized control approaches may not be able to respond in a timely manner.

In order to obtain the accurate state information of the power systems in real-time, the deployment of measuring equipment in the smart grid will continue to increase, resulting in a sharp increase in the amount of data that the control center needs to collect, and the limited communication resources, resulting in an increase in data transmission delay and a high communication cost. The traditional centralized control methods usually subject to poor scalability and cannot meet the requirements of accurate control of the smart grid with satisfying dynamic performance in a real-time manner. With the expansion of power system scale and the increasing number of controllable objects, distributed control gradually shows its superiority in terms of robustness and low control cost.

Compared with centralized control, distributed control has the following advantages: (i) Global optimal or near-optimal system objective is achieved based on a point-to-point sparse communication network, which reduces communication burden, (ii) parallel data processing and calculation is done without the central controller, which lower the computation cost significantly, (iii) the "plug and play" function is supported, which also facilitates scalable application to large systems, and (iv) control decision is made by a local control unit, which improves reliability. Therefore, distributed energy management approaches are more suitable for the cooperation of large-scale distributed intelligent equipment in vast geographical areas to ensure safe, stable, and economic operation of the smart grid, which has received extensive attention from worldwide researchers and scholars.

1.3 Existing Distributed Control Approaches

In recent years, distributed control and distributed optimization methods have been widely applied to solve the control and optimization of power systems with a high penetration level of distributed energy resources. Most of the existing distributed control methods can be classified as the following three typical approaches.

The first type of distributed energy management approach is based on dual decomposition. The main idea is to decompose the optimization problem into multiple suboptimization problems, which are coupled by a certain global variable or a uniform Lagrangian multiplier corresponding to the energy balance constraint, such as local marginal price. The control center interacts with all distributed units and also the information exchange among multiple distributed energy units is required to update global information. This type of energy management method does not need the control center to collect all the information from all participants, and the participants can obtain essential and necessary information about the power systems global state. Considering the economic dispatch problem with multiple distributed energy resource units, the objective of minimizing the total cost of power generation should be achieved while satisfying the constraints of power supply and demand balance, the upper and lower bounds of distributed energy resource generation output. Through the dual-decomposition method, the power utility company calculates the global Lagrangian multiplier, and each distributed generation unit solves the local optimization problem and calculates the local output power according to the dual-variable information broadcasted by the power utility company.

Next, a multiple time-scale energy dispatching problem with traditional power generation units, controllable load units, distributed energy storage units, and renewable power generation to minimize the energy transaction cost minus the load utility benefit function is considered. In order to solve this problem, a distributed energy dispatch and demand response algorithm is designed based on the dual-decomposition method and applied to electric vehicle charging scenarios. The cost of traditional power generation is modeled as a quadratic function, a penalty function is imposed to reduce renewable power generation curtailment, and the user preferences are characterized by their willingness to pay for services, which can be seen as benefit functions of users. Users with small energy consumption need to broadcast their aggregated load to the power utility company. The power utility company updates the electricity price based on the deviation of the power generation and load demand. Last, DC-optimal power flow problem considering the upper and lower bounds of generation units power output, transmission line physical limits, power generation, and load demand balance constraints to minimize the total cost of all power generation units is investigated. Each unit updates its power supply or demand and broadcasts it to neighboring units based on the Lagrangian multiplier estimates of the neighboring units' net power. After receiving the transmission line congestion information of the power system, each unit updates the local power supply or demand in an average manner and then updates the Lagrangian multiplier associated with the power supply and demand constraints. The drawback of this type of energy management method is the requirement of a control center. Thus, these are not the fully distributed protocols, and the robustness, privacy protection, and scalability are relatively limited.

The second type of distributed energy management method is based on game theory. The main idea is based on the potential energy game, which can guarantee the existence of Nash equilibrium. Considering a large number of end users cooperate to decide the scheduling of household electrical equipment or load in the next day, power utility company needs to adopt a fixed electricity pricing method to provide guidance for the reasonable energy consumption and proper electricity use time to achieve the Nash equilibrium with the minimum power generation costs. However, this type of approach requires all users to know the power generation cost function information, and the user needs to update the energy schedule asynchronously, so its algorithm is less scalable. Moreover, for the strategy-based potential energy game method, there are special requirements for the network topology, which only apply to fully connected networks, and has no advantage in dealing with complex coupling constraints among units.

The third type of distributed energy management is based on the consensus algorithm, which can achieve global goals through local information exchange. Therefore, this type of distributed energy management has high flexibility, strong robustness, and decent scalability and is fully distributed. As an important branch of distributed computing with minimum communication, consensus algorithms have been widely used in the economic dispatching, demand response, and topology identification problems in smart grids. For the economic scheduling problem of multiple power generation units, the power generation and supply equation constraints are described in the objective function according to the Lagrangian multipliers, and the problem is decomposed into multiple suboptimal problems. Multiple Lagrangian multipliers in the optimal objective function need to converge to the same value, which can be regarded as the optimal marginal cost of the power generation units. Each unit updates the Lagrangian multiplier based on the estimation of the neighboring units' power supply and demand deviation using the consensus algorithm. The updated Lagrangian multiplier is then used to calculate the desired local power generation or load demand.

The distributed energy management methods have the advantages of simplicity, high convergence precision, strong robustness, and decent scalability. Consensus algorithm-based energy management approaches proposed in recent literature are more practical than traditional distributed algorithms. The convergence and the optimality of the consensus algorithm-based energy management methods can be rigorously proved. However, existing consensus algorithm-based energy management methods still have shortcomings in problem modeling, robustness analysis in communication nonideal situations, complex constraints handling, and so on. Therefore, there are still many problems in the consensus algorithm-based energy management methods that are challenging and worth studying.

As a combination of distributed control and artificial intelligence, the multi-agent system (MAS) can decompose large and complex problems into multiple small local problems and realize the local optimal control decision making through the cooperative operations of each agent and other units. Therefore, the MAS is fully applicable to the power systems' collaborative control with well-structured and complex operational objectives because of the following merits: (i) The autonomy of the agent corresponds to the autonomous decision-making ability of the distributed unit in the power systems; (ii) the sociability and teamwork spirit of the agent corresponding to the communication interaction between distributed energy resource units; and (iii) the initiative and adaptability of the agent accommodates the topology change scenario of the power systems with high level of distributed energy resources, such as plug and play, reconfiguration, restoration, etc. As one of the most popular distributed control approaches, a well-designed MAS is flexible, reliable, and less expensive to implement, and it has a better chance of surviving single-point failures. Recently, MAS-based approaches have been applied to various power system applications. However, most of the existing methods are mainly rule-based and lack rigorous stability analysis. The potential applications of MAS for the power grid need to be further explored. Because consensus algorithm can provide the fundamental support, therefore, a MAS-based approaches using consensus algorithm promising have promising applications to address various problems in smart grids with a high penetration level of distributed energy resource units, which is the main focus of this book.

2

Algorithm Evaluation

Before the implementation of the control algorithm or solution, the off-line evaluations are always required. In this chapter, we will discuss two aspects for off-line algorithm evaluation. The first aspect we would like to discuss is the communication network topology configuration for the distributed solution. Then, we will present our existing ideas and experiences for the real-time digital simulation regarding the algorithm evaluation.

2.1 Communication Network Topology Configuration

As for the typical distributed algorithm, the centralized processor or controller is no longer required. The computational efforts are distributed among scattered processors or controllers, which are commonly referred to as agents or intelligent agents. In order to cooperate to achieve the predetermined goal or optimal strategy, these agents need to exchange necessary information with each other. Therefore, the communication network should be appropriately designed to fulfill the functionality of essential information exchange, depending on the applications.

The design of the communication network for distributed control and optimization is case to case. However, in this chapter, we will discuss the common issues and matters that should be considered for distributed algorithms, and especially, we will use the consensus algorithm for demonstration.

2.1.1 Communication Network Design for Distributed Applications

The communication topology of a network of n agents can be represented using a directed graph $G = (V, E)$, with the set of nodes $V = v_1, v_2, \dots\ v_n$ denoting the agents and edges $E \subseteq V \times V$ representing the communication channels. The agents can communicate with an agent, say, agent i is defined as the communication neighbors of this agent, and they are denoted by

Distributed Energy Management of Electrical Power Systems, First Edition.
Yinliang Xu, Wei Zhang, Wenxin Liu, and Wen Yu.

Published 2021 by John Wiley & Sons, Inc.

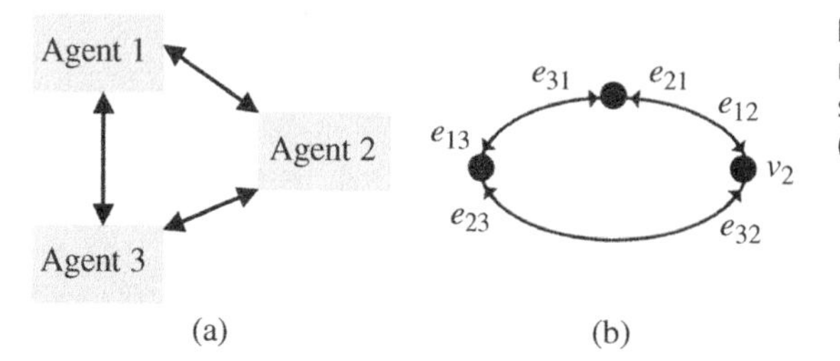

Figure 2.1 Communication network graphs of two scenarios: (a) Scenario 1 and (b) Scenario 2.

$N_i = \{j \in V : (i,j) \in E\}$. Figure 2.1a is an example of a three-agent system and its corresponding communication network graph.

The adjacency matrix $\boldsymbol{A}$ is used to denote the connectivity of the communication network, and the element of $\boldsymbol{A}$, a_{ij}, is defined as:

$$a_{ij} = \begin{cases} 1 & if e_{ij} \neq \emptyset \\ 0 & if e_{ij} = \emptyset \end{cases}$$

For most of the distributed applications, the common practice is that information flow is bidirectional, that is to say, if $e_{ij} \neq \emptyset$, then $e_{ji} \neq \emptyset$, and vice versa. Therefore, the element of the adjacency matrix satisfies:

$$a_{ij} = a_{ji}$$

Hence, A is symmetrical. This type of communication network graph is referred to as balanced graph and more details can be found in [1]. In this book, we only consider the balanced graph because the communication among agents is always bidirectional for power system applications. Accordingly, the communication network graph discussed here can be represented with an undirected graph. In the following context, the communication network graph is assumed to be undirected graph unless it is specially mentioned as a directed graph.

For the fixed communication network topology, if a specific agent in the system needs to cooperate with another agent in a directed or undirected way, this agent needs to comprehend certain kind of information of the other agents through a communication network and vice versa. From another perspective, the information regarding an agent, let us say agent i, needs to be spread to another agent, say agent x, through the communication network. Figure 2.1b demonstrates the information spreading process.

As shown in Figure 2.2, agent i and agent x cannot communicate with each other directly. Yet, agent i can communicate with agent j, and through this communication, the agent j can obtain certain information that can represent the agent i. Through this way, the information of agent actually spreads on the network, and after n rounds of information exchange, it finally reaches agent x. Note that in order to make this process work properly, there must exist a path that connects

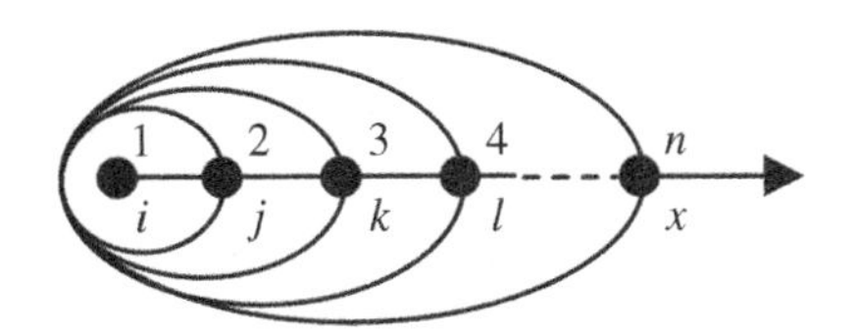

Figure 2.2 Information spreading process.

agent i and agent x. Consequently, for any two agents in the system, there should be a communication chain that makes the cooperation possible. The graph corresponding to a communication network should be connected.

2.1.2 $N-1$ Rule for Communication Network Design

There are many ways to design a connected graph, hence the communication network. For example, Figure 2.3a,b shows two graphs with six agents. As shown in Figure 2.3a,b, both types of the graphs with six nodes have five edges. This kind of connection is the simple tree structure, and obviously, it is connected. Note that if any of the edges are missing, these graphs are no longer connected. Under this circumstance, a node is isolated from the other, which make the cooperation of this node with other nodes impossible. It is not recommended to adopt a tree structure for the communication network configuration of the distributed algorithm or optimization because a single communication channel failure will jeopardize the entire process.

The $N-1$ rule is usually enforced at the phase of communication network design to improve the robustness of the communication network against single communication channel failure. The $N-1$ rule states that any two nodes of a graph under consideration are still connected directly if any one of the edges of this graph is disabled. To this end, the original graph must contain at least one loop that connects all of the nodes in the graph. Because the tree graph is defined as the graph without any loop, it surely does not satisfies the $N-1$ rule. We can modify the structures given in Figure 2.3a,b to make them satisfy the $N-1$ rule, and possible modifications are shown in Figure 2.4a,b. For the graphs illustrated

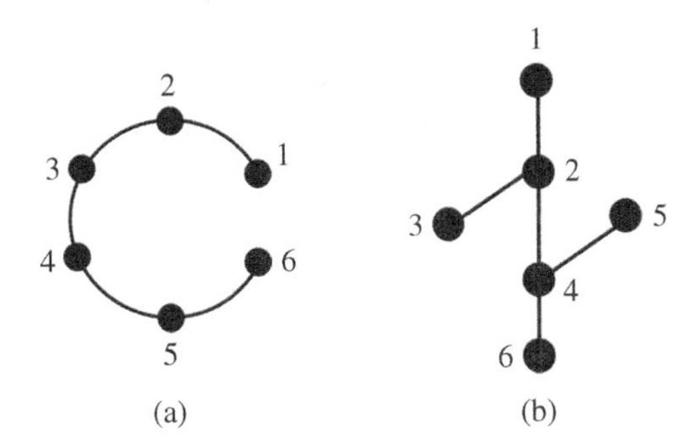

Figure 2.3 Two types of tree graphs for six-agent-system: (a) type 1 (b) type 2.

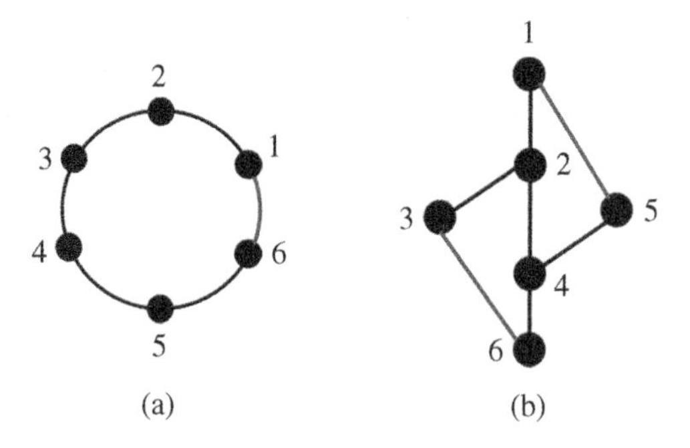

Figure 2.4 Two types of graphs satisfy the $N-1$ rule: (a) type 1 (b) type 2.

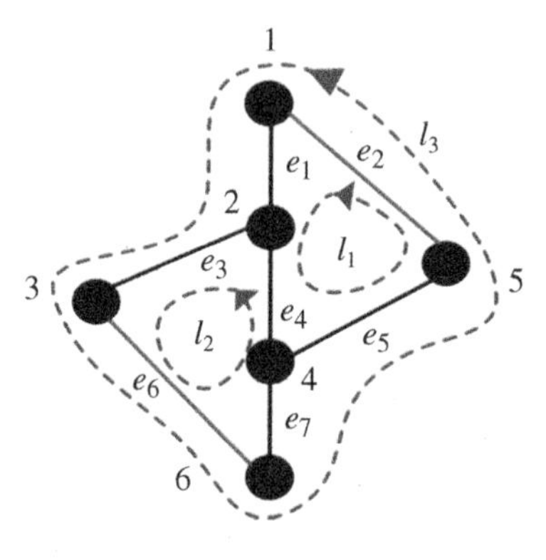

Figure 2.5 Graph with three loops.

in Figure 2.4a, there is only one loop, which indeed contains all the nodes of the system. Accordingly, the corresponding communication network design satisfies the $N-1$ rule. We redraw Figure 2.4b as Figure 2.5. It can be noticed that this graph has three loops, l_1, l_2, and l_3. Loops l_1 and l_2 contain only part of nodes of the system, whereas loop l_3 encircles all the nodes of the system. Therefore, the corresponding communication network design of this graph also follows the $N-1$ rule. One can verify that disabling any one of the edges of graphs in Figure 2.3a,b will not isolate any of the nodes from the remaining nodes. In general, the communication network satisfies the $N-1$ rule if and only if the complete loop matrix (CLM) of the corresponding graph satisfies the following condition.

$$\max_{i=1\ldots n_l} \sum_{j=1}^{n_e} C_{ij} = n \tag{2.1}$$

where n_l and n_e are the total number of loops and edges of the graph, respectively, and n is the total number of nodes. C_{ij} is the element of the CLM C and it is defined as:

$$C_{ij} = \begin{cases} 1 & \text{if loop } i \text{ contains edge } j \\ 0 & \text{otherwise} \end{cases} \tag{2.2}$$

The CLM of the graph given in Figure 2.5 is shown as follows.

$$\begin{array}{c} \quad e_1\; e_2\; e_3\; e_4\; e_5\; e_6\; e_7 \Big|\; \sum_{j=1}^{7} C_{ij} \\ \mathbf{C}_1 = \begin{matrix} l_1 \\ l_2 \\ l_3 \end{matrix} \begin{pmatrix} 1 & 1 & 0 & 1 & 1 & 0 & 0 \\ 0 & 0 & 1 & 1 & 0 & 1 & 1 \\ 1 & 1 & 1 & 0 & 1 & 1 & 1 \end{pmatrix} \Bigg| \begin{pmatrix} 4 \\ 4 \\ 6 \end{pmatrix} \end{array} \tag{2.3}$$

Because $\max_{i=1\ldots3} \sum_{j=1}^{7} C_{ij} = 6$, which is the number of nodes in the network, the communication network satisfies the $N-1$ rule. If e_3 is disabled, the CLM of the graph is:

$$\begin{array}{c} \quad e_1\; e_2\; e_3\; e_4\; e_5\; e_6\; e_7 \Big|\; \sum_{j=1}^{7} C_{ij} \\ \mathbf{C}_2 = l_3 \begin{pmatrix} 1 & 1 & 1 & 1 & 1 & 1 & 1 \end{pmatrix} \Big| (6) \end{array} \tag{2.4}$$

Because $\sum_{j=1,j\in l_3}^{7} C_j = 6$, the network still satisfies the $N-1$ rule. However, if e_5 is disabled, the CLM of the graph becomes:

$$\begin{array}{c} \quad e_1\; e_2\; e_3\; e_4\; e_5\; e_6\; e_7 \Big|\; \sum_{j=1}^{7} C_{ij} \\ \mathbf{C}_3 = l_2 \begin{pmatrix} 0 & 0 & 1 & 1 & 0 & 1 & 1 \end{pmatrix} \Big| (4) \end{array} \tag{2.5}$$

Because $\sum_{j=1,j\in l_2}^{7} C_j = 4 < 6$, the network no longer satisfies the $N-1$ rule. Yet, it is easy to verify that the graph is still connected.

If we design the communication network for distributed applications by following the $N-1$ rule, the malfunction of any one of the communication channels will not cause the malfunction of the others, which can improve the robustness of the overall system. The essence of the $N-1$ rule is to design a loop that encircles all nodes, which does not necessarily make the graph much denser. As can be seen in Figure 2.4a, we merely add only one communication channel to acquire the communication network satisfying the $N-1$ rule. Yet, the density of the communication network does affect the performance of the distributed algorithm that is implemented on it, and we can discuss this matter in the subsequential.

2.1.3 Convergence of Distributed Algorithms with Variant Communication Network Typologies

As discussed before, the communication network of the distributed algorithm is configured for information or knowledge spreading. Of course, a higher density of the corresponding graph of the communication network indicates that the knowledge can spread in a faster way. Thus, it is intuitive to the premise that the denser graph enables the distributed algorithms to converge faster. However, a graph with higher density usually bounds to larger investment and communication cost, and that is a trade-off which must be made during the design phase for the communication network.

Let us take a look at how the convergence of consensus algorithm is affected by the density of the corresponding graph of the communication network. Discrete form of the consensus algorithm can be written as:

$$\mathbf{x}\,[k+1] = \mathbf{W}\mathbf{x}\,[k] \tag{2.6}$$

Here, $\mathbf{W}$ is the weight matrix, which is defined as:

$$\mathbf{W} = \mathbf{I} - \Delta t\mathbf{L} \tag{2.7}$$

where $\mathbf{I}$ is the identity matrix, and Δt is the sampling time of the discrete system. $\mathbf{L}$ is known as the Graph Laplacian of the graph $\mathbf{G}$ [2], and its element is defined as:

$$l_{ij} = \begin{cases} -1 & j \in idx_i \\ N_i & j = i \\ 0 & \text{else} \end{cases} \tag{2.8}$$

where N_i is the degree of node i. According to the definition of $\mathbf{L}$, $\mathbf{W}$ becomes a Perron matrix, which has the following properties: (i) $\mathbf{W}$ is a doubly stochastic matrix; (ii) All eigenvalues of $\mathbf{W}$ are in a unit circle if the sampling time $\Delta t < 1/\Delta$, where $\Delta = \max\{l_{ii}\} i = 1, 2 \cdots n$. Based on the Perron–Frobenius lemma given in [3]

$$\lim_{k\to\infty} \mathbf{W}^k = \frac{1}{n}\mathbf{e}\mathbf{e}^T = \mathbf{J} \tag{2.9}$$

According to Eqs. (2.6) and (2.9), the system will reach consensus as k approaches infinity and is represented as

$$\lim_{k\to\infty} \mathbf{x}[k] = \lim_{k\to\infty} \mathbf{W}^k \mathbf{x}[0] = \frac{ee^T}{n}\mathbf{x}[0] \tag{2.10}$$

As discussed in [4], the number of iterations needed for convergence of the consensus algorithm is estimated as

$$K = \frac{-1}{\log_E\left(\frac{1}{\lambda_2(\mathbf{W})}\right)} \tag{2.11}$$

where $\lambda_2(\mathbf{W})$ is the second largest eigenvalue of $\mathbf{W}$. As can be seen in Eq. (2.11), the smaller $\lambda_2(\mathbf{W})$ yields faster convergence of the consensus algorithm. Because $\mathbf{W}$ is determined in Eq. (2.7), then

$$\lambda_2(\mathbf{W}) = 1 - \Delta t\lambda_2(\mathbf{L}) \tag{2.12}$$

where $\Delta t\lambda_2$ is the second smallest eigenvalue of $\mathbf{L}$, which is also called Fiedler eigenvalue of the graph. A well-known observation regarding the Fiedler eigenvalue of a graph is that for dense graphs, $\lambda_2(\mathbf{L})$ is relatively large, whereas for sparse graphs, $\lambda_2(\mathbf{L})$ is relatively small [5] Accordingly, $\lambda_2(\mathbf{L})$ is also defined as the algebraic connectivity of graph [6]

Now, we can see that the convergence speed of the consensus algorithm depends on the algebraic connectivity of the corresponding graph of the communication network. Here, we provide an example with 16 agents to demonstrate how the consensus algorithm responds to the algebraic connectivity. We initialize the $\mathbf{X}$ with $X_0 = [1, 2, 3.., 16]^T$, and set the $\mathbf{W}$ by using Eq. (2.7). Since the sampling time should $\Delta t < 1/\Delta$, we set it as follows:

$$\Delta t = 1/(\Delta + 1) \tag{2.13}$$

Six communication network graphs shown in Figure 2.6 are tested, and the test results are provided in Figure 2.7. It can be observed that as the graph grows denser, i.e. the algebraic connectivity grows larger, the number of iterations needed for convergence decreases. Note that there is a sharp drop in the number of iterations when we increase the number of edges to 32. This implies that the performance of the distributed algorithm has increased significantly with a reasonable increase in communication cost, which is the trade-off that we attempt to achieve. In practice, we can always carry out off-line studies to find this trade-off between achieving high performance and maintaining a low communication cost [7]

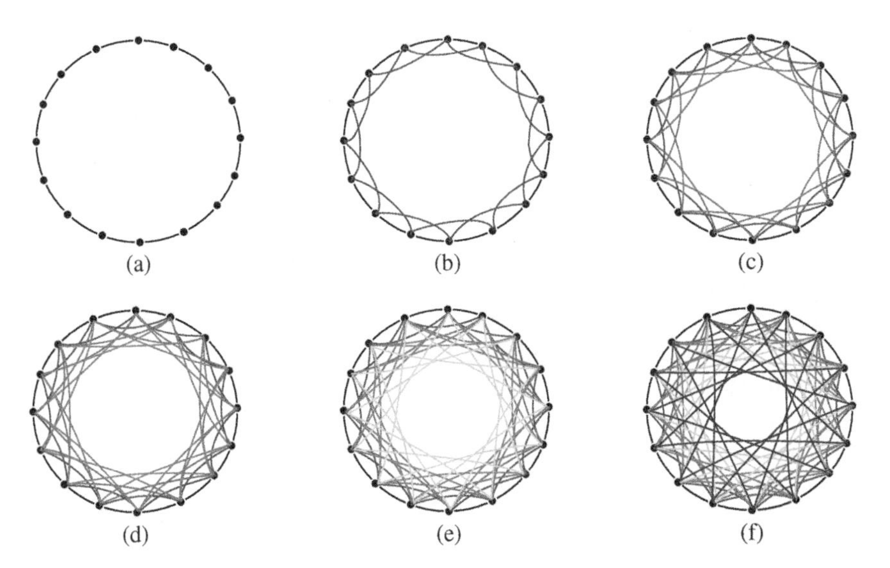

Figure 2.6 Variable communication network configurations for 16-agent systems. (a) $N_e = 16$, (b) $N_e = 32$, (c) $N_e = 48$, (d) $N_e = 64$, (e) $N_e = 80$, and (f) $N_e = 96$.

2.2 Real-Time Digital Simulation

In this book, most of the introduced algorithms will be evaluated through simulations. According to the concept of technology readiness level (TRL), further steps should be technology development, technology demonstration, system development, then system test, and field operation. In this chapter, the MAS platform we developed for distributed applications is introduced, which is suitable for MAS-based control technology development. By integrating the MAS with the real-time digital simulator, a MAS-based real-time simulation platform for testing distributed control solution is also introduced.

2.2.1 Develop MAS Platform Using JADE

We develop the MAS platform based on the platform known as JADE (Java Agent Development), which is a software platform developed using JAVA that can provide basic middle-ware layer functionalities independent of the specific application. JADE can provide programmers with ready-to-use and easy-to-customize core functionalities.

Started by Telecom Italia in late 1998, JADE went open source in 2000 and was distributed by Telecom Italia under the LGPL (Library Gnu Public License). It can simplify the realization of distributed applications via the usage of the abstraction called "agent." Because the object-oriented language, JAVA, is used for the implementation of the JADE platform, it can provide simple and friendly application programming interfaces (APIs).

The MAS developed using JADE has the following features [8]: (i) An agent is autonomous and proactive: an agent makes its own decisions; each agent maintains its own thread of execution. This feature enables the easy plug-and-play operation of the agent. (ii) The agents are loosely coupled: agents in a JADE-based platform communicate with each other via an asynchronous communication protocol. Temporal dependency between sender and receivers is avoided: a receiver may be unavailable when the sender initiates sending a message. This feature facilitates the handling of communication issues such as communication delay, single-point failures, etc. (iii) The overall system consists of multiple agents developed using JADE is Peer-to-Peer: each agent is identified by a globally unique name, known as the agent identifier or AID. A specific agent can join and leave a host platform anytime and can discover other agents via white-page and yellow-page service. The main elements of a JADE platform are shown in Figure 2.7. A JADE platform is composed of agent containers that are distributed over networks. Agents, who live in containers, are the Java processes, providing the JADE run-time and all the services needed for hosting and executing agents. When the main container is launched, two special agents are automatically

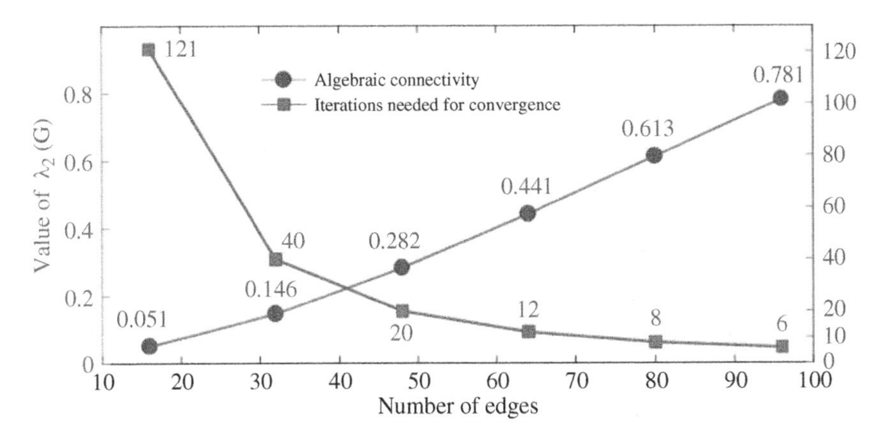

Figure 2.7 Test results of the consensus algorithm with variable communication network graphs.

initiated and started by JADE: (i) The agent management system (AMS) is the agent that supervises the entire platform. Every agent is required to register with the AMS to receive a valid AID. (ii) The directory facilitator (DF) is the agent that implements the yellow pages service, used by any agent wishing to register its services or search for other available services. Multiple DFs can be started concurrently to distribute the yellow pages service across several domains. Each agent can provide services to other agents (no matter on the same platform or not) with the help of DF. The agents can be created on different platforms, as shown in Figure 2.8. A unique AID distinguishes each agent. The agent can be configured to either send information to other agents or receive information from other agents,

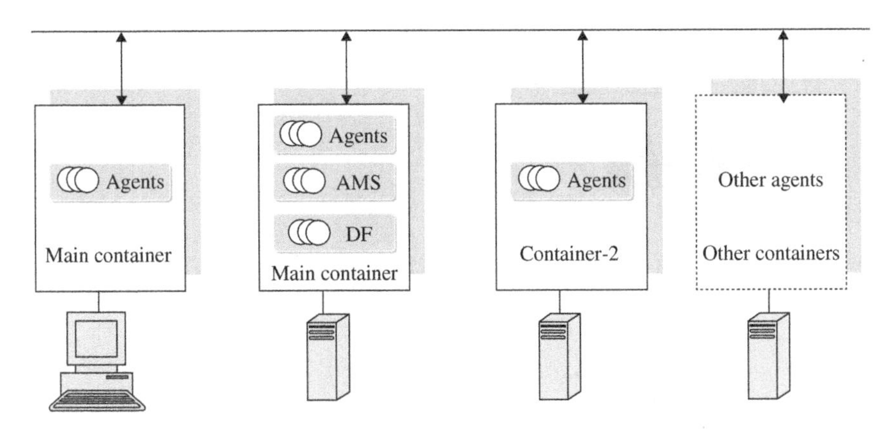

Figure 2.8 Illustration of the JADE environment.

which indicates that the virtual communication network among agents can be easily adjusted. Because of the outstanding features of the JADE, it is used to develop MAS to evaluate the distributed solutions in this book. In the following, we provide the test results with one of the types of distributed algorithms – consensus algorithm to demonstrate the implementation of distributed solutions using MAS.

2.2.2 Test-Distributed Algorithms Using MAS

2.2.2.1 Three-Agent System on the Same Platform

We first deploy three agents on the same platform, which is built using book-size PCs/NanoPCs (dual-core, 1.86 GHz). We choose this book-size PC because it is configured with wireless communication functionality. Besides, the book-size PC is relatively cheap and is more energy-saving compared with the traditional PCs. Figure 2.9 shows the developed platform with logic agents being also demonstrated.

Here, we test an example with consensus algorithm on this platform. Example: $x_1 = 15, x_2 = 20,$ *and* $x_3 = 35$, and the communication network is designed the same as in Figure 2.10, $\mathbf{W}$ is set according to Eq. (2.7) with Δt being set to 0.2, and the weight matrix $\mathbf{W}$ is calculated as:

$$\mathbf{W} = \begin{bmatrix} 0.8 & 0.2 & 0 \\ 0.2 & 0.6 & 0.2 \\ 0 & 0.2 & 0.8 \end{bmatrix} \tag{2.14}$$

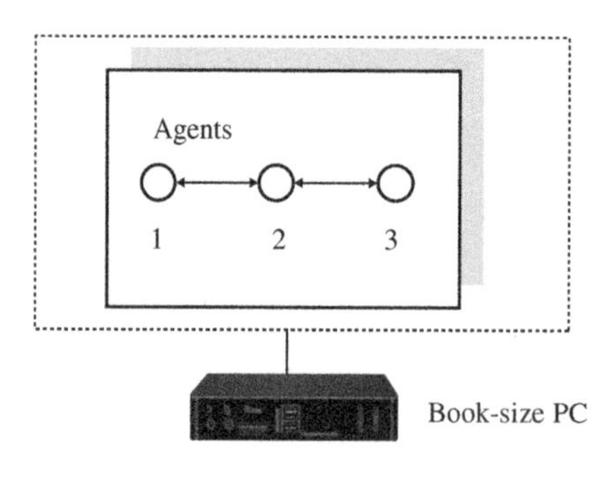

Figure 2.9 Platform with three-agent system.

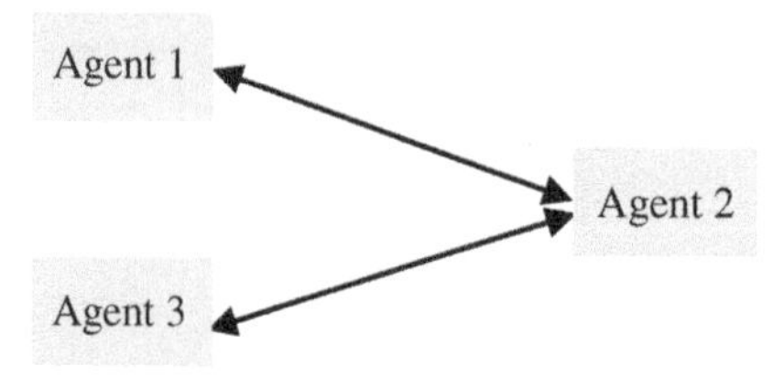

Figure 2.10 Communication topology configuration of the three-agent system.

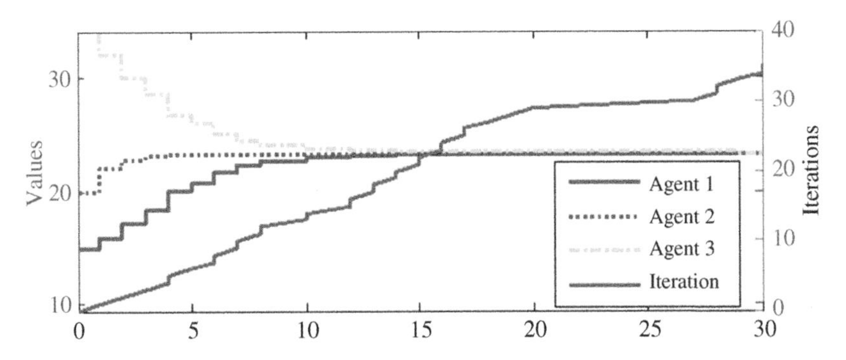

Figure 2.11 Tests of consensus algorithm in JADE platform.

The test results of this example are shown in Figure 2.11. It can be seen that the implemented consensus algorithm on this platform takes 30 ms for 36 iterations, on an average, 0.84 ms for one iteration. Note that the time consumed for the algorithm during the test is random to some extent and can vary depending on the hardware used (different computers) and software implementation (coding and compilers). However, considering all these variables, the test results here still provide some insights into implementing the MAS-based distributed control solutions.

2.2.2.2 Two-Agent System with Different Platforms

The two-agent system we used for the test is shown in Figure 2.12. Here, we deploy these two logic agents on two different book-size computers. The communication between these two computers is based on TCP/IP protocol through the internet with a bandwidth of 100 Mb/s.

The update process of the consensus algorithm on a two-computer platform is shown in Figure 2.13. The initial values of agent-1 and agent-2 are set to 10 and 30, respectively. As can be seen in Figure 2.13, for 15 iterations, the implemented algorithm takes about 40 ms, i.e., about 2.7 ms per iteration. Here, we should point out that the test setup provided is just a special case. The performance of the algorithm

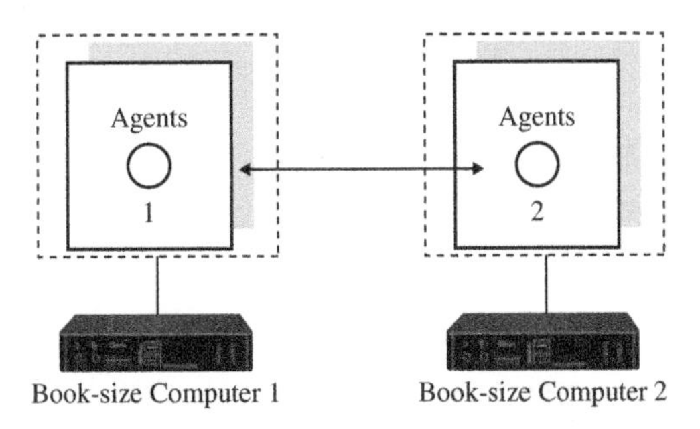

Figure 2.12 Two-agent system configuration.

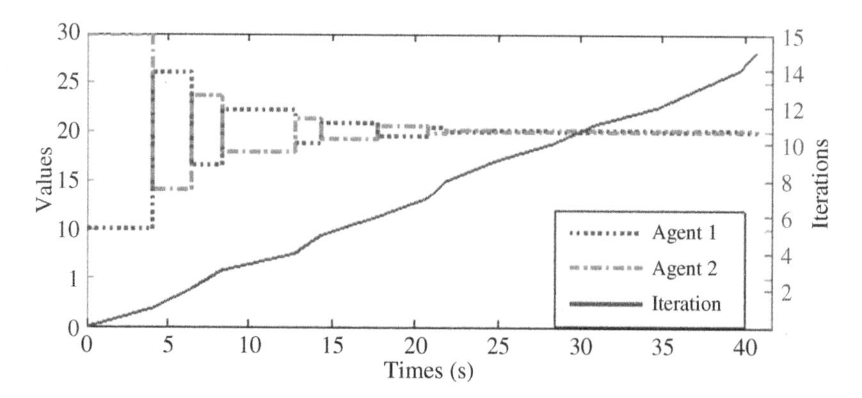

Figure 2.13 Convergence information of the two-agent system in JADE platform.

may vary if implemented with different computers, communication networks, or protocols or with different communication traffics.

2.2.3 MAS-Based Real-Time Simulation Platform

To advance the system development of the distributed control solution to a further step, we developed a MAS-based real-time simulation platform. This platform consists of a host computer, a real-time digital simulator, and an MAS implemented with multiple NanoPCs, as shown in Figure 2.14. The host computer

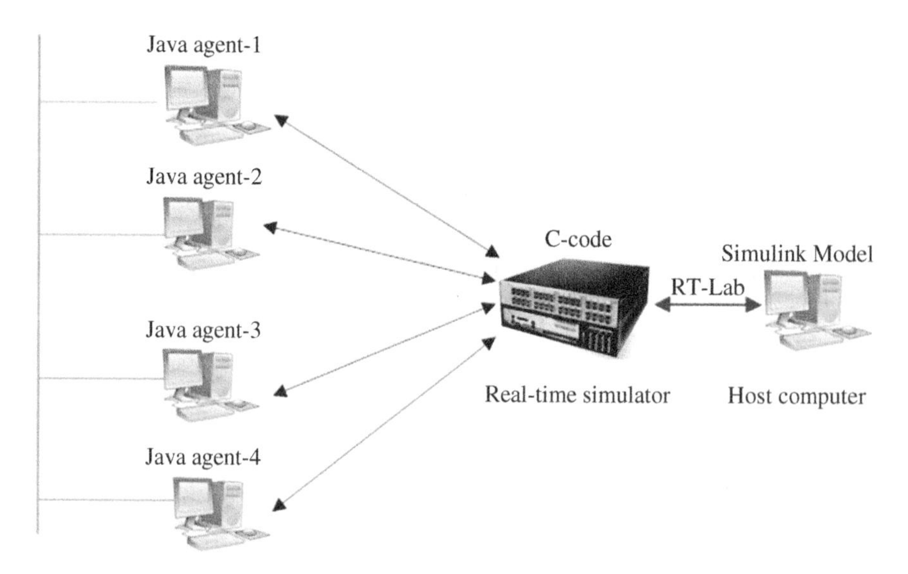

Figure 2.14 Real-time simulation platform with MAS.

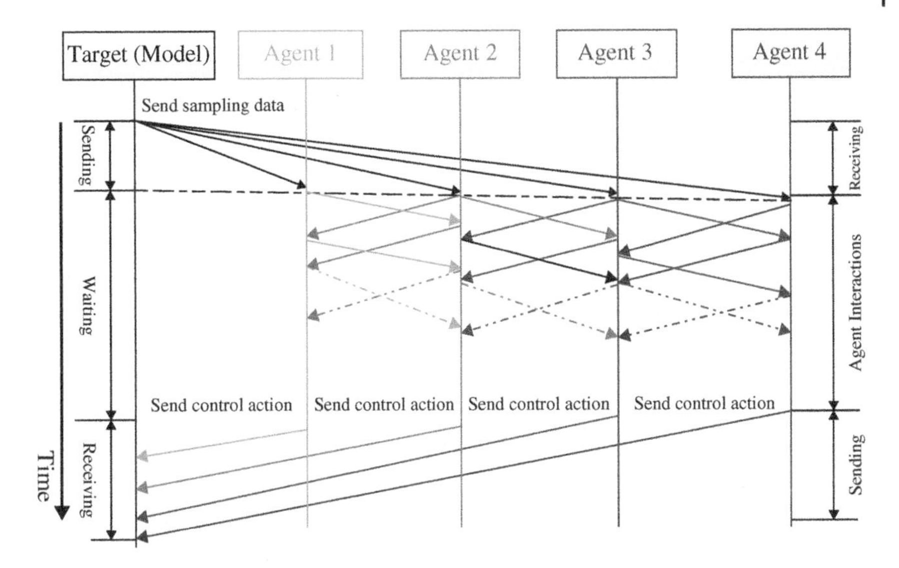

Figure 2.15 Agent interactions among four-agent system.

is used for compiling the developed power system model into C-code and then uploads the code to the real-time digital simulator for simulation. The host computer is also used to monitor the states of the running model. The real-time simulator is also called a target computer, which is manufactured by OPAL-RToR. The MAS is implemented using NanoPcs. As can be seen in Figure 2.14, agents interact with each other on the JADE platform, and the communication between the simulator and the agent is realized by our developed Java-C interface.

Because the developed models to describe actual power systems are running in real time in the simulator, this platform can be used to evaluate the control procedures such as data sampling, communication, the act of control signal, etc. The interactions among agents during this real-time simulation process are demonstrated in Figure 2.15. It shows that the agent directly communicates with the target to access operation data of the power system models that are running in the simulator, which acts similarly to the data sampling in real power systems. The algorithms for control and optimization are integrated into the interactions process of agents. In this way, each agent is equivalent to the distributed controller of actual power systems.

In this book, we will constantly use this MAS-based real-time simulation platform to test our distributed control solutions, including the algorithm convergences, communication performances, and robustness against abnormalities, etc.

References

1 Olfati-Saber, R., Fax, J.A., and Murray, R.M. (2007). Consensus and cooperation in networked multi-agent systems. *Proceedings of the IEEE* 95 (1): 215–233.

2 Doob, M., Cvetković, D.M., and Sachs, H. (1998). *Spectra of Graphs: Theory and Applications*, 3rd Revised and Enlarged Edition. New York: Wiley.

3 Horn, R.A., Horn, R.A., and Johnson, C.R. (1990). *Matrix Analysis*. Cambridge University Press.

4 Xu, Y. and Liu, W. (2011). Novel multiagent based load restoration algorithm for microgrids. *IEEE Transactions on Smart Grid* 2 (1): 152–161.

5 Rota, G. C. (1976). *Algebraic Graph Theory, Graduate Texts in Mathematics*, vol. 20, 415. Springer.

6 Fiedler, M. (1973). Algebraic connectivity of graphs. *Czechoslovak Mathematical Journal* 23: 298–305.

7 Olfati-Saber, R. and Murray, R.M. (2004). Consensus problems in networks of agents with switching topology and time-delays. *IEEE Transactions on Automatic Control* 49 (9): 1520–1533.

8 Bellifemine, F.L., Caire, G., and Greenwood, D. (2007). *Developing Multi-Agent Systems with JADE*, vol. 7. Wiley.

3

Distributed Active Power Control

Proper control of active power of the power system is crucial for frequency stability. Although the frequency of the power system is generally considered to be a global indicator of system performance, the active power of the entities of the system all will impact this indicator. Because of the integration of the renewable technologies such as distributed wind and solar generation, as well the variant demand response programs, the number of active components in the power system increases. Therefore, the control of active power of the power system calls for new solutions for enhancement of reliability and improved cost-benefit effectiveness. The distributed solution has many merits that nicely fit to these demands, and it is investigated in this chapter. We will discuss three distributed control methods/solutions for active power control, where the applications of these solutions differ depending on the control targets, control objectives, and available resources. The first control solution introduced is subgradient-based active power sharing, which aims at maintaining the supply-demand balance in a microgrid. The second control method is used for the economic dispatch (ED) of the smart grids, which is not limited to the microgrid. The last control solution aims at integrating the ED with the frequency control. From the perspective of control implementation, all of these three methods can be implemented in a distributed manner, which not only improve the reliability of the control system but can also improve the frequency response performance of the power systems.

3.1 Subgradient-Based Active Power Sharing

For an autonomous microgrid with high penetration of renewable energy, the supply-demand balance of active power bears vital significance. However, for the maximum peak power tracking (MPPT) algorithm, which is rather popular, it may cause a supply-demand imbalance when the power generation exceeds the load

Distributed Energy Management of Electrical Power Systems, First Edition.
Yinliang Xu, Wei Zhang, Wenxin Liu, and Wen Yu.

Published 2021 by John Wiley & Sons, Inc.

demand. At present, droop control is a widely used decentralized approach for distributed generators (DGs) in active and reactive power sharing. Nevertheless, the traditional droop control approaches suffer from some drawbacks, such as slow dynamic response, oscillation, and steady-state deviations. To address these issues, a fully distributed subgradient-based approach for an autonomous microgrid is proposed in this chapter, which can realize coordinate operations of multiple distributed renewable generators (RGs). By adjusting the utilization level of distributed renewable energy generators, the dynamic performance of the microgrid is considerably improved while maintaining the supply-demand balance. The effectiveness of the proposed fully distributed subgradient-based coordination approach is validated by simulation studies using a 6-bus microgrid.

3.1.1 Introduction

A microgrid has two operation modes, islanded and grid-connected, which is serviced by a distribution system and composed of DGs, energy storage systems, and different kinds of loads. A microgrid can be used for many different applications such as the housing estates, commercial buildings, industrial factory, and municipal regions [1]. The advantages of microgrid technology include improving system flexibility and reliability, enhancing energy efficiency, and promoting the utilization of renewable energy sources [2].

Wind power and photovoltaic (PV) is developing rapidly in recent years because of their clean, pollution-free, and abundant nature. However, the intermittency and volatility of renewable energy sources bring significant challenges to the control and operation of the microgrid. For distributed renewable energy sources, the power reference control under different load conditions is an important issue. The most popular method to enhance the energy usage efficiency of renewable energy sources is the MPPT-based algorithm [3, 4]. Nevertheless, for this method, it may cause a supply-demand imbalance when the power generation exceeds the load demand. The energy storage device is required to overcome this problem, but the capacity may be insufficient, and the utilization of renewable energy sources is also limited by the charging and discharging rate of energy storage devices [5].

To maintain the supply-demand balance and regulate the system frequency, the control issues for autonomous microgrid are very similar to the large-scale power system. Thus, methods used in the traditional power system can be directly introduced to the autonomous small-scale microgrids. Reference [6] proposed an integrated control system of wind farms, which is based on two-level control schemes. In this scheme, the active and reactive power set points for DGs are decided by the supervisory control level, and the set points are ensured by the machine control level. An optimal control strategy for wind power is presented in [7]. According to the requirement of the system's central operator, the active and reactive power of all doubly fed induction generators (DFIGs) is regulated automatically. In [8], the author proposed a control method for wind farms, which can provide a sufficient generation reserve according to the requirements of supervisors. A coordinated control scheme for PV generation using fuzzy reasoning to generate the central leveling generation commands is proposed in [9], which can reduce the frequency deviation effectively.

All the methods discussed above are centralized, which requires a powerful central controller to process a complex problem and collect complicated system information [10]. Therefore, this kind of centralized approach is costly to implement and require high data reliability. Besides, because of the intermittency and volatility of renewable energy sources, these centralized approaches may fail to respond in a timely manner when the operating conditions of the system change rapidly.

Focusing on the self-contained, medium voltage, and small-scale microgrid, which consists of multiple renewable energy sources, the purpose of this chapter is to maintain the supply-demand balance of active and reactive power and regulate the voltage magnitude and system frequency to the desired values. The control approach for the DGs in a microgrid should be of high efficiency and low cost because the DGs are always diverse and distributed. Because of the fast, reliable, robust, and scalable characteristics of distributed control approaches, they have a promising application value to be implemented in microgrids [10].

Multi-agent systems (MASs) are also very popular in recent research, as in [11–14]. MAS has been widely used in various problems of microgrid, such as the active and reactive power management and dispatch [12, 13], distributed control, and optimization [14]. However, the existing MAS are generally rule-based and lack rigorous stability analysis. Recent literature focus on the development of consensus and cooperative control to improve the stability and applicability of MAS-based approaches, which have been implemented in practical systems, as in [15–17]. In this chapter, a MAS-based fully distributed control approach for renewable generators' controller consisting of two control levels is proposed. In the upper control level, by adjusting local utilization levels of renewable energy according to the local frequency and available renewable energy prediction, a subgradient-based optimization algorithm is implemented to maintain the supply-demand balance of microgrid. In the lower control level, after the utilization level is updated, the reference value of renewable energy generators can be determined and deployed. Besides, the DGs can be controlled in the reactive power control mode or voltage regulation mode according to operating conditions. The major contributions of this chapter are summarized as follows:

(1) The coordination of multiple renewable energy sources within a microgrid is modeled as a convex optimization problem and solved by a distributed subgradient-based algorithm introduced in [18–20].
(2) According to the near-optimal and stable coefficients setting algorithm mentioned in [15] and [16], the convergence of the proposed fully distributed MAS-based algorithm is analyzed and guaranteed.
(3) Using the control law according to system frequency dynamics in [21], only predicted available renewable energy generation, local frequency measurement, and neighboring DGs' utilization levels are required under the presented distributed subgradient-based coordination approach.

3.1.2 Preliminaries - Conventional Droop Control Approach

Microgrids have two operating modes: islanded and grid-connected. In grid-connected mode, a microgrid can exchange power with the main grid. In this scenario, the main grid can be assumed as the slack bus to maintain the supply-demand balance of the system, and all of the renewable energy generators can be controlled using MPPT algorithms. However, when a microgrid is operated in an islanded mode, all of the components within the microgrid should be operated coordinately to maintain the supply-demand power balance autonomously.

To control the DGs in a microgrid, droop control approaches are applied coordinately to determine the active power sharing originating from the control of the synchronous generator (SG) in large-scale power systems [22–26]. For readers' convenience, the principle of a conventional droop control approach is briefly introduced in the following statement.

The active power outputs of each renewable energy generator should be adjusted to maintain the system frequency according to the P–f droop property predefined in [27] as follows:

$$P_i = P_i^0 + k_{f,i}(f^* - f_i) \tag{3.1}$$

where f_i and f^* are the local and desired operating frequencies, respectively. $k_{f,i}$ is the frequency droop coefficient of renewable energy generator i. P_i^0 is the initial active power that corresponds to f^* of renewable generator i, P_i is the output active power generation reference value of renewable energy generator i.

In the same way, according to the predefined Q–V droop property, the reactive power of each renewable energy generator can be adjusted as follows:

$$Q_i = Q_i^0 + k_{V,i}(V_i^* - V_i) \tag{3.2}$$

where V_i^* and V_i are the measured and rated voltage magnitudes, respectively, and $k_{V,i}$ is the voltage droop coefficient of renewable generator i. Q_i^0 is the initial reactive power that corresponds to V_i^* of renewable generator i. Q_i is the output reactive power reference value of renewable generator i.

Droop control has an advantage that it does not need to communicate with other DGs in the microgrid. Nevertheless, for the conventional droop controller, it has several shortcomings. For instance, there exists voltage and frequency deviations and inaccurate active power sharing, as noticed in [24, 27–33]. $k_{f,i}$ and $k_{V,i}$ are the predefined parameters, which cannot be adjusted online. For the microgrid with small inertia and lack of grid support, because the maximum generations of renewable generators change constantly, the updated P_i^0 may not be achievable, causing inaccurate power sharing and reducing system stability. In summary, for the droop control approach, on the positive side, it can eliminate the mismatch of power supply and demand. On the negative side, it may cause frequency and voltage deviations.

For microgrids, the system's static stability and dynamic performance are very critical because the operating conditions are always changing. Therefore, a fast, reliable, robust, and scalable controller for a microgrid with high penetration of intermittent and volatile renewable energy is desperately required.

3.1.3 Proposed Subgradient-Based Control Approach

In this section, a fully distributed algorithm used to achieve the system power balance is proposed.

3.1.3.1 Introduction of Utilization Level-Based Coordination

P_D is the overall active power demand of a microgrid, which can be expressed as follows:

$$P_D = \sum_{i=1}^{n} P_{L,i} + P_{\text{Loss}} \tag{3.3}$$

where P_D is the overall active power demand of a microgrid, $P_{L,i}$ is the load demand of bus i, P_{Loss} is the active power loss and n is the bus number of the microgrid.

In the same way, the overall available renewable energy generation can be expressed as follows:

$$P_G^{\max} = \sum_{i=1}^{m} P_{G,i}^{\max} \tag{3.4}$$

where $P_G^{\max}$ is the maximum overall available renewable energy generation, $P_{G,i}^{\max}$ is the maximum available generation of renewable generator i, and m is the number of renewable energy generators.

For an autonomous microgrid, if P_D is larger than P_G^{max}, all of the renewable generators will be operated in MPPT mode. Otherwise, if P_D is less than P_G^{max}, the MPPT method is invalid. An effective deloading strategy of sharing the load demand among the renewable generators is to control the utilization levels of renewable generators to a common value as follows:

$$u^* = \min \left\{ \frac{P_D}{P_G^{max}}, 1 \right\} \tag{3.5}$$

where u^* is the common utilization level for all renewable generators.

$P_{G,i}^{ref}$ is the reference active power generation of renewable generator i calculated as follows:

$$P_{G,i}^{ref} = u^* \cdot P_{G,i}^{max} \tag{3.6}$$

Based on Eq. (3.5) and Eq. (3.6), it can be easily validated that the system power balance can be maintained when the generation is larger than load demand, as:

$$\sum_{i=1}^{m} P_{G,i}^{ref} = \sum_{i=1}^{m} u^* \cdot P_{G,i}^{max} = \frac{P_D}{P_G^{max}} \sum_{i=1}^{m} P_{G,i}^{max} = P_D \tag{3.7}$$

3.1.3.2 Fully Distributed Subgradient-Based Generation Coordination Algorithm

For the autonomous microgrid, the supply-demand balance should be maintained to guarantee the static stability. Thus, for multiple renewable generator coordination, the objective function is formulated as follows:

$$\min\ H(u_i[k]) = \frac{1}{2}\left(\sum_{i=1}^{m} u_i[k] P_{G,i}^{max} - P_D \right)^2 \tag{3.8}$$

where $u_i[k]$ is the utilization level of renewable generator i at step k.

The problem formulated in Eq. (3.8), is a convex optimization problem and can be solved by traditional distributed subgradient algorithms. $u_i[k]$ can be obtained using the following equation according to [18, 24, 25].

$$u_i[k+1] = \sum_{j=1}^{m} a_{ij} u_j[k] - d_i \frac{\partial H(u_i[k])}{\partial u_i[k]} \tag{3.9}$$

where d_i is the step size during the iteration and a_{ij} is the communication coefficient between renewable generator i and j.

In Eq. (3.9), $\frac{\partial H(u_i[k])}{\partial u_i[k]}$ can be formulated as follows:

$$\frac{\partial H(u_i[k])}{\partial u_i[k]} = P_{G,i}^{max} \left(\sum_{i=1}^{m} u_i[k] P_{G,i}^{max} - P_D \right) \tag{3.10}$$

If every renewable generator is regarded as an independent agent, the communication links in a microgrid form an undirected graph, i.e. $a_{ij} = a_{ji}$. According to our previous research, a_{ij} can influence the converging speeds of the distributed algorithm [15]. The mean metropolis algorithm presented in Eq. (3.11) is a fully distributed algorithm and has the advantages of being adaptive to changes in communication network topology and providing satisfactory converging speed. Thus, the mean metropolis algorithm is adopted in this chapter.

$$a_{ij} = \begin{cases} 2/(n_i + n_j + 1) & j \in N_i \\ 1 - \sum_{j \in N_i} 2/(n_i + n_j + 1) & i = j \\ 0 & \text{otherwise} \end{cases} \tag{3.11}$$

where N_i is the set of the agents, which have communication links with agent i. n_i and n_j are the agents' number that are connected to agent i and j, respectively.

According to Eqs. (3.10), (3.9) can be rewritten as follows:

$$u_i[k+1] = \sum_{j=1}^{m} a_{ij} u_j[k] - P_{G,i}^{\max} d_i \left(\sum_{i=1}^{m} u_i[k] P_{G,i}^{\max} - P_D \right) \tag{3.12}$$

Defining a communication coefficient matrix $\mathbf{A}$ of m dimensional, which is composed of a_{ij}. Then, the Eq. (3.14) can be represented by the following matrix form:

$$\mathbf{U}[k+1] = \mathbf{A} \cdot \mathbf{U}[k] - \left(\sum_{i=1}^{m} u_i[k] P_{G,i}^{\max} - P_D \right) \cdot \mathbf{D} \tag{3.13}$$

where $\mathbf{U}[k] = [u_1[k], \cdots, u_i[k], \cdots, u_m[k]]^T$, $\mathbf{D} = [P_{G,1}^{\max} d_1, \cdots, P_{G,i}^{\max} d_i, \cdots, P_{G,m}^{\max} d_m]^T$

Using the mean metropolis algorithm, the communication coefficient matrix $\mathbf{A}$ has the following properties, which is a doubly stochastic matrix.

(1) The eigenvalues of matrix $\mathbf{A}$ are less than or equal to 1.
(2) The following relationship is satisfied: $\mathbf{A} * \mathbf{v} = \mathbf{v}$, $\mathbf{v}^T * \mathbf{A} = \mathbf{v}^T$ and $\mathbf{v}^T * \mathbf{v} = \mathbf{1}$, where $\mathbf{v} = \frac{1}{\sqrt{m}}\mathbf{1}$, $\mathbf{1} = [1, \ldots, 1]^T$.
(3) Based on Perron–Frobenius theorem [34], the communication coefficient matrix $\mathbf{A}$ satisfies the following relationship: $\lim_{k\to\infty} A^k = \mathbf{v} * \mathbf{v}^T = \frac{1}{m}\mathbf{1} \cdot \mathbf{1}^T$.

According to [19], a fully distributed subgradient algorithm can converge when satisfies the following two conditions: (i) The communication coefficient matrix $\mathbf{A}$ should satisfy: $\lim_{k\to\infty} A^k = \frac{1}{m}\mathbf{1} \cdot \mathbf{1}^T$. (ii) The iteration step sizes d_i should be sufficiently small.

According to the above analysis, condition (i) is satisfied automatically. d_i should be adjusted small enough for algorithm implementation to guarantee the convergence of the proposed algorithm.

For Eq. (3.13), the equilibrium point can be calculated by summing up Eq. (3.12). Let $u_i[k+1] = u_i[k] = u_i^*, i \in [1, ..., m]$, then, the Eq. (3.12) can be represented by the following matrix form:

$$\sum_{i=1}^{m} u_i^* = \sum_{i=1}^{m}\sum_{j=1}^{m}(a_{ij}u_j^*) - \left(\sum_{i=1}^{m} u_i^* P_{G,i}^{\max} - P_D\right)\sum_{i=1}^{m}(P_{G,i}^{\max} d_i) \tag{3.14}$$

Since matrix **A** is symmetrical and the summation of each column equals to 1. For Eq. (3.14), the first term of right-hand side can be obtained as follows:

$$\sum_{i=1}^{m}\sum_{j=1}^{m}(a_{ij}u_j^*) = \sum_{i=1}^{m}\sum_{j=1}^{m}(a_{ji}u_i^*) = \sum_{i=1}^{m}(u_i^* \sum_{j=1}^{m} a_{ji}) = \sum_{i=1}^{m} u_i^* \tag{3.15}$$

Because of $\sum_{i=1}^{m} P_{G,i}^{\max} d_i \neq 0$, therefore:

$$\sum_{i=1}^{m} u_i^* P_{G,i}^{\max} - P_D = 0 \tag{3.16}$$

Since $U[k+1] = U[k] = U^*$, according to Eqs. (3.16), (3.13) can be rewritten as follows:

$$U^* = AU^* \tag{3.17}$$

where $U^* = [u_1^*, ..., u_m^*]^T$.

The solution of Eq. (3.17) has the following properties based on [35]:

$$U^* = u^* \cdot \mathbf{1} \tag{3.18}$$

Substituting Eq. (3.18) into Eq. (3.16), when the algorithm is converged to a steady state, the utilization level u^* is obtained as follows:

$$u^* = P_D \Big/ \sum_{i=1}^{m} P_{G,i}^{\max} = \frac{P_D}{P_G^{\max}} \tag{3.19}$$

Thus, according to Eqs. (3.16) and (3.19), the system power supply-demand balance can be guaranteed in a microgrid.

However, it is rather difficult to measure the overall load and estimate the system power loss accurately. Because the system frequency can reflect the supply-demand imbalance, it is convenient to use the system frequency deviation to address these difficulties.

According to [21], the dynamic frequency response can be represented by the following equation:

$$\frac{df}{dt} = \frac{f_0}{2\omega_{kin0}}\left(\sum_{i=1}^{m} u_i P_{G,i}^{\max} - P_D\right) \tag{3.20}$$

where f_0 is the reference system frequency and ω_{kin0} is the rated kinetic energy of the generator.

Since $\frac{df}{dt} \approx \frac{f[k]-f[k-1]}{\Delta t} = \frac{\Delta f[k]}{\Delta t}$, Eq. (3.20) can be discretized as follows:

$$\Delta f[k] = \frac{f_0 \Delta t}{2\omega_{kin0}} \left(\sum_{i=1}^{m} u_i[k] P_{G,i}^{\max} - P_D \right) \tag{3.21}$$

Thus:

$$\sum_{i=1}^{m} u_i[k] P_{G,i}^{\max} - P_D = \frac{2\omega_{kin0}}{f_0 \Delta t} \Delta f[k] \tag{3.22}$$

Substitute Eq. (3.22) into Eq. (3.14), the updating law of utilization level u_i can be proposed as follows:

$$u_i[k+1] = \sum_{j=1}^{m} a_{ij} u_j[k] - \alpha_i \Delta f[k] \tag{3.23}$$

where $\alpha_i = \frac{2P_{G,i}^{\max} \omega_{kin0} d_i}{f_0 \Delta t}$.

According to [21], ω_{kin0} relies on the system capacity. Notice that because the influence of ω_{kin0} and d_i is absorbed by α_i, ω_{kin0} does not need to be estimated accurately. Thus, only the value $\frac{2\omega_{kin0} d_i}{f_0 \Delta t}$ should be confirmed and not every parameter should be identified.

When all of the renewable generators are working at the rated condition, $P_{G,i}^{\max}$ is a constant and $\frac{2\omega_{kin0} d_i}{f_0 \Delta t}$ can be obtained using the following trials. With the increasing value of α_i, the convergence speed will be increased, but it may cause system instability and dynamic response oscillations. Firstly, α_i is assigned by a small value of 10^{-1} and increased by a small step size of 0.02 until the system dynamic response diverges. Then, when the system dynamic response occurs continuous oscillation, which is also called stability margin, 1/2 of the largest α_i is selected and the value of $\frac{2\omega_{kin0} d_i}{f_0 \Delta t}$ can be obtained by $\frac{\alpha_i}{P_{G,i}^{\max}}$.

3.1.3.3 Application of the Proposed Algorithm

The topology of the proposed control approach is presented in Figure 3.1, consisting of renewable generators, synchronous generator, and loads. The number of renewable generator is m and the number of load is n.

For every renewable generator agent, it should predict its maximum available power generation, measure the frequency, and exchange its local information with neighboring renewable generator agents. Notice that for a MAS, the information communication network can be independent to the system topology. This means that a complex system may correspond to a simple communication network based on the agent location, operation cost, information communication convenience, and so on. For each synchronous generator agent, a synchronous generator is assigned. The synchronous generator agent is not involved in the utilization-level

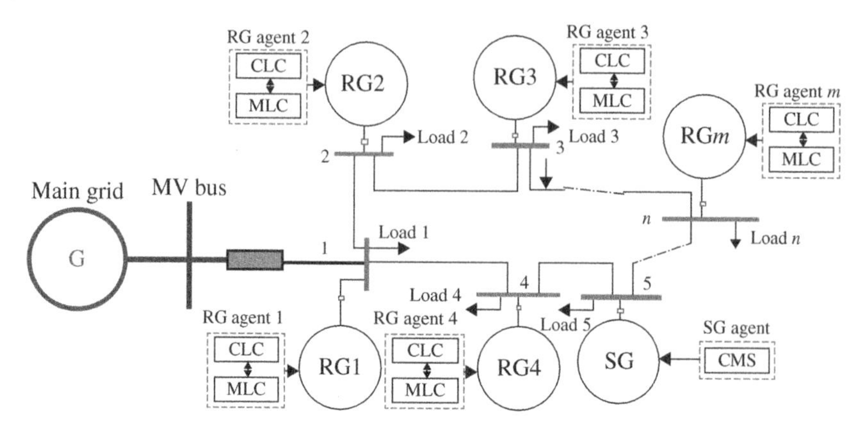

Figure 3.1 Proposed control approach topology in a microgrid.

iteration process. The synchronous generator agent will decide the control mode of the synchronous generator corresponding to it. More details can be found in Synchronous generator control approach.

For a specific renewable generator agent, the operation diagram is presented in Figure 3.2. A two-level control approach is applied to each renewable generator agent. In the upper cooperative-level control (CLC), the desired utilization level and reference active power generation for each renewable generator agent are obtained. For the upper CLC, it has the following modules: (i) system frequency measurement module, (ii) maximum available renewable power generation forecasting module, (iii) information communication module that can exchange the utilization-level information with the neighboring renewable generator agents, (iv) utilization-level updating module. According to the system frequency deviation and utilization-level information of the last iteration, the utilization level can be updated using Eq. (3.23) and the active power generation reference can be obtained. In the lower machine-level control (MLC), active power tracking is realized while considering the reactive power and terminal voltage regulation requirements.

Using the proposed fully distributed algorithm, the global overall load condition and active power transmission loss information are not required. Because of the fact that the power supply-demand imbalance will cause frequency deviation, the utilization level of each renewable generator can be regulated based on the measured system frequency deviation, as in Eq. (3.23). Thus, the amount of measurements is considerably reduced. Besides, the complexity and communication cost are also reduced.

According to the wind speed, the maximum available power generation of DFIGs can be obtained [36]. In addition, based on the weather conditions such

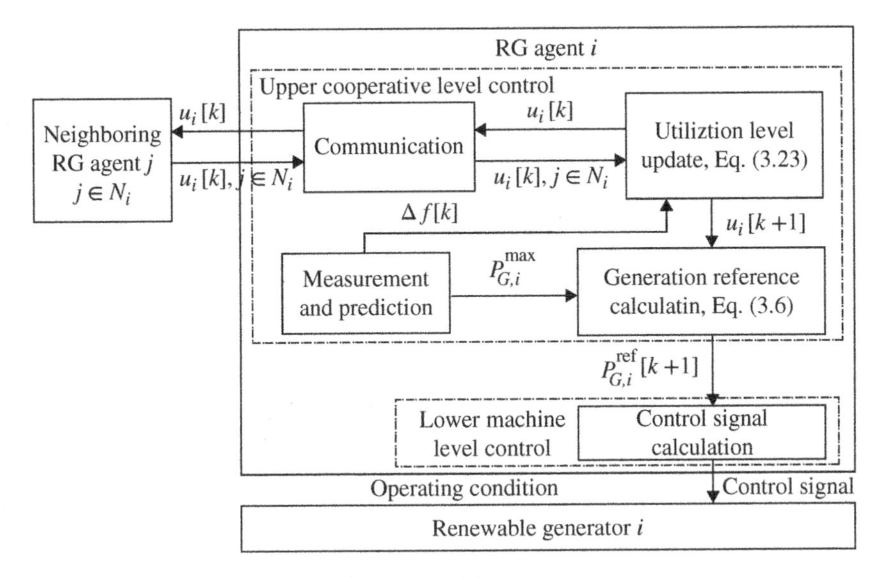

Figure 3.2 Operation diagram of the renewable generator agent.

as the solar radiation condition, the maximum available power generation of PV generators can be obtained [37]. Besides, many kinds of MPPT forecasting algorithms for wind power generators and PV generators, such as fuzzy logic control and neural network, have been widely used, as in [3, 38].

However, there always exists a forecasting error for the MPPT forecasting algorithm, which causes the forecasted maximum available power unachievable [39]. Thus, to make the forecasted maximum available power achievable, the renewable generators can be controlled conservatively to generate power, which is less than the MPPT forecasting. Thus, rather than controlling renewable generators in traditional MPPT mode, the authors decide to control the utilization level of renewable generators in Eq. (3.5) less than 1.

3.1.4 Control of Multiple Distributed Generators

3.1.4.1 DFIG Control Approach

In the lower MLC, all DFIGs need to realize the active power tracking according to the reference profile from the upper level. Besides, the reactive power terminal voltage regulation and DC-link voltage regulation requirements should be considered. Thus, the lower MLC of each DFIG is composed of the mechanical control of pitch angle and electrical control of two converters, as shown in Figure 3.3. A DFIG model is applied in this chapter according to [40]. By adjusting the rotor speed or tuning pitch angle, the active power generation of each DFIG can be regulated

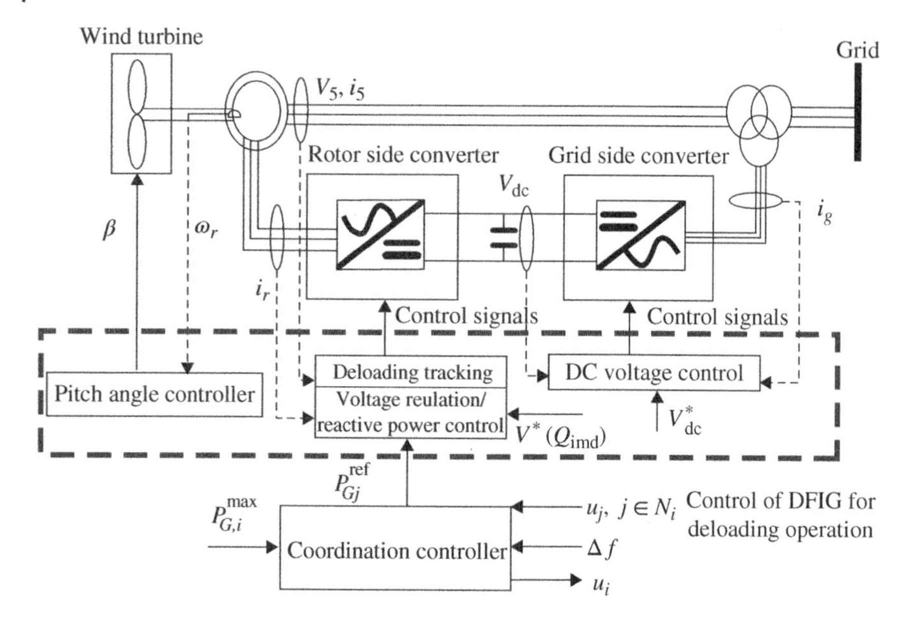

Figure 3.3 Diagram of lower machine-level control for DFIG.

[38]. This method has the following two advantages: (i) the rotor speed of DFIG is controlled by the converter control using power electronics, which offers a faster response speed than the mechanical pitch angle control and (ii) this kind of electrical control method can reduce the wear and tear and increase the service life of the equipment. Nevertheless, when the rotor speed reaches the upper limit, pitch angle tuning needs to be activated. The detailed implementation of this method can refer to [17].

3.1.4.2 Converter Control Approach

In this chapter, the DFIG is controlled by AC–DC–AC converters. According to the decoupled control strategy given in [41], the active and reactive power of DFIG can be controlled by the rotor-side converter. As presented in Figure 3.4, the active power can be adjusted by controlling d-axis rotor current i_{dr} and the reactive power can be adjusted by controlling q-axis rotor current i_{qr}. For a specific DFIG, the deviation between the reference power output $P^{ref}_{G,i}$ and actual active power output $P_{G,i}$ is managed by a PI controller to generate a reference value of rotor current i^*_{dr}. Besides, using another PI controller, the difference between the actual rotor current i_{dr} and reference rotor current i^*_{dr} is used to regulate the voltage amplitude.

For reactive power control, there are two operation modes. One is the voltage regulation mode and the another is the reactive power regulation mode. For

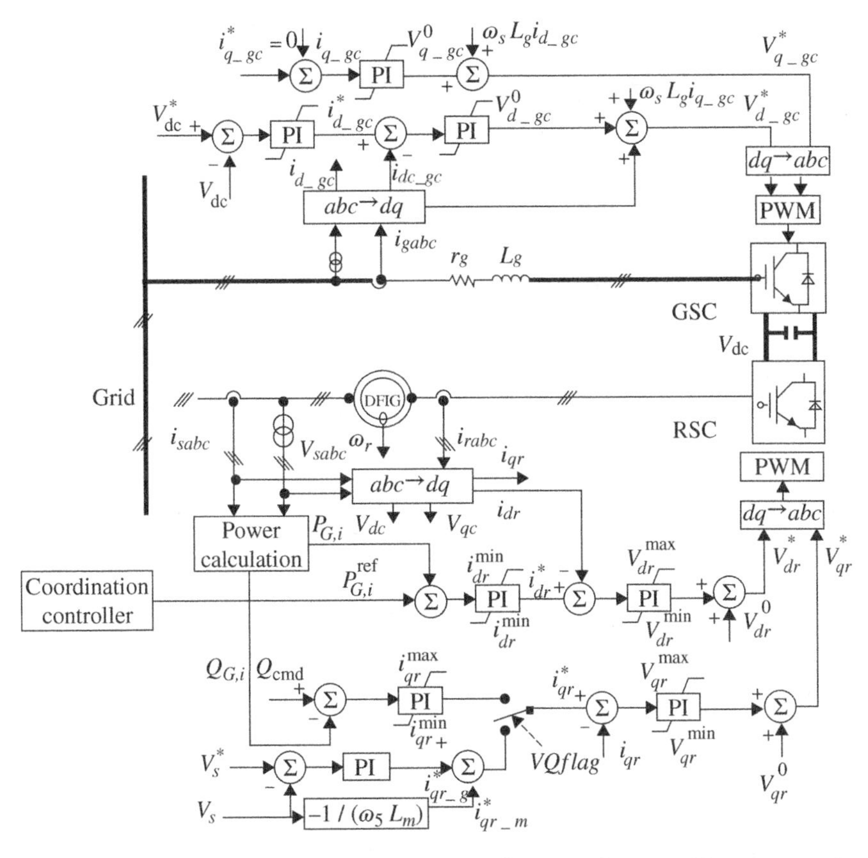

Figure 3.4 Block diagram of the decoupled control strategy for rotor-side converter of a DFIG.

reactive power regulation mode, the deviation between the reactive command Q_{cmd} and the power output Q_w is used to generate reference rotor current i_{qr}^{*} using a PI controller. For voltage regulation mode, i_{qr} is used to reduce the voltage fluctuation, as introduced in [64].

In this chapter, the grid-side converter (GSC) is used to maintain the stability of the DC-link voltage. See more details in [23, 43].

3.1.4.3 Pitch Angle Control Approach

As presented in Figure 3.5, the pitch angle control approach is composed of a pitch angle actuator and a PI controller. Set the threshold angular speed to 1.3 p.u and $\beta_{\min}$ to 0. The maximum changing rate of pitch angle $\frac{d\beta}{dt}$ has certain limits.

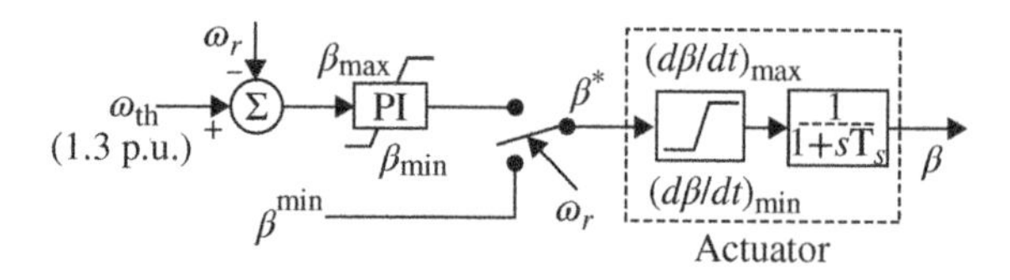

Figure 3.5 Block diagram of pitch angle control.

3.1.4.4 PV Generation Control Approach

The PV power generation model presented in [44] is applied in this chapter. The solar array current and voltage are represented by I and V, respectively. The local bus voltage and current are represented by V_{abc} and I_{abc}, respectively. In this chapter, the unit power factor mode is adopted for PV control. During the application process, the maximum available power generation $P_{G,i}^{max}$ of PV can be estimated off-line according to a look-up table [45]:

$$P_{G,i}^{max} = P_{STC}\frac{G_{ING}}{G_{STC}}[1 + k_{pv}(T_c - T_r)] \tag{3.24}$$

where P_{STC} is the module maximum power generation under standard test conditions. G_{ING} and G_{STC} are the actual incident irradiance and standard incident irradiance, respectively. T_c and T_r are the actual cell temperature and reference temperature, respectively. k_{pv} is the temperature coefficient.

To estimate the maximum available power generation $P_{G,i}^{max}$, every PV generator agent is allocated with radiation and temperature sensors. The reference PV generation value is calculated according to utilization level u_i and $P_{G,i}^{max}$ using Eq. 3.19 After that, a PI controller is implemented to control the DC–AC converter to output desired power generation, as presented in Figure 3.6.

3.1.4.5 Synchronous Generator Control Approach

The synchronous generator in a microgrid has the following functions. When the renewable generation exceeds the load, the synchronous generator does not need

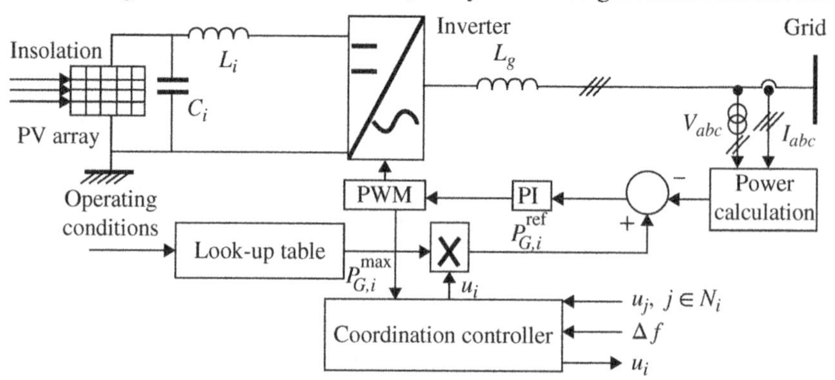

Figure 3.6 Diagram of the control approach for PV system.

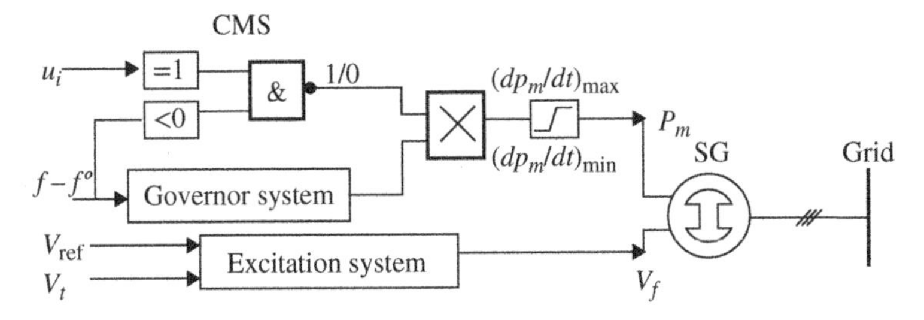

Figure 3.7 Control diagram of synchronous generator.

to generate active power and only need voltage regulation. When the renewable generation is less than the overall system load, the synchronous generator needs to compensate for the active power vacancy apart from the voltage regulation. Notice that the synchronous generator does not participate in utilization-level control. To identify the control mode of a synchronous generator, it can regulate the instantaneous utilization level of a neighboring renewable generator using the control mode selector presented in Figure 3.7.

When the system frequency deviation is less than 0 and the utilization level is equal to 1, the synchronous generator is operated to generate active power to compensate for the power vacancy. Otherwise, the synchronous generator only needs to provide reactive power control to regulate the voltage output to the desired value.

Moreover, the ramp rate of synchronous generator can be modeled by a rate limiter (represented by $\frac{dPm}{dt}$) and the governor system is modeled by a PI controller, as shown in [46]. The excitation system adopts DC exciter as described in [47].

3.1.5 Simulation Analyses

A 6-bus microgrid model is formulated in the MATLAB/SIMULINK platform to test the proposed fully distributed cooperative control approach. The simulation system is composed of three DFIGs, two PVs, one SG, and six loads, as given in Figure 3.8. According to the control mode presented in Synchronous generator control approach, DFIG 1 is controlled in reactive power regulation mode and DFIG 4 and 5 are controlled in voltage regulation mode. PV 2 and 3 are controlled in unit power factor mode. The ramp constraint of the synchronous generator is set to 400 kW/s. Besides, the topology of supporting the communication network for the test system is presented in Figure 3.9.

Taking technical feasibility and system performance into account, the time step size for utilization-level update is set to 0.1 seconds. Then, the proposed control approach is implemented in two cases of different operating conditions. In case 1, load and maximum available renewable generation are set to constant values.

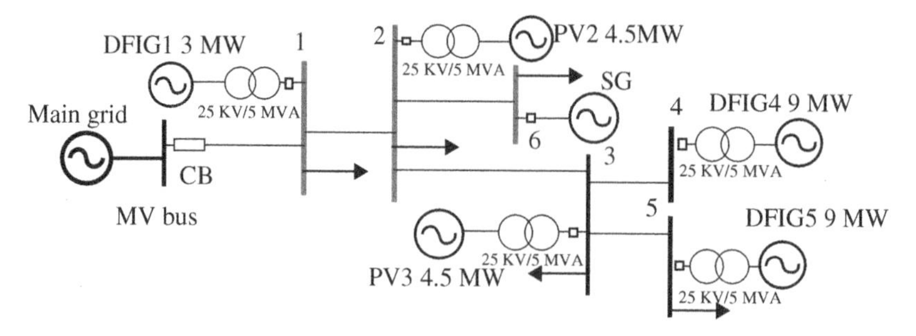

Figure 3.8 A 6-bus microgrid test system.

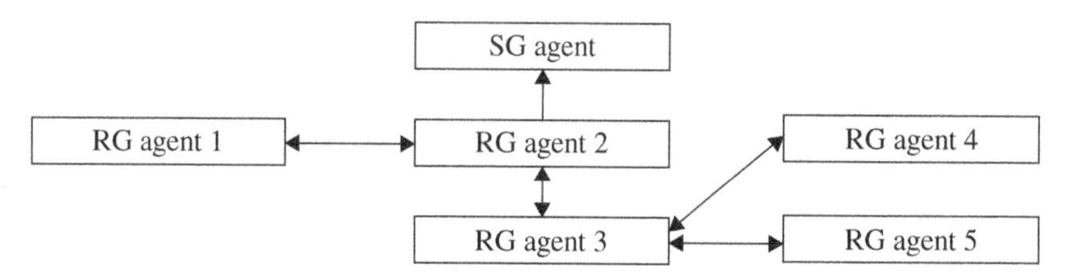

Figure 3.9 Topology of supporting communication network for the 6-bus microgrid.

In case 2, load and maximum available renewable generation are variable, corresponding to realistic conditions.

3.1.5.1 Case 1 – Constant Maximum Available Renewable Generation and Load

In case 1, the load demands are set to be a constant value and the renewable generations are set as follows. The wind speed of DFIGs 1, 4, and 5 are set to 11, 14, and 14 m/s, respectively. The solar radiation of PV 2 and PV 3 are set to 900 and 1000 w/m^2, respectively. The utilization-level updating profile of the five renewable generators is shown in Figure 3.10. The islanding event for microgrid occurs at the 60 seconds. To test the performance of the proposed fully distributed control algorithm before islanding, all generators are operated under MPPT mode. Synchronous generator generates 2 MW active power to tackle disturbances. At the moment of the islanding event, the maximum available renewable energy generation exceeds the system's overall load demand, which causes the system frequency to increase at this instant, as presented in Figure 3.12. The utilization level is reduced using the proposed control approach to maintain the system power supply-demand balance.

The dynamic performances of five renewable generators are shown in Figure 3.11. Under the proposed control approach, after the islanding event, the renewable generation of all the DGs converges to a value that is less than the

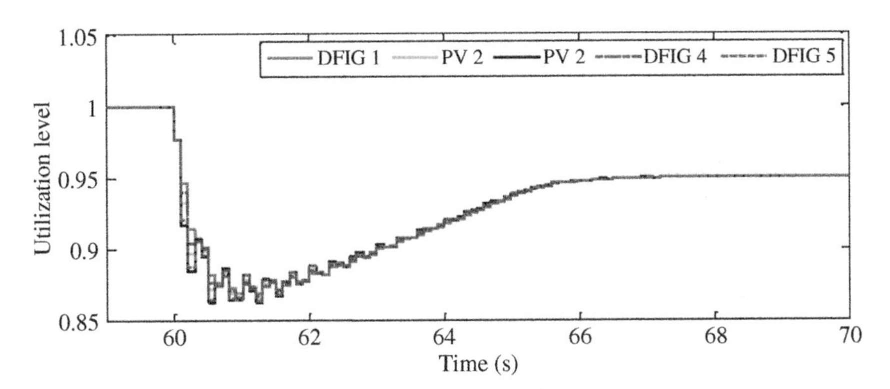

Figure 3.10 Utilization-level updating profile of the five renewable generators (Case 1).

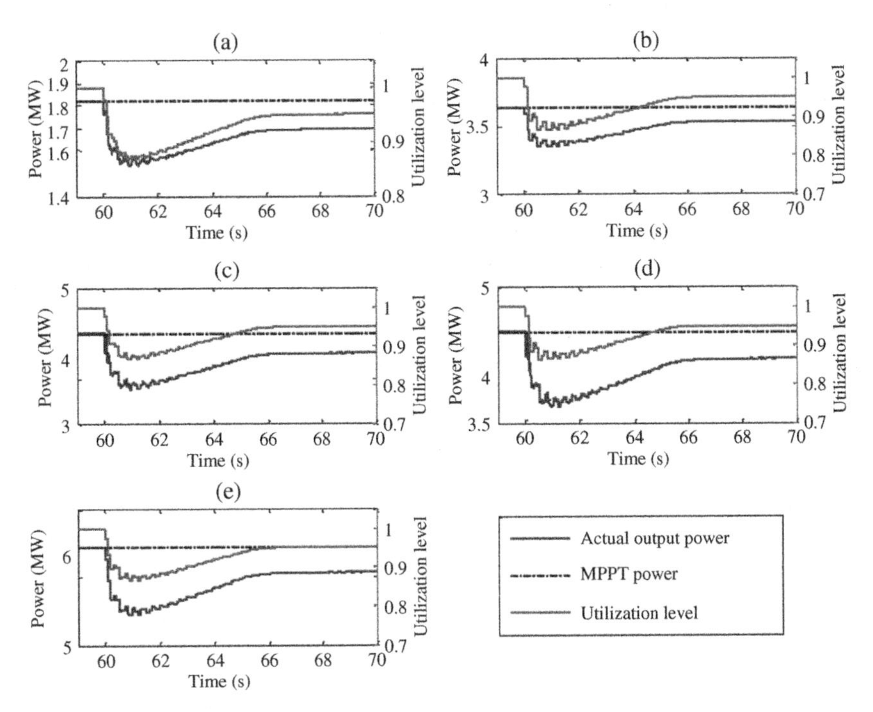

Figure 3.11 Dynamic performance of five renewable generators. (a) DFIG-1 active power output, (b) PV-2 active power output, (c) PV-3 active power output, (d) DFIG-4 active power output, and (e) DFIG-5 active power output.

maximum available power generation. At this time, the utilization levels of all the renewable generators are the same with the ration of actual active power output to the maximum available power generation. In this chapter, to illustrate the superiority of the proposed approach, a traditional droop control approach with automatic generation control (AGC) for secondary control is conducted for comparison. The Droop-AGC approach adopted in this section is applied hierarchically. Because of that, all of the maximum available renewable generations are constant in this case, P_i^0 and $k_{f,i}$ need to be identified off-line. All of the DFIGs and PVs are controlled in a deloaded mode to realize the adopted droop control approach. For a specific PV generator, the deload approach is achieved by operating the converters' current output and PV arrays' voltage output [44]. For a specific DFIG, the deload approach is achieved by operating the pitch angle and rotor speed [17]. For the AGC, the interval is set to five seconds since the time triggered and the interval of AGC in a microgrid is relatively small [48–55].

The system's dynamic performance, including frequency response and terminal voltage response using the proposed control approach and the droop-AGC approach, is presented in Figures 3.12 and 3.13 respectively. Comparing the two approaches, using the proposed control approach, the frequency response is converged within six seconds and the overshoot is less than 0.19 Hz, which offers a better frequency dynamic performance than the traditional droop-AGC approach (the converge time is nearly 30 seconds and overshoot is about 0.32 Hz). Similarly, the terminal voltage dynamic performance under the proposed control approach is better than that under the traditional droop-AGC approach. Therefore, the proposed approach offers a better frequency and voltage dynamic performance than the droop-AGC approach. The proposed control approach regulates the renewable generators automatically according to the system condition rather than P–f and Q–V properties, which can greatly improve the system dynamic performance and

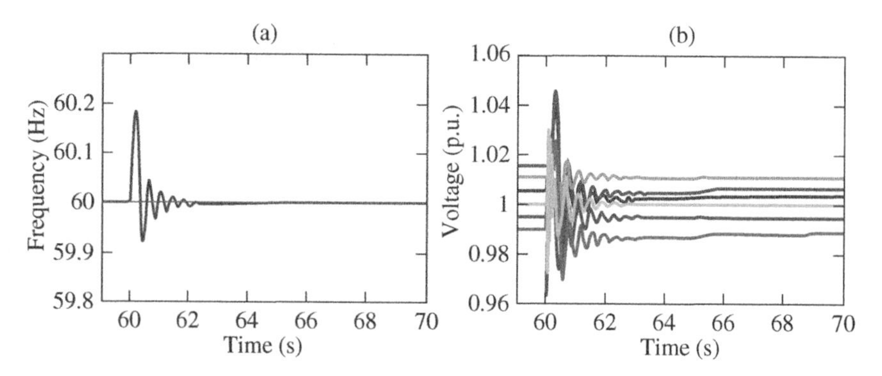

Figure 3.12 System frequency and voltage dynamic performance under the proposed control approach. (a) Frequency response and (b) Terminal voltages of DGs.

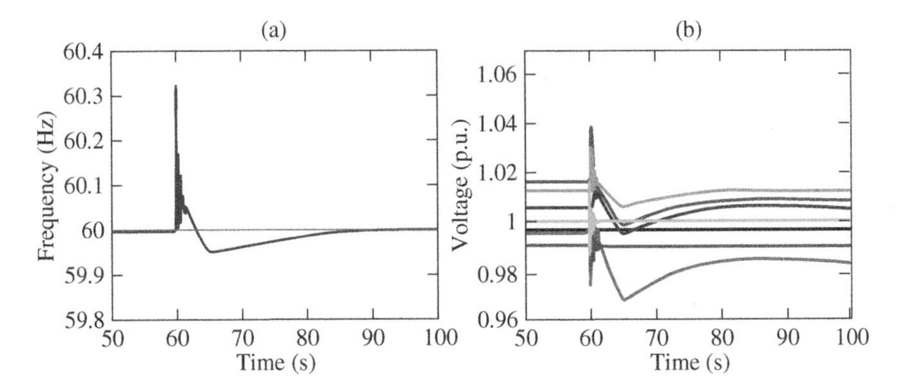

Figure 3.13 System frequency and voltage dynamic performance under the droop-AGC approach. (a) Frequency response and (b) Terminal voltages of DGs.

reduce the steady-state deviations. Note that when power generations of all the renewable generators are assumed constant, renewable generators act like traditional synchronous generators. Thus, the proposed control approach can also be implemented to coordinate multiple synchronous generators.

The simulation study of case 1 is unpractical because maximum generations of renewable generators are assumed to be constant values. At this time, P_i^0 and Q_i^0 or $k_{f,i}$ and $k_{V,i}$ in Eqs. (3.1) and (3.2) are possible to be identified off-line. Nevertheless, renewable energy sources have properties of unreliability, intermittency, and volatility. It is hard to regulate these parameters automatically. Thus, the traditional droop-AGC approach is not used for comparative analysis.

3.1.5.2 Case 2 – Variable Maximum Available Renewable Generation and Load

As presented in Figure 3.14, the wind speed for DFIGs and solar radiation for PVs change continuously. In this case study, the initial active power output of the synchronous generator is set to 4 MW. The islanding event of microgrid happens at 60 seconds. At 150 seconds, 2 MW load is shed, and at 200 seconds, the 2 MW load is restored. According to the previous analysis, we can conclude that the utilization levels of renewable generators are coordinated to the same value under constant wind speed and solar radiation. When the wind speed and solar radiation change continually, the utilization level also changes, as shown in Figure 3.15. At 60 seconds, the microgrid changes to the islanded mode and the utilization-level drop immediately, which causes the active power production of the synchronous generator to decrease gradually from 4 to 0 MW. At 150 seconds, 2 MW load is shed and the utilization-level drops so that the system power supply-demand balance is maintained. In addition, at 200 seconds, 2 MW load is restored and the utilization-level rises. When the maximum available renewable generation is less

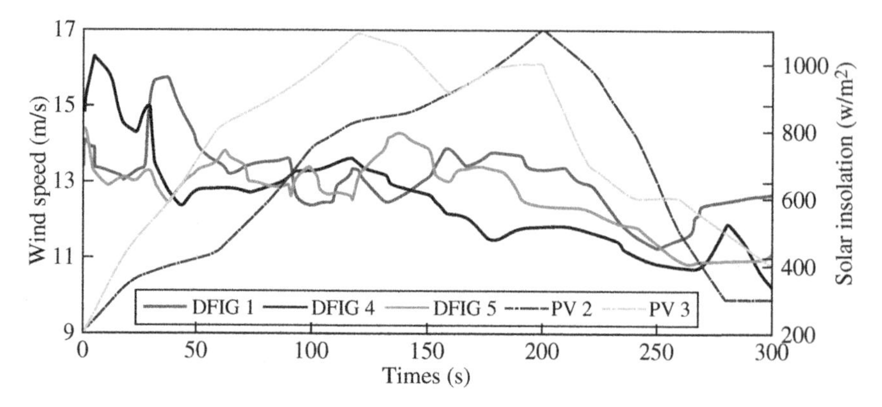

Figure 3.14 Wind speed and solar irradiation profiles for renewable generators.

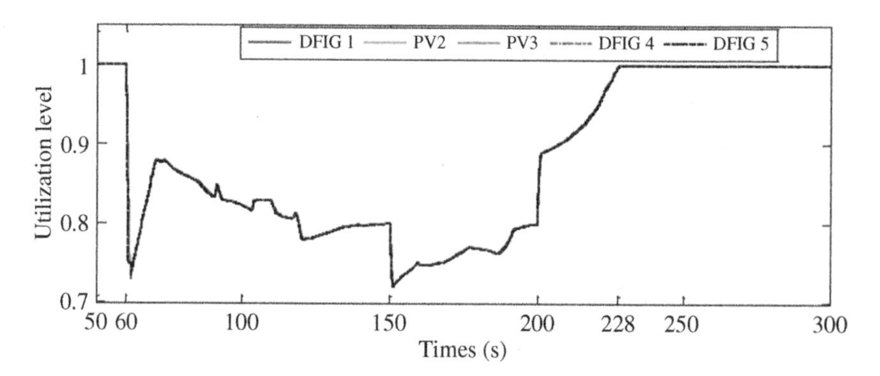

Figure 3.15 Utilization-level variation profiles of five renewable generators (case 2).

than the load demand (228 seconds < t < 300 seconds), the utilization level reaches the upper limit of 1. At this time, all of the renewable generators are controlled in MPPT mode and the synchronous generator is operated to produce active power to compensate active power shortage, as shown in Figures 3.17 and 3.19.

The maximum renewable power generation, utilization level, and system overall load are presented together in Figure 3.16. Notice that the values of serial RLC loads will oscillate because the load value is related to the system frequency and terminal voltage fluctuations. The dynamic response of system frequency and terminal voltage are shown in Figure 3.20.

Taking DFIG 4 as an example, the actual active power output, maximum available renewable generation forecast, and utilization-level variation profile of DFIG 4 are shown in Figure 3.17. When the maximum available renewable generation is larger than the system load demand (60 seconds < t < 228 seconds), DFIG 4 is controlled in deloading mode. When the maximum available renewable

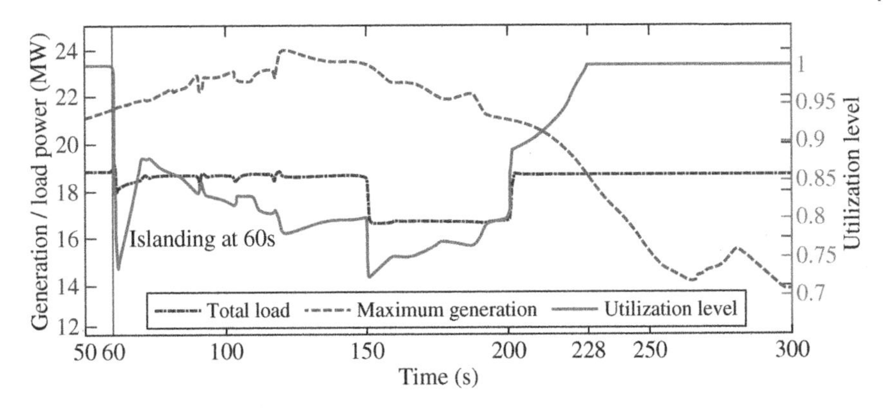

Figure 3.16 Maximum renewable power generation, utilization level, and system overall load demand variation profile.

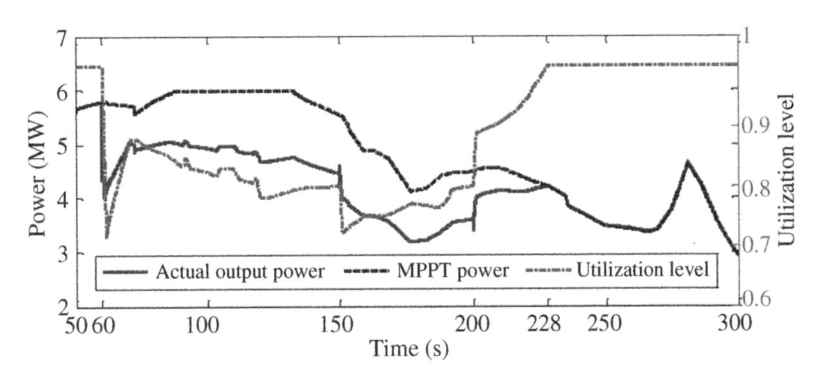

Figure 3.17 Active power tracking of DFIG 4.

generation is less than the system load demand (228 seconds < t <300 seconds), DFIG 4 is controlled in MPPT mode. The pitch angle and rotor speed response profile of DFIG 4 are presented in Figure 3.18. When the rotor speed is less than the threshold value (1.3 p.u.), the pitch angle control remains at the lower bound ($\beta_{\min}$). Otherwise, the pitch angle control is activated and controlled automatically. The active and reactive power output of the synchronous generator is shown in Figure 3.19. When the islanding event occurs (t=60 seconds), the active power generation of synchronous generator is gradually decreased to 0 and the decreasing speed depends on the ramping rate of the generator. The reason is that the available renewable generation is larger than the system's overall load demand at this time. In addition, when the renewable generation is insufficient (228 seconds < t < 300 seconds), the utilization level u_i is set to 1 and

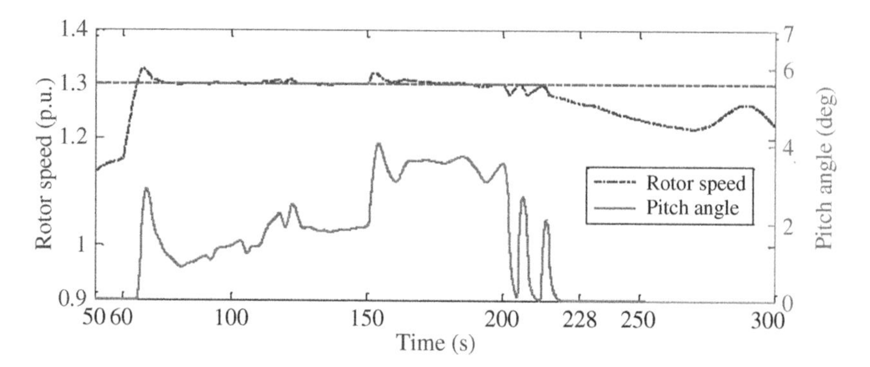

Figure 3.18 Pitch angle and rotor speed variation profile of DFIG 4.

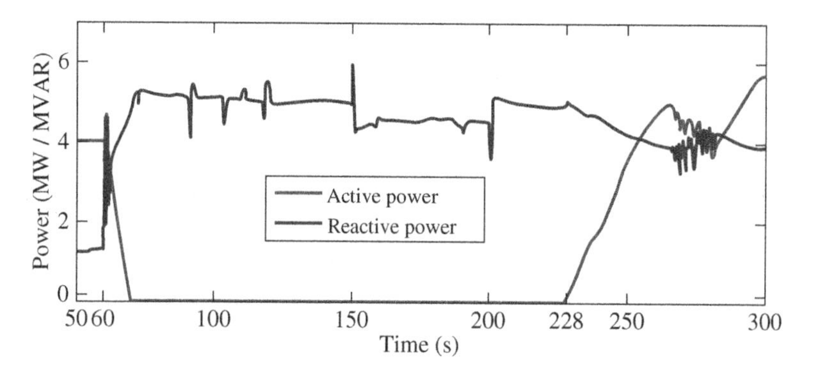

Figure 3.19 Active and reactive power output of the synchronous generator.

the synchronous generator is operated to produce active power to maintain the power supply-demand balance.

The frequency and terminal voltage responses are rather important to the end user. According to the IEEE standard, the frequency deviation should be controlled within ±0.05 Hz and the terminal voltage deviation should be maintained within ±0.1 p.u. [52]. From Figure 3.20, it can be indicated that the maximum frequency deviation is 0.4Hz and the maximum voltage deviation is controlled within ±0.05 p.u., which can meet the IEEE standards. In the simulation analysis, the DFIG 1, PV 2, and PV 3 are controlled in reactive power production mode while DFIG 4 and DFIG 5 are controlled in voltage regulation mode. The former mode is used to generate desired reactive power, so the corresponding terminal voltages have some fluctuations. Thus, the voltage response performances of the former mode are inferior to that of the latter mode.

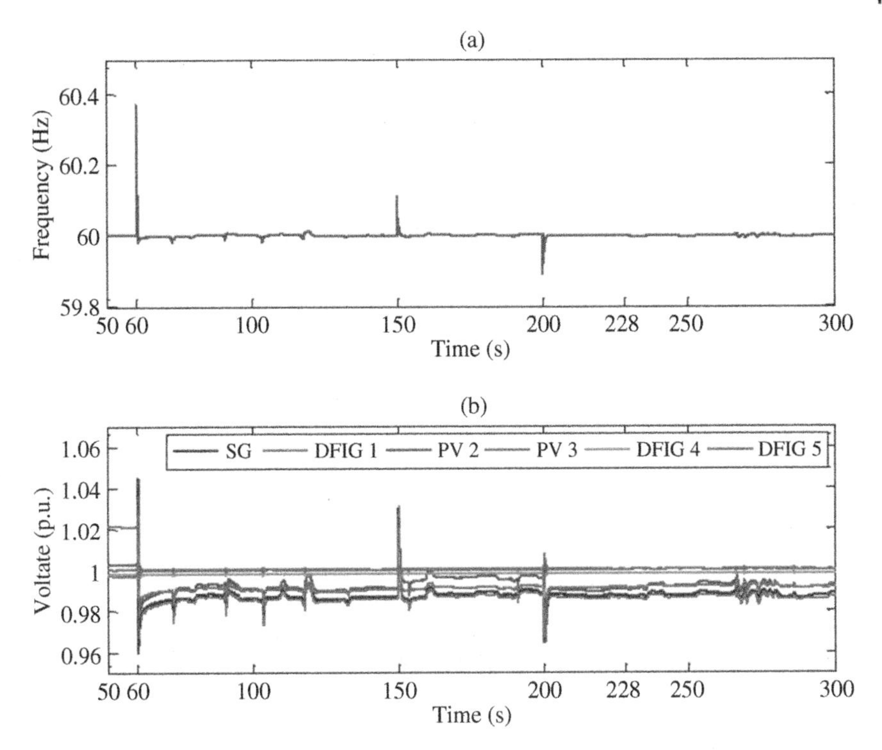

Figure 3.20 System dynamic frequency and terminal voltage responses of case 2. (a) System frequency response and (b) Terminal voltages of DGs.

3.1.6 Conclusion

In this chapter, the coordination control of multiple renewable generators in an autonomous microgrid is investigated. The motivation of our research is listed as follows: Firstly, the traditional MPPT control approach may cause system active power supply-demand imbalance when the generation exceeds the load demand. Secondly, the precise foresting for MPPT control approach is difficult to achieve. Controlling the utilization levels of renewable generation to a common value can effectively mitigate the impact of imprecise renewable generation forecasting. The proposed fully distributed control approach can effectively maintain the supply-demand balance of an islanding microgrid in different situations. In summary, the proposed method has the following four merits:

(1) The proposed MAS-based fully distributed approach is simple and reduces the information exchange (IE) time, which offers a lower communication cost than a traditional centralized approach.

(2) The proposed approach does not need to measure the loading conditions directly.
(3) The system dynamic performance is improved using the proposed control approach compared with the traditional droop-AGC approach.
(4) Using the proposed control approach, multiple DGs (including DFIG, PV, and synchronous generator) are operated coordinately within a microgrid, which can maintain the system power supply-demand balance effectively with satisfactory frequency and terminal voltage responses.

3.2 Distributed Dynamic Programming-Based Approach for Economic Dispatch in Smart Grids

In this section, the discrete ED problem is formulated as a knapsack problem. To allocate the total power demand optimally among different generation units, an effective distributed strategy based on a distributed dynamic programming (DDP) algorithm is proposed. In this algorithm, both the generation limits and ramping rate limits are considered. Based on a MAS framework, the proposed distributed strategy only requires local communication and computation among neighboring agents. As a result, the communication and computation burden is able to be shared with distributed agents. Besides, simpler implementation and faster convergence speed are obtained by implementing the proposed strategy with asynchronous communication. Finally, two cases, including a four-generator system and the IEEE 162-bus system, are simulated for the effective demonstration of this distributed strategy.

3.2.1 Introduction

The power grid is becoming increasingly efficient, flexible, reliable, and adaptive because of the integration of distributed generation (DG), advanced communication technology, and smart sensors and meters [56]. With the upgrading traditional power network, some new challenges in the electric grid emerge. Using the upcoming smart grid framework, some fundamental problems such as ED are able to be solved with better solutions [57]. ED is to find the most economical way to allocate the total power demand among multiple generation units and satisfy the component as well as system-level constraints at the same time, which is, therefore, an optimization problem. There are several traditional optimization methods to solve the ED problems including interior point method [51], gradient search [58], lambda iteration [59], Newton–Raphson method [60], etc. These methods, however, lead to global optimal solutions only when the formulated cost functions are convex and are sensitive to initial conditions. To solve ED problems with nonlinear and non-convex cost functions and deal with more stringent constraints, many intelligent techniques such as evolutionary programming [61], genetic algorithm [62], particle swarm optimization [63, 64], ant colony optimization [65], etc., have been explored.

The optimization methods tackling ED problem, despite the effectiveness and excellent performance, are mostly applied in a centralized way. However, some challenges in the existing smart grid applications cannot be ignored in centralized strategies when addressing the distributed features. Firstly, communication between every single component and the central controller in the system is required in the conventional centralized approaches. The potent central controller is used to manage the mass of data, while components are used to collect information globally [10]. As a result, these solutions are not preferable because of the expensive implementation cost and susceptibility to a single-point failure. Secondly, considering the possible subjection of the electric grid and the related communication network to topology variations in the future, the effectiveness of the centralized approaches will probably degrade.

To better address the features of smart grids and satisfying real-time application requirements, flexible, reliable, high-speed, and cost-effective distributed solutions are becoming increasingly needed [66]. Using distributed strategies, the sparse communication network can be exploited effectively to facilitate the cooperation of components that are located in the system disparately [57]. Recently, there are two main categories of distributed strategies to solve the ED problem. One strategy uses only local information and is called a fully decentralized strategy. This type of control strategy includes the cost-based droop scheme [67] and the frequency deviation based method [68]. It is less expensive because no communication is required when utilizing the available information locally. However, not all resources available in the network are utilized because of the lack of broader available information, which results in ineffectiveness [69]. The other is the distributed strategy, requiring local computation and exchanging information among several neighbor units by a local communication network only [15, 16, 54, 57, 66, 70–73]. Through distributed strategy, the communication and computation burden is able to be shared. Communication and computation are made through multiple distributed controllers, which enable them to work simultaneously. As a result, scalable, cheap, reliable, and flexible as the advantages, well-designed distributed strategies are regarded as promising options for the optimal control and management of smart grids.

Usually, the ED problem uses the incremental cost (IC), a Lagrangian multiplier, to characterize after formulating into a constrained problem. There are, however, a variety of consensus-based approaches used to calculate the optimal IC. The proposed consensus-based distributed strategies emphasize global equality constraints as well as local inequality constraints effectively [66]. They are able to accommodate to plug-and-play operation [66] and robust to the variations of communication topology [70]. However, the formulated objective function needs to be convex and continuous so that consensus-based approaches obtain the global optimality property. Moreover, in recent literature, most distributed approaches assume that information is updated and exchanged synchronously through all agents in the network. Therefore, a global clock is required for synchronization in these approaches, including distributed ED [54, 57, 66, 71, 72], smart microgrid control monitoring [73], supply/demand management [74], consensus-based PEV charging management [75], distributed decision making [76], and the consensus-based load management algorithm [15, 16]. Synchronous communication-based approaches, however, could have a disadvantage, that is, the significantly slowed down convergence speed because of the extra waiting time for the slowest one during each iteration.

To tackle the issues involved in consensus-based strategies, this chapter proposes a novel DDP-based strategy in which synchronous communication is not required. Taking the generation limits as well as ramp rate limits into consideration, this strategy is aimed at solving the discrete ED problem. There are several steps for implementation. Firstly, the discrete ED problem is reconstructed into a traditional knapsack problem, which could further be worked out by the DDP algorithm. Secondly, the proposed DDP-based strategy is used based on a MAS framework. In MAS, every generation unit in the system is entrusted with such an agent that communicates with its neighbor agents only and updates its local generation setting under fixed simple rules. This distributed approach is adaptive, robust, effective, and scalable. These advantages of effectiveness and scalability are verified by the simulation results of a four-generator system and the IEEE 162-bus system, respectively. There are several contributions of this proposed distributed approach:

- The reduced expense for supporting communication network compared with centralized strategies. Because local communication exclusively among neighbor agents are utilized in the MAS-based distributed approach.
- By reconstructing the discrete ED problem into a knapsack problem, this problem can further be worked out by implementing the DDP algorithm.
- Simpler implementation and faster convergence speed because synchronous communication is not required in the DDP algorithm.
- The verified scalability of the DDP algorithm using the modified IEEE 162-bus system.

The rest of this section is organized as follows. Section 3.2.2 presents the graph theory and dynamic programming (DP) briefly. Section 3.2.5 formulates the discrete ED problem as a knapsack problem. Section 3.2.8 gives an introduction to the proposed DDP algorithm. Section 3.2.11 shows simulation results with a four-generator system and the IEEE 162-bus system. Finally, Section 3.2.13.2 draws the conclusion.

3.2.2 Preliminary

This section presents the basic graph theory as well as DP.

3.2.3 Graph Theory

$G = (V, E)$ refers to a graph consisted of a set of vertices $V = 1, 2, \dots, n$ and a set of edges $E \subseteq V \times V$. The unordered and distinct pair $(i, j) \in E$ refers to an undirected edge from i to j. $N_i = \{j \in V | (i, j) \in E\}$ refers to the neighbor set of vertex i. When a sequence of vertices is connected by a sequence of edges, paths are produced. When the edges contain a path from i to j, two vertices are deemed to be connected. If and only if a path between any two vertices exists, the graph is connected. In this chapter, a vertex represents an agent in the power grid, and an edge represents the undirected communication link between two agents.

3.2.4 Dynamic Programming

Bertsekas et al. [77, 78] first introduced the abstract framework for DP. S is used to denote the state space and C to denote control space. x and u refer to states and controls and denote the elements in S and C, respectively. F is assumed to be the set of all extended real-valued functions $J : S \to [-\infty, +\infty]$ on S. The following notation is defined for arbitrary two functions $J_1, J_2 \in F$

$$\begin{cases} J_1 = J_2, & \text{if } J_1(x) = J_2(x), \forall x \in S \\ J_1 \leq J_2, & \text{if } J_1(x) \leq J_2(x), \forall x \in S \end{cases} \tag{3.25}$$

$H : S \times C \times F \to [-\infty, +\infty]$ is assumed to be a mapping that is monotonic and for any $x \in S$ the equation below is guaranteed:

$$H(x, u, J_1) \leq H(x, u, J_2), \forall J_1, J_2 \in F \text{ with } J_1 \leq J_2 \tag{3.26}$$

The goal of DP is obtaining a function $J^* \in F$ such that

$$J^*(x) = \inf_{x \in S} H(x, J^*), \quad \forall x \in S \tag{3.27}$$

The mapping $T : F \to F$ defined by Eq. (3.28) is considered.

$$T(J)(x) = \inf_{x \in S} H(x, J) T(J)(x) = \inf_{x \in S} H(x, J) \tag{3.28}$$

The DP problem is the same as finding the fixed point of T in F, such that

$$J^* = T(J^*) \tag{3.29}$$

3.2.5 Problem Formulation

In this section, the discrete form of the ED problem is reconstructed into a knapsack problem. In addition, the generation limits, as well as the ramping rate limits in the ED problem, are analyzed.

3.2.6 Economic Dispatch Problem

Considering the objective function, which is minimizing the total cost of power generation and constraints, including several equality and inequality constraints, the ED problem is formulated as follows.

$$\min \sum_{i=1}^{N} C_i(p_i[t]) \tag{3.30}$$

$$\text{s.t.} \quad p_i^{\min} \leq p_i[t] \leq p_i^{\max} \tag{3.31}$$

$$-p_i^{\text{ramp}} \leq p_i[t] - p_i[t-1] \leq p_i^{\text{ramp}} \tag{3.32}$$

$$\sum_{i=1}^{N} p_i[t] = P_L[t] \tag{3.33}$$

In the equations, t represents the time slot, which is non-negative integers. $p_i[t]$ and $P_L[t]$ represent the power produced by generator i and total load demand during time slot t, respectively. $p_i^{\min}/p_i^{\max}$, and p_i^{ramp} are used to represent the minimum/maximum power produced by generator i and ramp rate limits of generator i, respectively. The following quadratic equation is often used to approximate the generation cost [54, 66, 71, 72, 79]:

$$C_i(p_i[t]) = a_i p_i^2[t] + b_i p_i[t] + c_i \tag{3.34}$$

where a_i, b_i, and c_i are the generation cost coefficients of generator i. The IC(r_i) of generator i is defined as follows. It is the partial derivative of C_i with respect to p_i.

$$r_i[t] = 2a_i p_i[t] + b_i \tag{3.35}$$

The equal IC incremental cost criterion mentioned in [80] is a famous solution to Eqs. (3.30)–(3.34). It is shown below.

$$\begin{cases} 2a_i p_i[t] + b_i = r^*[t], & p_i^{\min} < p_i[t] < p_i^{\max} \\ 2a_i p_i[t] + b_i < r^*[t], & p_i[t] = p_i^{\max} \\ 2a_i p_i[t] + b_i > r^*[t], & p_i[t] = p_i^{\min} \end{cases} \tag{3.36}$$

where $r^*[t]$ stands for the optimal equal IC.

3.2.7 Discrete Economic Dispatch Problem

At time slot $t + 1$, there are adjustments for the power generated by all the generators in order to satisfy total load demand variation. The adjustment amount of power generated by generator i is assumed to be multiples of its minimum discrete incremental generation δp_i either for increase or decrease.

$$p_i[t+1] = p_i[t] + k_i[t]\Delta p_i \tag{3.37}$$

$$-K_i^{\max} \leq k_i[t] \leq K_i^{\max} \tag{3.38}$$

$$K_i^{\max} = p_i^{ramp} \Big/ \Delta p_i \tag{3.39}$$

$$\sum_{i=1}^{N} k_i[t]\Delta p_i = P_L[t+1] - P_L[t] \tag{3.40}$$

For ensuring tiny enough discrete minimum incremental generation so that the adjustment is achievable, this chapter assumes fast realization of generation adjustment and a slight difference in the total load demand between time slot t and time slot $t + 1$. After adjustment, the cost of the generator i for generating power $p_i[t] + k_i[t]\Delta p_i$ is given by

$$C_i(k_i[t]\Delta p_i, p_i[t]) = a_i(p_i[t] + k_i[t]\Delta p_i)^2 + b_i(p_i[t] + k_i[t]\Delta p_i) + c_i \tag{3.41}$$

When selling the generation at price R (constant value), the benefit function of the generator i is calculated as

$$J_i(k_i[t]\Delta p_i, p_i[t]) = R(p_i[t] + k_i[t]\Delta p_i) - C_i(k_i[t]\Delta p_i, p_i[t]) \tag{3.42}$$

where R is the generation price per unit.
The objective, maximizing the total benefit of every generator combined, is given by

$$\max \sum_{i=1}^{N} R(p_i[t] + k_i[t]\Delta p_i) - C_i(k_i[t]\Delta p_i, p_i[t]) \tag{3.43}$$

s.t. (3.31) - (3.33).

After finding the optimal combination of all generators' possible generation adjustments, problem (3.43) can be reconstructed into a knapsack problem. According to the introduction beforehand, every generator is assigned with one agent that calculates the corresponding generator's benefits of all available adjustment choice. The calculation equations are as follows.

$$\begin{aligned} \Delta J_i(k_i[t]\Delta p_i, p_i[t]) &= J_i(k_i[t]\Delta p_i, p_i[t]) - J_i((k_i[t]-1)\Delta p_i, p_i[t]) \\ &= (R - 2a_i p_i[t] - b_i)\Delta p_i - a_i(2k_i[t]-1)\Delta p_i^2 \\ \text{for } & -K_i^{\max} \leq k_i[t] \leq K_i^{\max} \end{aligned} \tag{3.44}$$

Every agent keeps an inconsistently updated benefit table of available choice by calculating Eq. (3.44). The objective can then be obtained through utilizing the DDP algorithm that is presented in Section 3.2.8.

3.2.8 Proposed Distributed Dynamic Programming Algorithm

The discrete ED problem has been reconstructed into a knapsack problem. Next, the DDP algorithm can be used to solve it. This section starts from the DDP algorithm and then introduces its implementation. In DDP algorithm, only local information and IE with neighbor agents are utilized by every agent.

3.2.9 Distributed Dynamic Programming Algorithm

The DDP algorithm contains calculations working in parallel with high possibility, which enables distributed computation implementation [81]. Divide the state space S into N sets $(S_1, \dots, S_N)$ that are disjoint and corresponding to N independent agents in the state spaces. Connecting by one edge, two agents that are also able to exchange information with each other are called neighbors.

In the DDP algorithm, every neighbor $j \in N_i$ of agent i has two buffers: J_{ij} and x_{ij} to keep the latest estimation of the optimal solutions J^* as well as the related states from agent j. Besides, there are another two buffers: J_{ii} and x_{ii} assigned to agent i. The agent i's estimation of the solutions J^* as well as the related states is stored in these two buffers.

During every iteration, the latest estimation of the optimal solutions, as well as related states, is firstly acquired through the communication between agent i and the neighbor agents. Then, the estimate of the optimal solution, as well as the associated state, is updated by agent i. The rules for updating are given below.

Step 1: Exchange information with neighbor agents

$$\begin{cases} J_{ij}(k+1) = J_{jj}(k), \\ x_{ij}(k+1) = x_{jj}(k), \end{cases} j \in N_i \tag{3.45}$$

Step 2: Process local information

$$\begin{cases} J_{ii}(k+1) = \{\inf_{x_i \in S_i} H(J_{ii}(k), J_{ij}(k+1), x_i)\} \\ x_{ii}(k+1) = \arg\{\inf_{x_i \in S_i} H(J_{ii}(k), J_{ij}(k+1), x_i)\} \end{cases} \tag{3.46}$$

In the equations above, the estimations of optimal solution estimates are represented by J_{ii} and J_{ij}. The related states that are obtained from agent i and the neighbor agent j are represented by x_{ii} and x_{ij}.

As analyzed in [81], Eq. (3.47) is used to express the values of J^* and x^* after convergence.

$$\begin{cases} \lim_{k\to\infty} J_{ij}(k) = \lim_{k\to\infty} J_{ii}(k) = J^* \\ \lim_{k\to\infty} x_{ij}(k) = \lim_{k\to\infty} x_{ii}(k) = x^* \end{cases} \tag{3.47}$$

According to the discussion in [81], DDP algorithm has to satisfy the following three conditions for convergence:

1) The number of communication time between agent i and the neighbor agents and the local information updating time are bigger than or equal to 1 during n continuous iterations, where n is a positive finite integer.
2) High-level connectivity of the communication network graph.

3) Two functions $\underline{J}$ and $\bar{J}$ exist. Equations (3.48) and Eq. (3.49) are assured $\forall J \in F$ $(\underline{J} \leq J \leq \bar{J})$.

$$\bar{J} \geq T(\bar{J}),\ T(\underline{J}) \geq \underline{J} \tag{3.48}$$

$$\begin{cases} \lim_{k\to\infty} T^k(\bar{J})(x) = J^*(x) \\ \lim_{k\to\infty} T^k(\underline{J})(x) = J^*(x) \end{cases} \tag{3.49}$$

Condition 1 implies that synchronous iterations are not required for all agents in the DDP algorithm. Because it is not necessarily waiting for the slowest agent for the response, implementation becomes simpler and converging speed becomes higher. By contrast, distributed algorithms requiring synchronous iterations converge slower. Condition 2 implies the existence of at least one path for any two agents, indicating that no agent is totally isolated. Condition 3 assures that there exists an optimal solution.

3.2.10 Algorithm Implementation

The control algorithm is carried out through a MAS framework. The MAS framework is developed in the Java Agent Development (JADE) platform, which consists of agent containers, distributing over communication networks. Accommodating agents, containers are the Java processes. They provide the JADE with a running time as well as services required in order to execute and host agents. The agent identifier (AID) is given to every agent for distinguishing purposes because AID is unique. Progressed by the Foundation for Intelligent Physical Agents (FIPAs), asynchronous communication protocols are the ways for bidirectional communications among agents after configuration. Based on the conditions of the applications, the virtual communication network between agents could be adjusted in an easy way.

The consensus-based algorithm mentioned in [16] is used to calculate the overall load demand during every time slot. The communication network in this algorithm enjoys equivalent to that in the DDP algorithm. To simplify the calculations, all the study cases assume it to be known in this chapter. There are two main functions for each agent: local information update (LIU) and IE with neighbor agents.

LIU calculates the benefit table and updates the state. The benefit table includes available adjustment choice, which is gained using Eq. (3.44). In addition, the update uses the benefit table and information gained from neighbor agents. Every agent is entrusted to a $K_i^{\max}$-dimensional state vector x_{ii}. The definition of the vector is given below.

$$x_{ii}(k_i) = \begin{cases} 1, & \text{if the } k_i\text{th incremental } \Delta p_i \text{ is chosen} \\ 0, & \text{otherwise} \end{cases} \tag{3.50}$$

Table 3.1 Distributed dynamic programming algorithm.

```
For time slot t
  Calculate the total load demand according to the algorithm introduced in [16].
  Update the benefit table for all available adjustment options based on (3.44).
    k = 1;
    while k < K (% K is the maximum iteration number)
  Each agent communicates with neighbor agents based on (3.45) and updates local
information based on (3.46).
    k = k + 1;
    end
Next control cycle, continue to time slot t + 1.
```

Using the state vector x_{ii} and Eq. 3.42 J_{ii} can be calculated. IE is in charge of IE with neighbor agents. Only if the designed communication graph has high-level connectivity, the topology of the communication network for supporting purpose and the power network could be either independent or identical. Table 3.1 summarizes the working process of the DDP algorithm.

3.2.11 Simulation Studies

In this section, a four-generator system is simulated to verify the effectiveness of the DDP algorithm under both synchronous and asynchronous communication circumstances. In addition, the modified IEEE 162-bus system is simulated to examine the scalability.

3.2.12 Four-generator System: Synchronous Iteration

Table 3.2 lists the coefficients of generation expenses and generation limits. Figure 3.21 shows the supporting communication topology.

Set the constant price R to be 10. 460 MW is the starting total load demand assumption. 110, 175, 90, and 85 MW are the starting generation assumptions. Initialization is nonoptimal because the associated IC of the four generators are according to Eq. (3.2)., i.e. 7.20, 7.25, 7.00, and 7.40, respectively.

3.2.12.1 Minimum Generation Adjustment $\Delta p_i = 2.5$ MW

In the first time slot, the increment of total load demand is 30 MW. The benefit table size for agent i is obtained according to Eq. (3.44). The benefit table of agent i, containing all generation adjustments available, is obtained using Eq. (3.44).

Table 3.2 Generators' parameters.

Unit	a_i	b_i	c_i	p_i^{ramp} (MW)	$[p_i^{min}, p_i^{min}]$ (MW)
G1	0.010	5.0	100	15.0	[80, 200]
G2	0.015	2.0	150	12.5	[120, 250]
G3	0.025	2.5	90	10.0	[50, 150]
G4	0.020	4.0	75	10.0	[40, 140]

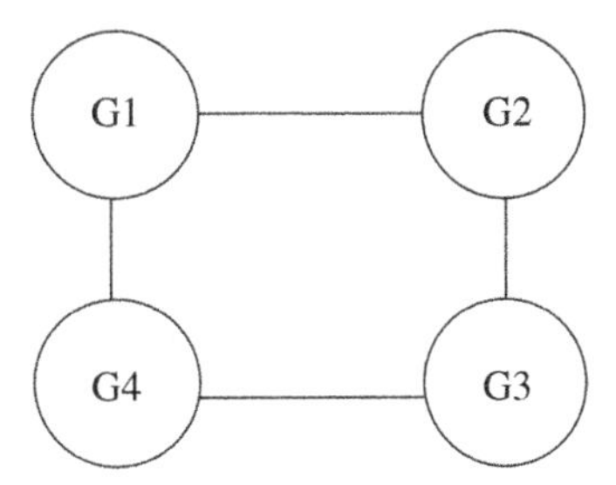

Figure 3.21 Four-generator system.

Table 3.3 Initialization (based on Eq. (3.44)).

	G1	G2	G3	G4
$\Delta J_i(\Delta p_i, p_i[0])$	6.9375	6.7813	7.3438	6.3750
$\Delta J_i(2\Delta p_i, p_i[0])$	6.8125	6.5938	7.0313	6.1250
$\Delta J_i(3\Delta p_i, p_i[0])$	6.6875	6.4063	6.7188	5.8750
$\Delta J_i(4\Delta p_i, p_i[0])$	6.5625	6.2188	6.4063	5.6250
$\Delta J_i(5\Delta p_i, p_i[0])$	6.4375	6.0313	—	—
$\Delta J_i(6\Delta p_i, p_i[0])$	6.3125	—	—	—

Table 3.3 shows it. Here, set the maximum number of iteration to be 30. Table parameters are (R = 10, $[p_1[0], p_2[0], p_3[0], p_4[0]] = [110, 175, 90, 85]$, and $\delta p_i =$ 2.5 MW)

As presented in Figures 3.22–3.24, the results converge in less than 15 iterations using the DDP algorithm. Figure 3.23 shows that during every time slot after convergence, the total generation and the load demand are met with each other. Figure 4 shows the total generation benefit update and proves that the DDP algorithm is monotonic.

IC, described in Eq. (3.2.6), is utilized to examine whether or not the DDP algorithm is optimal. Figure 3.25 shows the convergence of the ICs of every generation

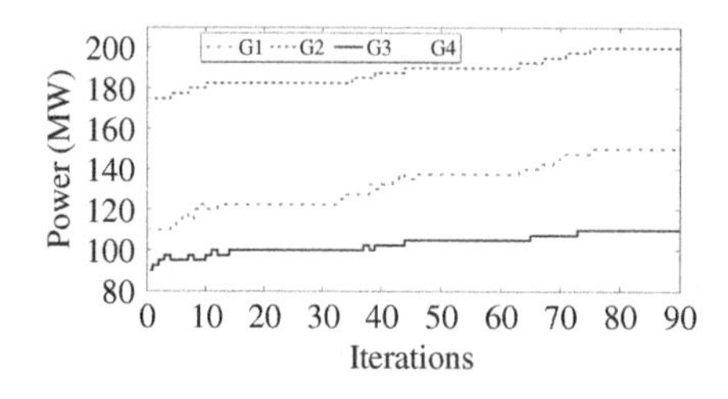

Figure 3.22 Generation update when $\Delta p_i = 2.5$ MW.

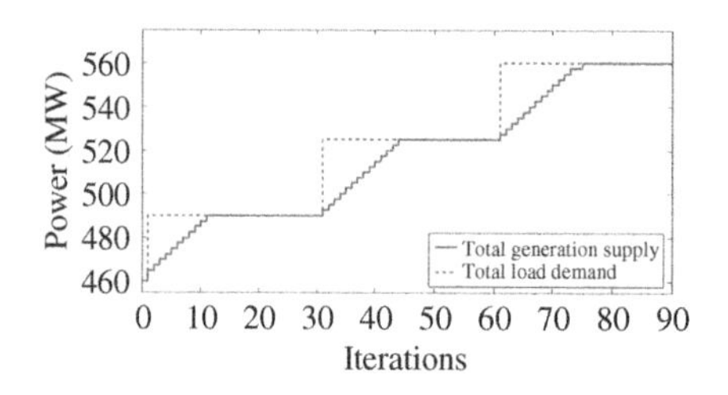

Figure 3.23 Total supply-demand update when $\Delta p_i = 2.5$ MW.

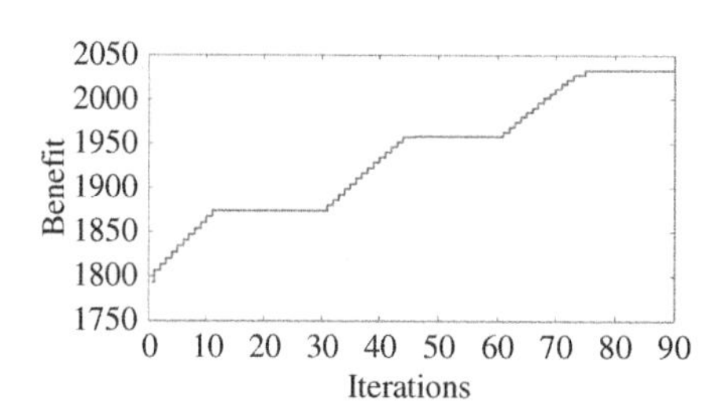

Figure 3.24 Benefit update when $\Delta p_i = 2.5$ MW.

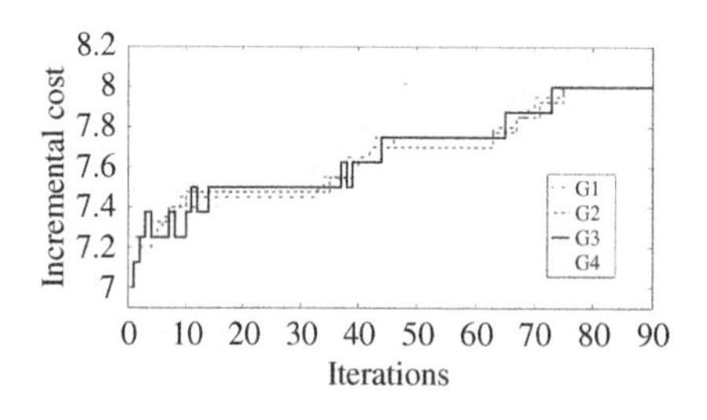

Figure 3.25 IC update when $\Delta p_i = 2.5$ MW.

combined at time slot 3, indicating that the optimal solution provided by the proposed DDP algorithm is the same as that by the centralized equal incremental cost (CEIC) algorithm. The result converges to the common optimal value. However, if the solutions for formulated discrete ED problems are different from those for continuous ED problems, this will not happen. In the first and the second time slot, the IC of the generation does not converge to the optimal value because of the discretization error. By decreasing the minimum generator adjustment Δp_i, the error can be reduced. The next test examines this generator adjustment.

3.2.12.2 Minimum Generation Adjustment $\Delta p_i = 1.25$ MW

This case sets the minimum generation adjustment to be 1.25 MW. Table 3.4 shows the doubled size of the benefits table because of the unchanged ramp rates of every generator combined. For the increased available adjustment options, more iterations are needed for the convergence of the DDP algorithm. This case assumes the maximum iteration to be 40. Table parameters: ($R = 10$, $[p_1[0], p_2[0], p_3[0], p_4[0]] = [110, 175, 90, 85]$, $\delta p_i = 1.25$ MW). Figure 3.26 shows that the DDP algorithm converges in less than 30 iterations. The fast convergence shows that the significant efficiency of DDP algorithm, when considering the large search space ($13 \times 11 \times 9 \times 9 = 11538$). Figure 3.27 shows that the total power generation satisfies the total load demand for every time slot. Also, that the DDP algorithm is monotonic is proved in Figure 3.28. Figure 3.29 shows that the convergence value, which is optimal, is identical to that in the previous case at the third time slot. In

Table 3.4 Initialization (based on Eq. (3.44)).

	G1	G2	G3	G4
$\Delta J_i(\Delta p_i, p_i[0])$	3.4844	3.4141	3.7109	3.2188
$\Delta J_i(2\Delta p_i, p_i[0])$	3.4531	3.3672	3.6328	3.1563
$\Delta J_i(3\Delta p_i, p_i[0])$	3.4219	3.3203	3.5547	3.0938
$\Delta J_i(4\Delta p_i, p_i[0])$	3.3906	3.2734	3.4766	3.0313
$\Delta J_i(5\Delta p_i, p_i[0])$	3.3594	3.2266	3.3984	2.9688
$\Delta J_i(6\Delta p_i, p_i[0])$	3.3281	3.1797	3.3203	2.9063
$\Delta J_i(7\Delta p_i, p_i[0])$	3.2969	3.1328	3.2422	2.8438
$\Delta J_i(8\Delta p_i, p_i[0])$	3.2656	3.0859	3.1641	2.7813
$\Delta J_i(9\Delta p_i, p_i[0])$	3.2344	3.0391	—	—
$\Delta J_i(10\Delta p_i, p_i[0])$	3.2031	2.9922	—	—
$\Delta J_i(11\Delta p_i, p_i[0])$	3.1719	—	—	—
$\Delta J_i(12\Delta p_i, p_i[0])$	3.1406	—	—	—

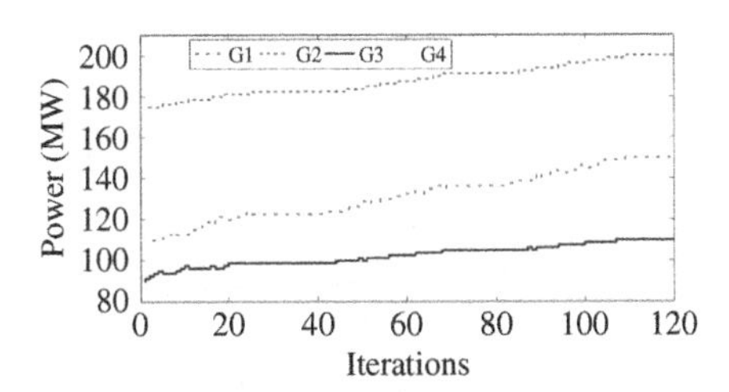

Figure 3.26 Generation update with $\Delta p_i = 1.25$ MW.

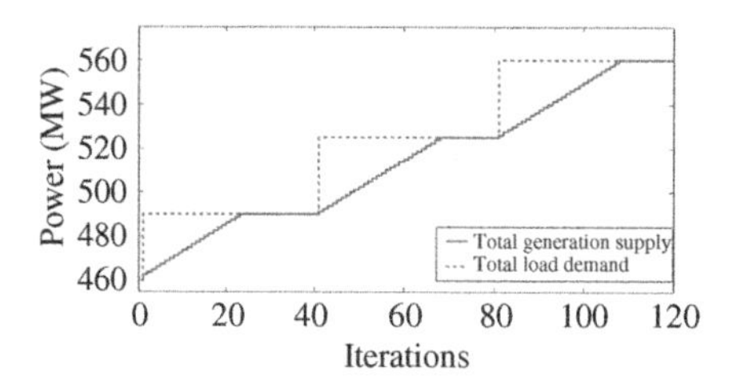

Figure 3.27 Total supply-demand update with $\Delta p_i = 1.25$ MW.

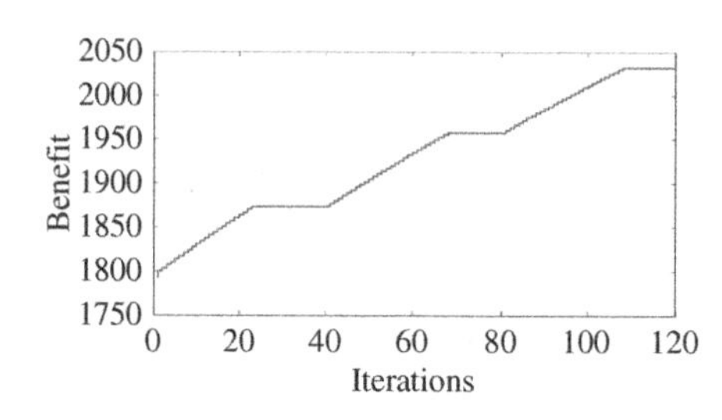

Figure 3.28 Benefit update with $\Delta p_i = 1.25$ MW.

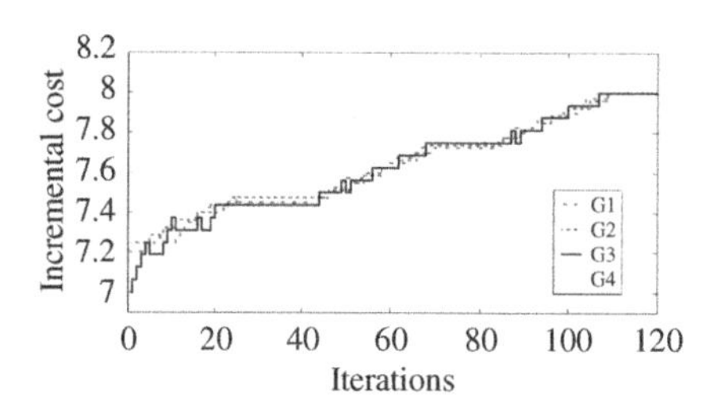

Figure 3.29 IC update with $\Delta p_i = 1.25$ MW.

addition, in the first and second time slot, there are slighter differences in the ICs between this case and the previous one. It indicates that when using CEIC algorithm, the difference with the optimal solution becomes smaller. For expedience, Tables 3.5 and 3.6 compare the DDP algorithm and the CEIC algorithm in first and second time slot, respectively. The results include the DDP algorithm when minimum generation adjustments are 2.5 MW and 1.25 MW and the CEIC algorithm for continuous ED problem.

Tables 3.5 and 3.6 show that competent results can be obtained by implementing the DDP algorithm. Results are in close proximity of the results from the CEIC algorithm for continuous ED problem. By decreasing the minimum generation adjustment, the small deviation from the optimal solution can be reduced at the expense of becoming more time-consuming.

Table 3.5 Comparisons of the proposal DDP algorithm with different Δp_i and the CEIC algorithm at time slot 1.

	$\Delta p_i = 2.5$ MW		$\Delta p_i = 1.25$ MW		Continuous ED	
	p_i (MW)	r_i	p_i (MW)	r_i	p_i (MW)	r_i
G1	122.50	7.450	122.50	7.450	122.73	7.455
G2	182.50	7.475	182.50	7.475	181.82	7.455
G3	100.00	7.500	98.75	7. 438	99.09	7.455
G4	85.00	7.400	86.25	7.450	86.36	7.455
Benefit	1873.344		1873.398		1873.409	

Table 3.6 Comparisons of the proposal DDP algorithm with different Δp_i and the CEIC algorithm at time slot 2.

	$\Delta p_i = 2.5$ MW		$\Delta p_i = 1.25$ MW		Continuous ED	
	p_i (MW)	r_i	p_i (MW)	r_i	p_i (MW)	r_i
G1	137.50	7.750	136.25	7.725	136.36	7.727
G2	190.00	7.700	191.25	7.475	190.91	7.727
G3	105.00	7.700	105.00	7.750	104.55	7.727
G4	92.50	7.700	92.50	7.700	93.18	7.727
Benefit	1957.688		1957.711		1957.727	

3.2.13 Four-Generator System: Asynchronous Iteration

No requirement on synchronous iterations is one great benefit of the DDP algorithm. Since there can be some iterations interspace for the estimation of optimal solution and the update of associated state of several agents, requiring no synchronous communication protocol makes it easier to implement. Based on two different asynchronous communication circumstances, the performances of the DDP algorithm is examined in this case.

3.2.13.1 Missing Communication with Probability

The simulation is performed based on the assumption that the probability of synchronous IE between agent 3 and others is 0.4 at every iteration. Figure 3.30 shows that the frequency of LIU by the agent 3 update is reduced when compared either to other agents or to its own performance in Figure 3.25. The DDP algorithm can still obtain the same optimal solution as in case Four-generator system: synchronous iteration under asynchronous communication. This is verified by the

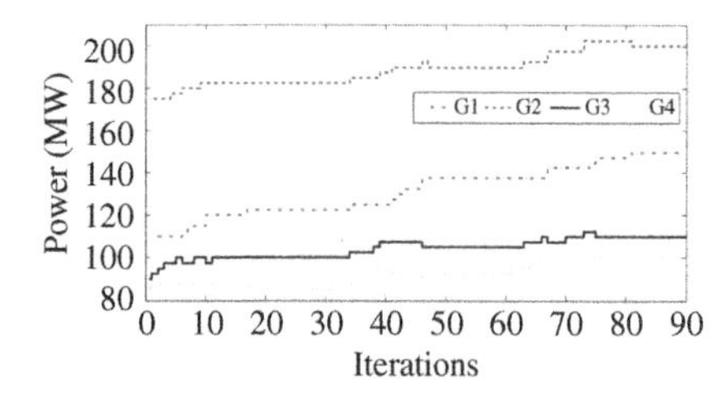

Figure 3.30 Generation update under intermittent communication.

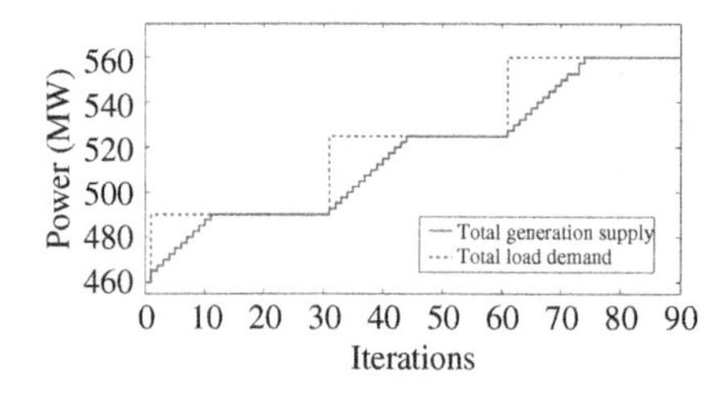

Figure 3.31 Total supply-demand update under intermittent communication.

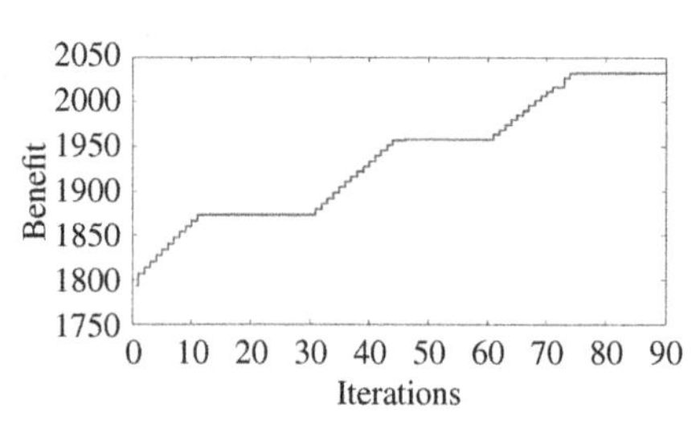

Figure 3.32 Benefit update under intermittent communication.

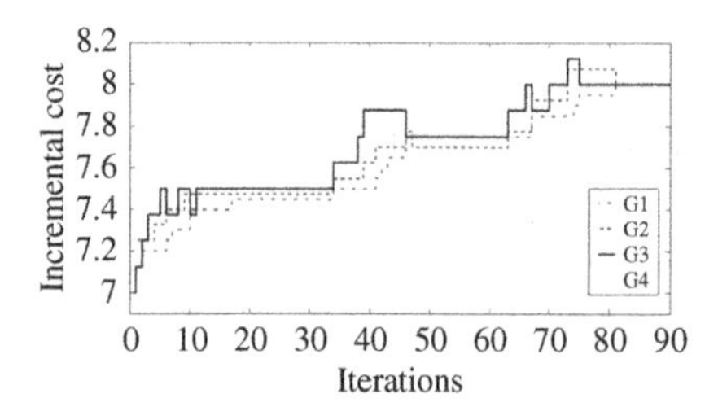

Figure 3.33 IC update under intermittent communication.

comparison of Figures 3.30–3.33 and Figures 3.22–3.25, respectively. The DDP algorithm, therefore, is significantly advantageous through asynchronous implementation for the response, requiring no waiting for the slowest agent.

3.2.13.2 Gossip Communication

The assumption in this case is that at every iteration, the number of neighbor agents that every agent can communicate with is less than 2. This is referred to as the gossip communication protocol. The global clock synchronization constraint is relieved in this protocol. Besides, it increases the robustness against packet loss.

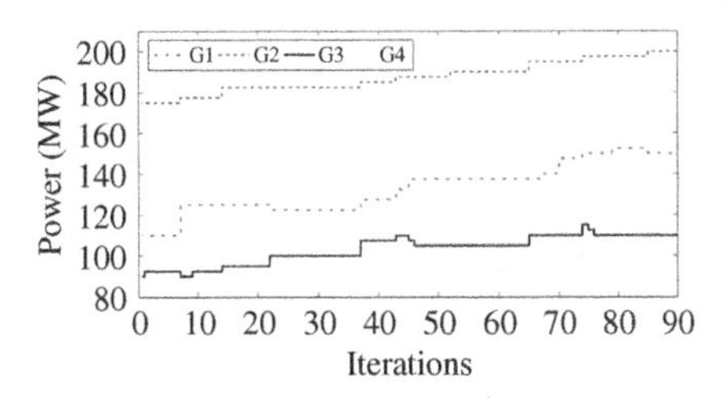

Figure 3.34 Generation update under gossip communication.

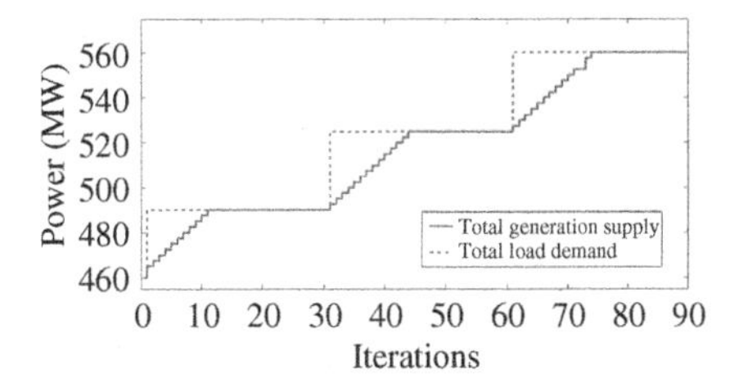

Figure 3.35 IC update under gossip communication.

This case assumes that every set of agents is one of the four combinations: 1 and 2, 2 and 3, 3 and 4, 4 and 1. In addition, they have the same probability of 0.25 to communicate for one iteration. As presented in Figures 3.34 and 3.35, under gossip communication protocol, the DDP algorithm converges to the same optimal solution as case 3.2.12.1. Because only two agents communicate with each other in every iteration, the average time consumed by one iteration can be reduced significantly. Therefore, the total time needed for convergence is shortened, although with more iterations to converge compared with case 3.2.12.1.

3.2.14 IEEE 162-Bus System

In this case, the modified IEEE 162-bus system, which has 17 generators, is studied. The DDP algorithm is implemented in this case to examine its scalability [82, 83]. The communication network and the physical bus connections can be designed to be independent. It is assumed that the communication order between an agent and the four adjacent neighbors follows the index number. The designed communication graph has a high-level connectivity (34 edges in total). Also, there are possibly 136 maximum edges in this system. After knowing the connecting edges and maximum edges, the density of the graph [72] can be calculated: $34/136 = 0.25$.

Originally, the total load demand in system is 16 046 MW. Next, there is an increase of 320.9, 237.8, and 320.2 MW, respectively, at 0, 150th, and 300th iteration in the load demand of the system. The minimum incremental adjustments are 2–5 MW. The limits for ramp rate are restricted with the range of 20–50 MW, Every agent has a benefit table with size 10. Figures 3.36-3.39 show that the convergence

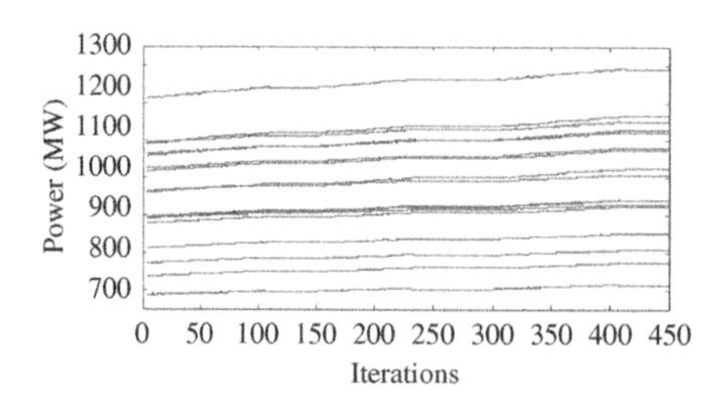

Figure 3.36 Generation update with 17-generator system.

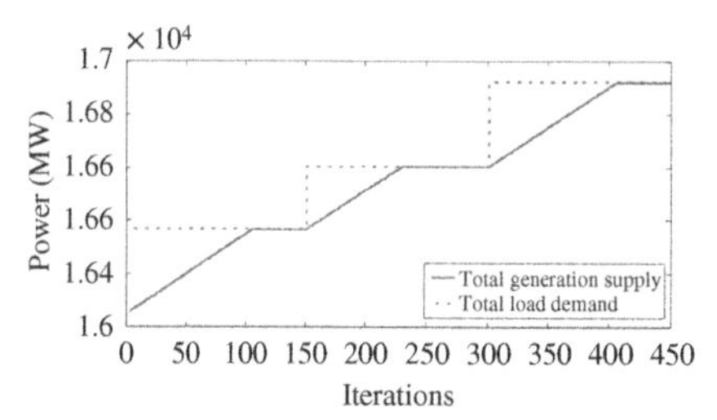

Figure 3.37 Total supply-demand update with 17-generator system.

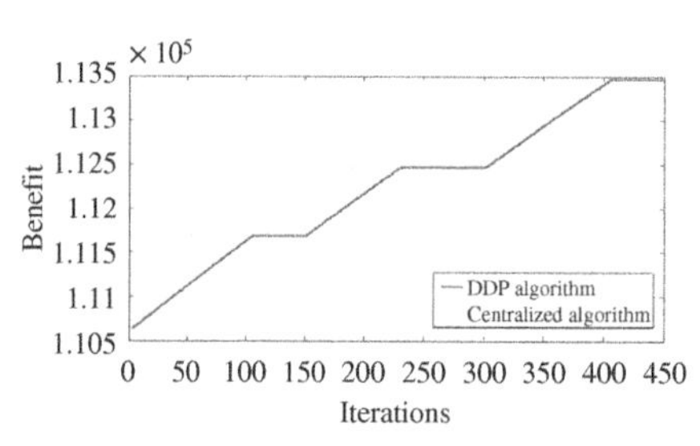

Figure 3.38 Benefit update with 17-generator system.

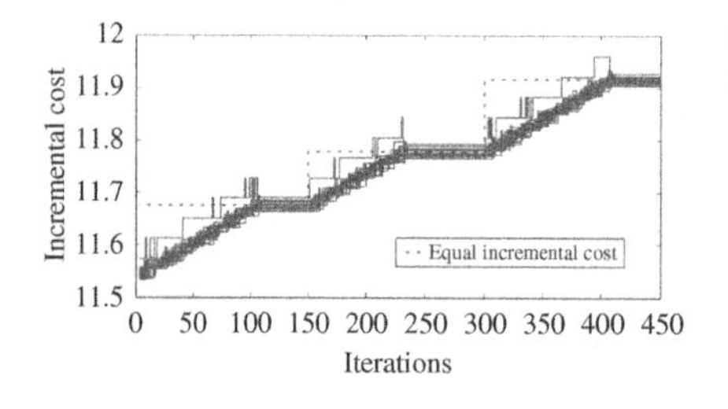

Figure 3.39 IC update with 17-generator system.

of the results obtained using the DDP algorithm happens in less than 120 iterations. Besides, the solutions provided by the DDP algorithm and the CEIC algorithm are similar. Table 3.7 provides with detailed comparisons. The graph density of the communication network is then increased in this case to examine the impact of communication network on how fast the DDP algorithm will converge. It is now assumed that the communication order between every agent and the eight adja-

Table 3.7 The DDP algorithm and the CEIC algorithm comparison.

	DDP		CEIC	
	r_i	Benefit	r_i	Benefit
Time slot 1	[11.6634, 11.6891]	111680.48	11.6757	111690.24
Time slot 2	[11.7650, 11.7924]	112466.21	11.7780	112468.23
Time slot 3	[11.9048, 11.9267]	113469.14	11.9159	113478.22

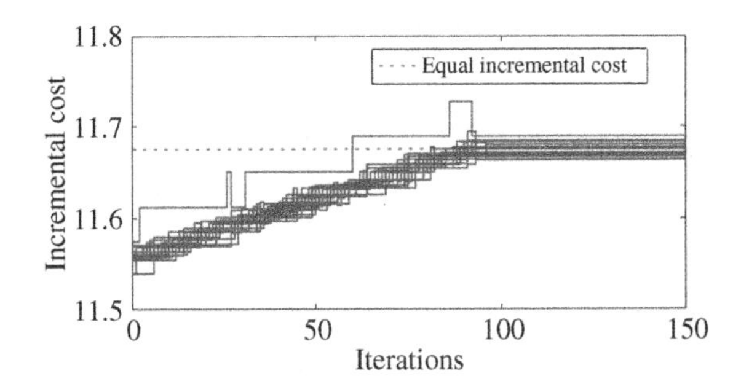

Figure 3.40 IC update with communication graph density of 0.5.

cent neighbors following index number $i-4, \ldots, i-1$ and $i+1, \ldots, i+4$, and the communication graph has a high-level connectivity (68 edges). Thus, the graph density is increased to $68/136 = 0.5$. As presented in Figure 3.40, 95 iterations are needed for convergence while when the graph density is 0.25, 110 iterations are required in previous case. In conclusion, by increasing the graph density, there will be an increase in the convergence speed at the expense of higher communication.

3.2.15 Hardware Implementation

PCs (dual core, 1.86 GHz) have various advantages including energy saving, small size, cheap price, and wireless communication features; therefore, they are used to build hardware platform. Figure 3.41 shows the schematic diagram of the hardware platform for this four-generator system. The following equation can calculate the maximum required communication bandwidth [84]:

$$\text{Bandwidth} = \underset{\text{consumed channels}}{\text{Number of communication}} \times \text{Frequency} \times \underset{\text{size}}{\text{Message}} \tag{3.51}$$

The number of successful communication times per second is used to define frequency. In this case, information exchanging among agents are happened in the

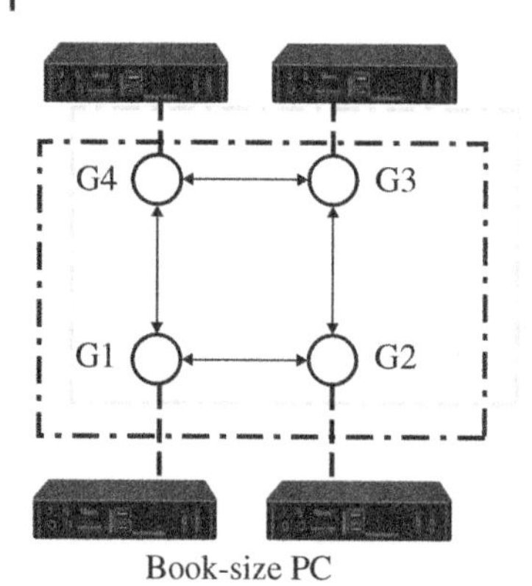

Figure 3.41 Hardware platform for four-generator system.

form of a vector. Vector sizes are minuscule: no more than several bytes. Generally, the update interval of ED schedules is more than five minutes [66], and Figure 3.40 shows that probably several hundred iterations are required for convergence of the algorithm as presented. Besides, the tests adopt 20 Hz for the IE frequency. According to Eq. (3.51), the 162-bus system, which has 17 generators, consumes the communication bandwidth maximally of approximately 100 kb/s and the consuming time is approximately $120/20 = 6$ seconds. The communication satisfies the bandwidth requirement because the speed of the internet it relies on is 10 Mb/s according to the TCP/IP protocol.

The difference in the use of hardware (computers) and software (coding, compilers) results in the various consuming time for every iteration. Therefore, this chapter uses the number of iterations to compare different communication protocols instead of the time consumed for simplicity.

3.2.16 Conclusion

This chapter considers the constraints of generation as well as ramp rate limits and proposes a DDP algorithm-based approach in order to solve the discrete ED problem. The simulation results and results obtained using the CEIC algorithm are similar, indicating that results from the DDP algorithm are competent. In addition, reducing the minimum generation adjustment could further enhance the optimality. There are three main merits of the proposed algorithm. The first is the reduction in the expense of the supporting communication network when compared with centralized strategies because it utilizes the sparsity property of the communication network and requires local communications among neighbor agents exclusively. The second is the feasible implementation with asynchronous communication, leading to faster convergence speed, easier implementation, robustness against packet loss, and increased flexibility. The third is that flexibility and scalability make it possible for large-scale power system applications in the future.

For the application of the DDP algorithm in practice, several practical issues, such as the constraints of transmission lines, the loss during transmission, etc., need to be further elaborated. In the future, the authors hope to improve the DDP algorithm further and publish state-of-the-art findings.

3.3 Constrained Distributed Optimal Active Power Dispatch

In the traditional power systems, ED and generation control are separately applied. Online generation adjustment is necessary to regulate generation reference for real-time control to realize the economic operation of power systems. Because most economical dispatch solutions are centralized, they are usually expensive to implement, susceptible to single-point failures, and inflexible. To address the abovementioned problems, this paper proposed a MAS-based distributed control solution that can realize optimal generation control. The solution is designed based on an improved distributed gradient algorithm, which can address both equality and inequality constraints. To improve the reliability of the MAS, the $N-1$ rule is introduced to design the communication network topology. Compared with centralized solutions, the distributed control solution can not only achieve comparable solutions but also respond timely when the system experiences a change of operating conditions. MAS-based real-time simulation results demonstrate the effectiveness of the proposed solution.

3.3.1 Introduction

The significance of ED has always been addressed by the power society. ED allocates the generation resource in the most economical way without violating the constraints of the power system [85]. Current methods for solving the problem for ED can be categorized into analytical algorithms (e.g. gradient search [59, 86] and lambda iteration [87]) and heuristic algorithms (e.g. Monte-Carlo [88], genetic algorithm [89], and particle swarm optimization [25, 90, 91]). The majority of the aforementioned methods are centralized. However, it is hard to achieve real-time optimal control by applying these centralized methods as they require to process large amounts of data and also need to handle the significant delay due to communication. The flexibility and stability of the power systems are limited with the centralized method, and the expense of using these centralized method-based control schemes can also be quite high.

It is not surprising that there exists a gap between real-time generation control and long time-scale ED, where the latter relies on the prediction of generation and load [49]. The prediction errors, operating condition variation, and optimization and communication delays require for the real-time generation adjustments. However, current methods by local frequency control and AGC do not consider optimization in the adjustments, thus degrading the efficiency [52].

Renewable energy and distributed energy resources (DERs) pose new challenges for the power system operation [50, 92]. DERs include distributed generation (DG), such as PV and wind, or from distributed storage (DS), such as pumped storage and flywheels [93]. However, these new resources are hard to control because of their intrinsic natures of intermittency and uncertainty.

The aforementioned gap between ED and generation control can be bridged by improving the solution speed of the ED problem. In addition, the difficulty of accommodating the changes in load composition and operating conditions calls for an upgrade of the solution speed.

Distributed control solutions can provide architectures that facilitate a much faster solution speed because they can allocate the computationally expensive task to multiple distributed controllers. By working in parallel, the solution speed increases with higher flexibility, improved robustness, and lower cost. Moreover, distributed control methods enable the integration of DERs and loads because they are both "distributed" [94]. Therefore, distributed methods are promising for optimal generation control.

There are some works on distributed optimal generation control. Current methods may cost a long time to achieve optimal dispatch [68, 95] or cause power mismatch and oscillation during optimization. A consensus algorithm is used in [70, 96, 97], where the authors proposed a two-level scheme, i.e. leader and follower level. However, the algorithm requires extra communication time as leaders need to collect global information. In addition, the optimality is not guaranteed as the power loss is not considered in this scheme.

In this chapter, we propose a fully distributed MAS-based optimal generation control solution. Based on our design, each generator is represented by an associated agent, and this agent can communicate with its neighboring agents that are within this agent's communication range. This structure requires no centralized agent. Moreover, the communication network is designed by applying the $N-1$ rule, which improves the robustness of the system against communication failures. Furthermore, we design a distributed gradient-based algorithm for online optimization of active power generation references, wherein the equality and inequality constraints of the optimization problem are handled by adjusting local generations and reconfiguring the virtual communication topology. At the end of this chapter, we provide real-time simulation results to demonstrate the performance of the proposed control solution.

The highlights of this chapter are listed as follows:

- We design a distributed algorithm that can handle both equality and inequality constraints.
- We utilize the $N-1$ rule in designing the topology of the communication network to improve reliability.
- We use a book-size PC-based MAS to achieve the proposed method and demonstrate the method performance by the MAS-based real-time simulation.

3.3.2 Problem Formulation

The optimal generation control problem is formulated as follows:

$$\min_{P_i} \quad \sum_{i}^{n} f_i(P_i) \tag{3.52a}$$

$$\text{subject to} \quad \sum_{i}^{n} f_i(P_i) = P_d, \tag{3.52b}$$

$$\underline{P_i} \leq P_i \leq \overline{P_i}, \tag{3.52c}$$

where P_i represents the active power generation of the ith generator, n is the total number of generators, $f_i(P_i)$ is the generation cost for the ith generator, P_d is the total active power demand in the power system, and $\underline{P_i}$ and $\overline{P_i}$ are the lower and upper bounds for generator i, respectively.

The control objective (3.52a) is to minimize the total generation costs. The generation cost of generator i, $f_i(P_i)$, can be approximated by a quadratic function [90]

$$f_i(P_i) = a_{i2}P_i^2 + a_{i1}P_i + a_{i0}. \tag{3.53}$$

Equality constraints (3.52b) indicate that the supply and demand should be balanced and inequality constraints (3.52c) denote that a generator's output should be controlled within its bounds. The Eq. (3.53) characterizes the input–output relationship of the generators [86], which can be obtained by the following methods:

- *Experiments*: The efficiency of the generating units.
- *Data*: Historic operation data of the generating units.
- *Design*: Design data from the manufacturer.

With different characteristics of production, different generators yield different parameters. Generally, the generation cost function is convex because of the positive parameter a_2. Therefore, the optimal generation control problem described in (3.52) is actually a convex optimization problem, and we can solve this optimization problem to obtain the optimal solution.

3.3.3 Distributed Gradient Algorithm

The problem with current distributed gradient algorithms for solving the convex optimization problem described in (3.52) is that they cannot handle problems with both equality and inequality constraints. Usually, they consider only equality constraints (3.52b) [18] or only inequality constraints (3.52c) [19, 20, 98, 99]. In this chapter, we propose an improved distributed gradient algorithm based on work [18] to handle both two types of constraints. We first introduce the algorithm to handle the equality constraint and we develop an improved algorithm that can hold both equality and inequality constraints.

3.3.4 Distributed Gradient Algorithm

If inequality constraints in (3.52c) are neglected, the optimal generation control problem (3.52) is simplified as

$$\min_{P_i} \quad \sum_{i}^{n} f_i(P_i) \tag{3.54a}$$

$$\text{subject to} \quad \sum_{i}^{n} f_i(P_i) = P_d, \tag{3.54b}$$

In the following, let $\mathbf{P} = [P_1, P_2, P_3, \cdots, P]^T \in \mathbf{R}^n$ denote the vector of power generation, $f(\mathbf{P}) = \sum_i^n f_i(P_i)$ denote the objective function, and $\nabla f(\mathbf{P}) = [\dot{f}_1(P_1), \cdots, \dot{f}_n(P_n)]^T$ denote the gradient vector of the cost functions, where $\dot{f}_i(P_i)$ denotes the derivative of $f_i(P_i)$ with respect to P_i. The convex problem (3.54) has a unique optimal solution $\mathbf{P}^*$ [18]. The conditions for optima are given as follows.

$$\mathbf{1}^T\mathbf{P}^* = P_d, \qquad \nabla f(\mathbf{P}^*) = \lambda^*\mathbf{1}, \tag{3.55}$$

where $\mathbf{1}$ is a column vector of ones and λ^* is the unique optimal Lagrange multiplier.

The key to designing a distributed optimization algorithm is to find $\mathbf{P}^*$ in a distributed way. The distributed gradient algorithm in [18] proposed that an agent pertains to a generator obtains the gradient of its cost function and its neighbors and then updates its own gradient by the weighted sum of the gradients in each iteration. The local updating rule is given by

$$P_i[k+1] = P_i[k] - W_{ii}\dot{f}_i(P_i[k]) - \sum_{j \in ix_i} W_{ij}\dot{f}_i(P_i[k]) \tag{3.56}$$

where ix_i is the indices of the neighboring agents of agent i, W_{ii} is the self-weight of agent i, and $W_{ij}(j \in ix_i)$ is the weight of agent j. The updating rule in the matrix form is given by

$$\mathbf{P}[k+1] = \mathbf{P}[k] - \mathbf{W}\nabla f(\mathbf{P}[k]). \tag{3.57}$$

With $\mathbf{W}$ being predetermined, the algorithm can lead to a distributed solution because each agent only requires local information to update its generation. The determination of $\mathbf{W}$ is introduced in the following [18].

- During update, $\mathbf{P}[k]$ is always feasible for all k, i.e.

$$\mathbf{1}^T\mathbf{P}[k+1] = \mathbf{1}^T(\mathbf{P}[k] - \mathbf{W}\nabla f(\mathbf{P}[k])) = \mathbf{1}^T\mathbf{P}[k] = P_d, \tag{3.58}$$

  which yields

$$\mathbf{1}^T\mathbf{W}\nabla f(\mathbf{P}[k]) = 0. \tag{3.59}$$

  For any $\nabla f(\mathbf{P}[k])$, the above equation must hold. Consequently, we have $\mathbf{1}^T\mathbf{W} = \mathbf{0}^T$, where $\mathbf{0}^T$ is a column vector of zeros.
- The Lagrange multiplier method requires the following condition when the algorithm converges.

$$\nabla f_1(P_1) = \nabla f_2(P_2) = \cdots \nabla f_i(P_i) = \cdots = \nabla f_n(P_n) = \lambda^*. \tag{3.60}$$

  By virtue of (3.57),

$$\mathbf{P}^* = \mathbf{P}^* - \mathbf{W}\nabla f(\mathbf{P}^*) = \mathbf{P}^* - \lambda^*\mathbf{W}\mathbf{1}, \tag{3.61}$$

  which yields $\mathbf{W}\mathbf{1} = \mathbf{0}$.

To summarize, $\mathbf{W}$ must have the following two properties:

$$\mathbf{1}^T\mathbf{W} = \mathbf{0}^T, \qquad \mathbf{W}\mathbf{1} = \mathbf{0}. \tag{3.62}$$

Note that if $\mathbf{W}$ is set to symmetric, satisfying either of the conditions (3.62) will satisfy the other. In this chapter, we design a dynamic weight matrix that changes with the topology of the communication network. In the following, we denote this dynamic weight matrix with $\mathbf{W}[k]$, instead of $\mathbf{W}$. We use the improved Metropolis method [15, 100] to $\mathbf{W}$ and its element, $W_{ij}[k]$, is calculated as follows:

$$W_{ij}[k] = \begin{cases} \frac{-2}{n_i[k]+n_j[k]}, & j \in ix_i[k],\ j \neq i,\ n_i \neq 0 \\ -\sum\limits_{j \in ix_i[k]} W_{ij}[k], & j = i \\ 0, & \text{otherwise} \end{cases}, \tag{3.63}$$

where $ix_i[k]$ is the set of indices of agents that communicate with agent i, $n_i[k]$ is the number of neighboring agents of agent i, and is decided by the topology of communication network.

We can use an undirected graph with n nodes to represent the communication network with n agents. The connectivity of the graph is described by an *adjacency matrix* $\mathbf{A}[k]$. The elements of $\mathbf{A}[k]$ are defined as

$$a_{ij}[k] = \begin{cases} 1, & \text{if } i \text{ and } j \text{ are connected} \\ 0, & \text{otherwise} \end{cases}. \tag{3.64}$$

By virtue of (3.60), $n_i(k)$ in (3.59) can be determined as

$$n_i[k] = \sum_{j=0}^{n} a_{ij}[k]. \tag{3.65}$$

Note that the problem described in (3.54) only takes account of equality constraints. Without considering inequality constraints, the results might become infeasible for the original problem. In the following, techniques for addressing inequality constraints are introduced to handle this problem.

3.3.5 Inequality Constraint Handling

To consider the inequality constraints, we modify the gradient algorithm as follows:

$$P_i[k+1] = \begin{cases} \overline{P_i} \text{or } \underline{P_i}, & i \in ix_b[k] \quad (a) \\ P_i[k] - \alpha[k]\sum_{j=1}^{n} W_{ij}[k]\dot{f}_j(P_j[k]), & i,j \in ix_b[k]^c \quad (b) \end{cases}, \tag{3.66}$$

where $P_i[k+1]$ is the update of $P_i[k]$, $\alpha[k]$ denotes the step size, $\dot{f}_j(P_j[k])$ is the derivative of the local generation cost, $\underline{P_i}$ and $\overline{P_i}$ are the lower and upper bounds of the generation for generator i, respectively, $i \in ix_b[k]$ is the indices of the generators that reach the bounds, and $ix_b[k]^c$ is the complement of $ix_b[k]$.

The modified algorithm is able to tackle both equality and inequality constraints. If the power generation of a generator does not exceed the bounds, its update continues by using (3.66a). Otherwise, the value of generation is fixed and excluded from further update. This operation can be realized by changing the values of the weight matrix $\mathbf{W}[k]$. Bacause the weight matrix is determined by the topology of the communication network, changing the value of $\mathbf{W}[k]$ can be understood as the "reconfiguration" of the topology.

Figure 3.42 shows the principle for reconfiguration of the virtual communication network. To consider the case of violating the bound, we consider a special case where agent #1 violates the lower bound. The influenced links are labeled in red related to agent #1 in Figure 3.42a. Next, the generation of agent #1 is fixed and prohibited from the further update. Note that this does not influence the IE of node #1. In Figure 3.42b, agent #1 still receives the data from a neighboring agent and sends it directly to all other neighboring agents that are originally connected with agent #1, as long as there is no direct communication link between these two neighboring agents. Therefore, this operation is similar to the case where the topology of a communication network is reconfigured, as shown in Figure 3.42c. It should be noted that the agent #1 still checks the value of the weighted gradients with the original weight matrix during the following update process. If the weighted sum becomes negative, agent #1 can rejoin its generation adjustment and

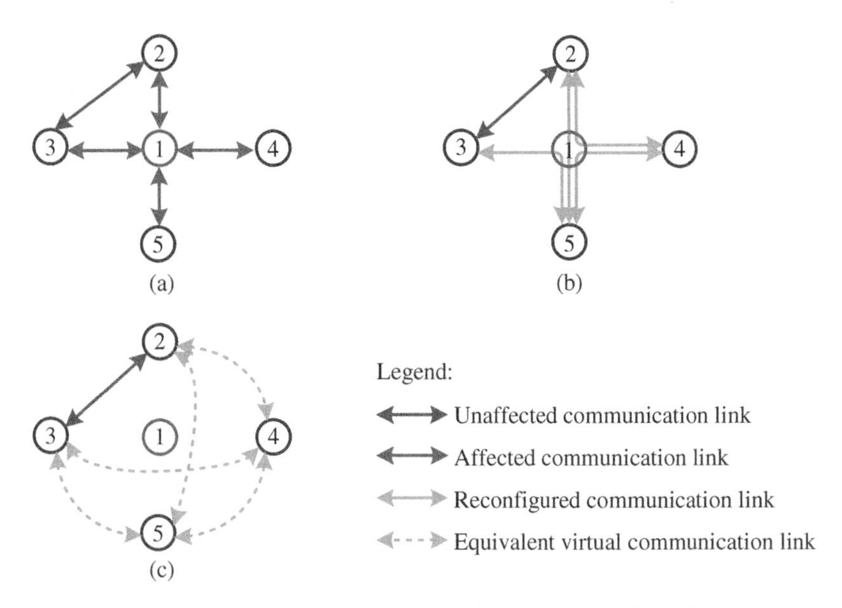

Figure 3.42 Construction of the virtual communication network. (a) Original communication network, (b) Reconfigured communication network, and (c) Equivalent virtual communication network.

the network topology is correspondingly restored. The case with an upper bound violation is handled in a similar manner.

The operation for virtual communication network reconfiguration can be formulated by updating a_{ij} associated with agent i, which is given by

$$a_{ij}[k+1] = \begin{cases} 0, & i \in ix_b[k] \text{ or } j \in ix_b[k] \\ a_{ib}[k] \ \varsigma a_{bj}[k] \mid a_{ij}[k], & b \in ix_b[k] \text{ and } i,j \in ix_b[k]^c \\ a_{ij}[k], & \text{otherwise} \end{cases}, \tag{3.67}$$

where"&" and "|" are logic operators "and" and "or", respectively.

During the bound violation, the boundary agent sends the indices of the neighboring agents to its neighboring agents and its neighboring agents then update their own a_{ij}, n_i, and W_{ij} consecutively based on (3.67), (3.65), and (3.63), respectively. This reconfigured virtual communication topology enables the boundary agent to transfer data for its neighboring agents that are not directly connected as per the original communication network. In such a way, operations defined in (3.67) can be realized, and thus, the inequality constraints can be handled in a distributed manner.

Note that the techniques for handling inequality constraints proposed here are applicable to algorithms introduced in [68, 70, 95–97]. Moreover, the cost function does not necessarily have to be quadratic; it only requires them to be convex.

3.3.6 Numerical Example

Here, we use a simple example to demonstrate the proposed approach. There are three generators that participate in the optimization. Each generator has a quadratic function in the form of (3.53). Table 3.8 shows the parameters for generation cost and boundary conditions. The total amount of demand is 40, which is initialized to 13, 14, and 13 for the generating units, respectively. Figure 3.43a portrays the original communication network. There are two test cases. Case 1 is designed with equality constraints only, while case 2 is designed to handle both equality and inequality constraints.

Note that although the parameters in the cost functions are not realistic data, they are useful in demonstrating our proposed algorithm.

3.3.6.1 Case 1

The adjacency matrix **A** and weight matrix **W** associated with the the communication network are

Table 3.8 Parameters for the numerical example.

	Cost function			Bounds of generators			
Generator	a_0	a_1	a_2	Case 1		Case 2	
1	12	30	4	0	20	0	20
2	20	40	3	0	20	0	14
3	15	16	5	0	20	0	20

Generating cost function: $f_i = a_{i0} + a_{i1}P_i + a_{i2}P_i^2$(\$/h).

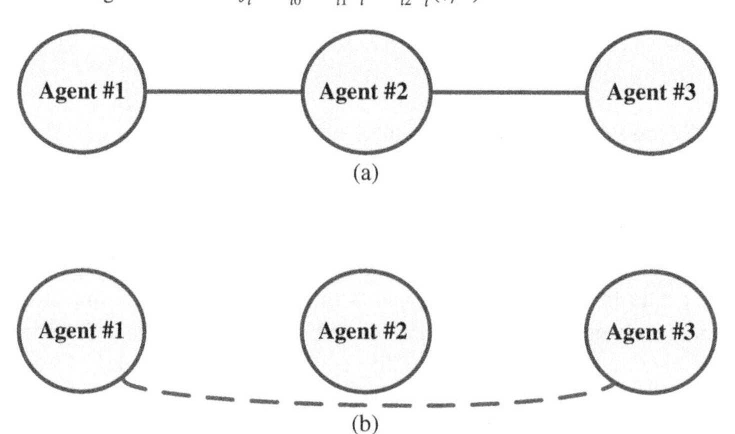

Figure 3.43 Topology of the designed communication network. (a) Original communication topology and (b) Reconfigured communication topology.

$$\mathbf{A} = \begin{bmatrix} 0 & 1 & 0 \\ 1 & 0 & 1 \\ 0 & 1 & 0 \end{bmatrix} \rightarrow \mathbf{W} = \begin{bmatrix} 2/3 & -2/3 & 0 \\ -2/3 & 4/3 & -2/3 \\ 0 & -2/3 & 2/3 \end{bmatrix}. \tag{3.68}$$

For case 1, we use a step size of $\alpha = 0.05$. Figure 3.44 provides the optimization results. As can be seen in the figure, none of the variables reach their bounds; accordingly, $\mathbf{W}$ is not changed during the whole optimization process.

Table 3.9 demonstrates that the optimal values obtained with the distributed algorithm obtain the same as that with the centralized method. Noted that none of the inequality constraints are violated during optimization.

In this case, the converging speed can be improved by setting the step size to a larger value. Figure 3.45 shows the optimization results with $\alpha[k] = 0.1$. It shows that only 10 iterations are needed to reach the optima. Compared with the previous one of 14 iterations, the increase of step size does improve the convergence speed.

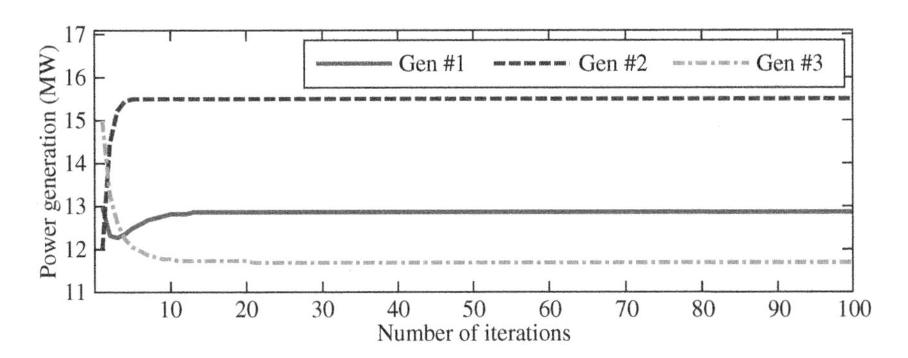

Figure 3.44 Optimization results of case 1 (step size = 0.05).

Table 3.9 The optimization results of numerical examples.

	Centralized		**Distributed**			
Case	**Cost ($/h)**	**Mismatch**	**Cost ($/hr)**		**Mismatch**	
1	3298.6	−0.001 MW	$\alpha = 0.05$, Iterations = 14		$\alpha = 0.10$, Iterations = 6	
			3298.7	0.00 MW	3298.7	0.00 MW
2	3307.4	−0.001 MW	$\alpha = 0.0035$, Iterations = 66		$\alpha = 0.01$, Iterations = 31	
			3307.2	−0.019 MW	3297.0	−0.091 MW

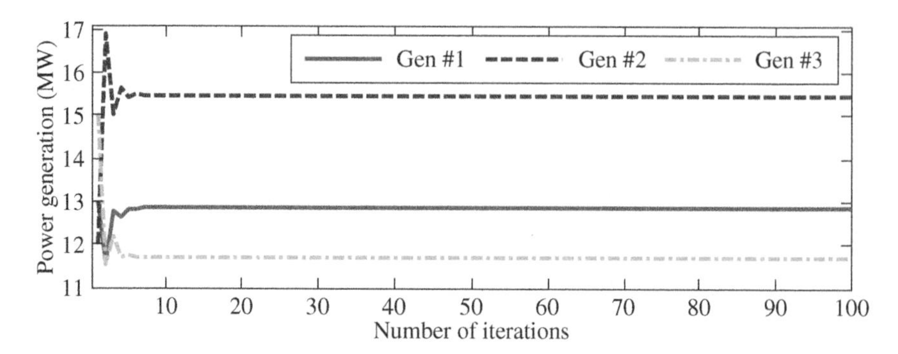

Figure 3.45 Optimization results of case 1 (step size = 0.1).

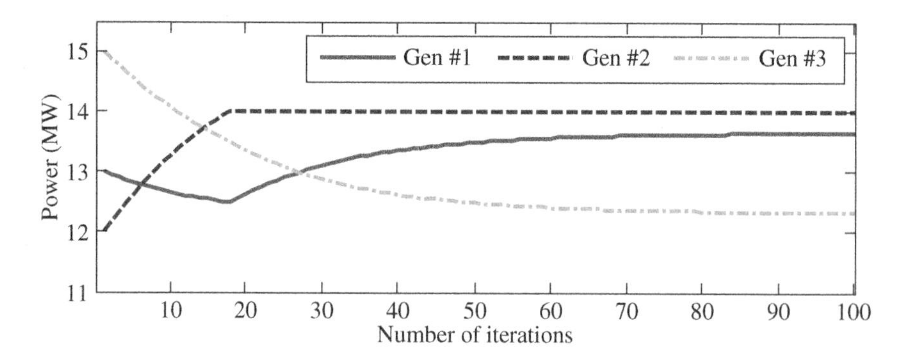

Figure 3.46 Optimization results for case 2 (step size = 0.0035).

3.3.6.2 Case 2

Figure 3.46 provides the simulation results with $\alpha = 0.0035$ in case 2. Gen #2 reaches its upper bound at the 14th step; consequently, the original network is reconfigured to form a virtual network at a 15th step, as shown in Figure 3.43b, and the adjacency matrix and weight matrix are correspondingly updated as follows:

$$\mathbf{A}(k+1) = \begin{bmatrix} 0 & 0 & 1 \\ 0 & 0 & 0 \\ 1 & 0 & 0 \end{bmatrix} \rightarrow \mathbf{W}(k+1) = \begin{bmatrix} 1 & 0 & -1 \\ 0 & 0 & 0 \\ -1 & 0 & 1 \end{bmatrix}. \tag{3.69}$$

The optimization results for this case are shown in Table 3.9. Here, the typical lambda iteration method introduced in [87] is utilized for comparison with the centralized method. In case 1, the distributed algorithm yields the exactly optimal solution because there is no boundary constraint violation. In case 2, the solution by a distributed algorithm with $\alpha[k] = 0.0035$ is very close to that with the centralized method even with boundary constraint violation. As can be seen from Table 3.9, the solution with $\alpha[k] = 0.01$ merely deviates slightly from the optimal values. Decreasing step size can decrease the gap between solutions of distributed gradient and that of the centralized algorithms. However, a smaller step size usually means a slower convergence speed. One can always test the settings of the step size through off-line studies to find the one that yields satisfactory performance.

Because the power system can be assumed to operate in a steady state or quasi-steady state, the current output of the generators can be taken as the initial conditions for them. Therefore, in the initialization, there is no need for the exchange of global information. As for the optimization, each agent updates its generation reference with information regarding current output and the gradients of its neighboring agents, and no central coordinator is required. Therefore, we can implement the proposed algorithm for optimization in a fully distributed manner.

In the following, we introduce the details on implementing the algorithm.

3.3.7 Control Implementation

Figure 3.47 demonstrates the architecture of the proposed control system. Each generator in this power system is assigned with an intelligent agent, and this agent can communicate with other agent(s) as per the topology of the designed communication network. An agent is designed to have the following functionalities: (i) acquiring the measurements of the corresponding generator; (ii) exchanging local information with its neighboring agents; (iii) participating in the generation reference optimization, and (iv) updating the generation reference for the generator. There is no direct connection between the design of the physical network and that of the communication network. Instead, the communication network is designed based on the $N-1$ rule, which is illustrated in the following.

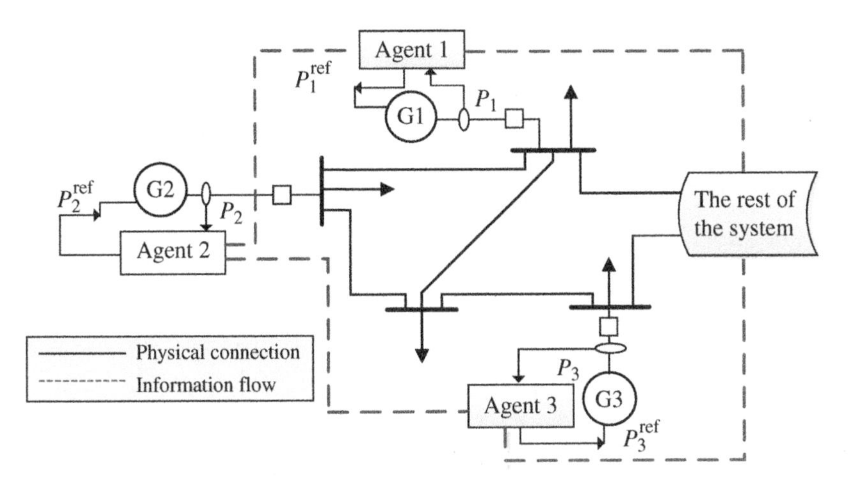

Figure 3.47 Control system architecture.

3.3.8 Communication Network Design

We use a graph to represent the structure of the communication network. The nodes in this graph represent the agents corresponding to generators, whereas the edges of the graph represent the communication channels among the agents. The reliability of the system is mainly determined by the edge connectivity of the graph. Higher connectivity signifies the more reliable network [34, 101]. In this chapter, the topology of the communication network of the control system is designed to comply with the $N-1$ rule introduced in Chapter 2. Following the $N-1$ rule in the communication network design can guarantee that the control system works properly with any single-channel failure in the communication network. This is important in the distributed optimization, which is illustrated in the later simulation studies.

3.3.9 Generator Control Implementation

Figure 3.48 provides the control block diagram for the generators. The inputs of agent i includes three parts: the output of the governor system $\Delta P_i[k]$, gradients of its neighboring agents $\dot{f}(P_j[k+1])$, $j \in ix_i$ and local active power generation $P_i[k]$. The output of agent i is the reference setting for the active power generation $P_i[k+1]$. We can denote the simplified governor control model as [102]

$$\Delta P_i[k] = -K_{Pi}\Delta f[k]. \tag{3.70}$$

Note that the assumption behind this type of governor is that the increase of power output brings about a frequency deviation at steady state [102].

By virtue of $\Delta P_i[k]$, Eq. (3.66) can be rewritten as

$$P_i[k+1] = \begin{cases} \overline{P_i} \text{ or } \underline{P_i}, & i \in ix_b[k] \\ P_i[k] - \alpha[k]\sum_{j=1}^{n} W_{ij}[k]\dot{f}_j(P_j[k]) - K_{Pi}\Delta f[k], & i,j \in ix_b[k]^c \end{cases}, \tag{3.71}$$

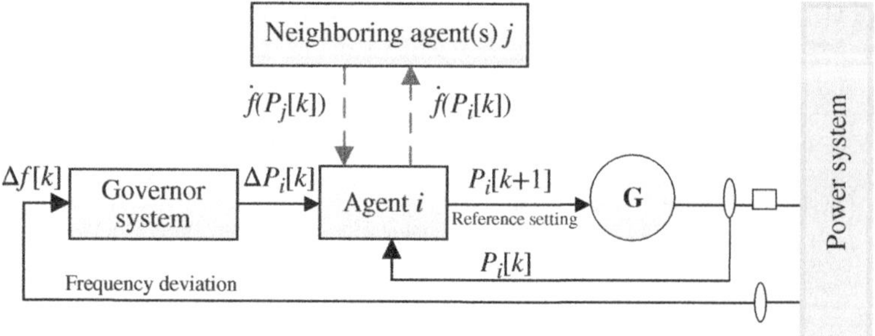

Figure 3.48 Control block-diagram of the generator.

The way to update the generation references by using (3.66) ensures that the total generation is not changed during the generation adjustment in the optimization process; in addition, it can satisfy both equality and inequality constraints in (3.52). The update rule of Eq. (3.66) is feasible only when the overall demand does not change. If the change of demand occurs, applying (3.66) will inevitably lead to the supply-demand imbalance. Therefore, we replace the original update rule of (3.66) with (3.71), by adding governor output term in it.

By using Eq. (3.71), the frequency deviation (supply-demand imbalance in (3.70)) can lead to the timely adjustment of the generation. The value of Δf is zero at steady state and tends to be nonzero where there is an operating condition change. The governor system of a generator estimates the additional generation adjustment that should be provided for this generator. In addition, by taking the output of the governor system into account, the generator can dynamically readjust its power output to maintain the balance between supply and demand while still optimizes its generation setting.

Note that the term $K_{Pi}\Delta f[k]$ acts as a role to make the physical system feedback enter the optimization process. It should be noted that the power balance constraint may not strictly hold during the optimization. However, our proposed algorithm drives the system toward the optimal point while keeping the constraints being satisfied. Moreover, the load change ΔP_L can result in a transient change of frequency Δf. Therefore, including the term $K_{Pi}\Delta f[k]$ in the control system can signify the change of load conditions. Furthermore, the feasibility and optimally is guaranteed by the term $K_{Pi}\Delta f[k]$, which actually brings the physical system feedback to the optimization process.

Equation (3.71) shows that both generation cost reduction and supply-demand balance require generation adjustment. In our proposed implementation, the two operations are designed to act simultaneously instead of in serial to improve the response speed. Because the optimization process is discrete and model-based, its converging speed is not influenced by the response speed of the physical system, and thus, it can be designed to be faster or slower by adjusting the step size α. However, the governor control-related operation is nonadjustable because of the physical constraint, and it dominates the overall response speed of the control system. Consequently, the speed of the optimization operation should be adjusted to match that of the governor system, which will be discussed in detail later.

3.3.10 Simulation Studies

In this section, we first introduce the real-time simulation platform we designed for testing the distribution solutions, and then we use this platform to test the IEEE 30-bus test system to demonstrate the control solution proposed in this chapter.

3.3.11 Real-Time Simulation Platform

Figure 3.49 portrays the simulation platform, which contains a MAS that is developed by using eight book-size PCs., a real-time digital simulator (also named the target computer), and a host computer. The host computer is capable of compiling and uploading the power system model to the real-time digital simulator, displaying real-time simulation results, and relaying communication between the simulator and agents. The real-time simulator is manufactured by *OPAL-RT*® and is also known as the target computer. The software package for running the agents is developed by using JADE Framework [103]. The MAS can interact with running power system models in the target computer through the relay of the host computer so as to realize the MAS-based real-time simulation.

3.3.12 IEEE 30-Bus System

The IEEE 30-bus system contains 6 generations and 30 buses. One can refer to [104] for more information. The parameters for generation cost functions and constraints of generators are displayed in Table 3.10. A turbine governor control system is installed for each generator and its model can be found in [47]. Other parameters regarding the generators and governor control system are provided in Appendix 3.A.

Figure 3.50 shows the topology of the communication network, which has six nodes (agents) and nine edges (communication channels). Obviously, this topology design follows the $N-1$ rule.

Figure 3.51 shows the implementation steps of the control method. The agents iteratively update the generation references until the convergence criterion is

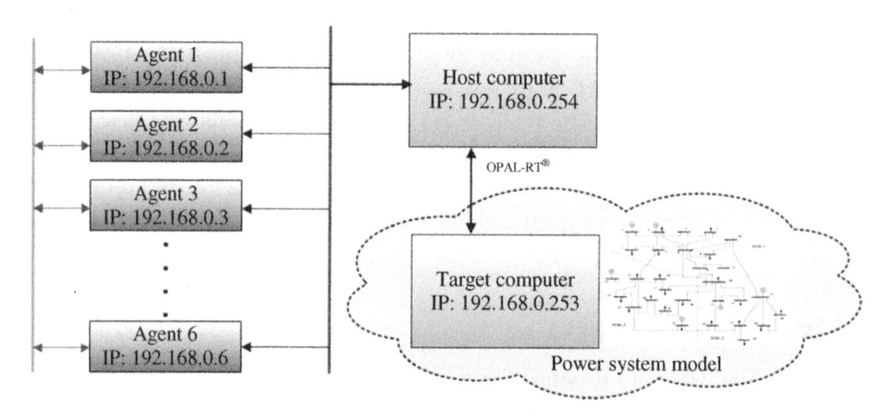

Figure 3.49 Developed real-time simulation platform for MAS.

Table 3.10 Parameters of the cost functions and constraints.

	Cost Function					
Generator	a	b	c	Lower Bound (MW)	Upper Bound (MW)	Initial Output (MW)
1 1	0	2.0	0.00375	50	200	157.36
2 2	0	1.75	0.0175	20	80	60
3 5	0	1.0	0.0625	15	50	30
4 8	0	3.25	0.00834	10	35	20
5 11	0	3.0	0.025	10	30	13
6 13	0	3.0	0.025	12	40	15

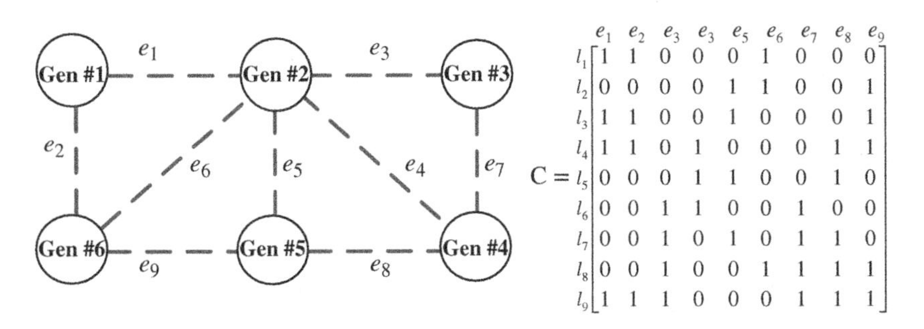

Figure 3.50 Communication network topology.

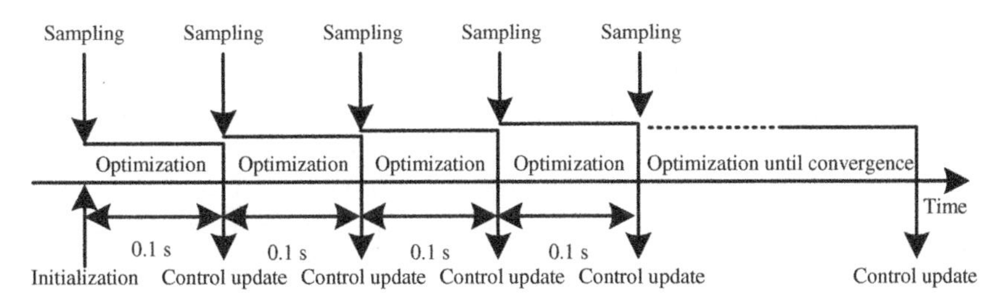

Figure 3.51 Implementation of the proposed distributed solutions.

reached. In each update interval (0.1 seconds), the agents are supposed to sample data, exchange data, calculate the reference, and update the control action.

We first test the performance of the developed MAS-based platform. Figure 3.52 shows that the distributed algorithm can converge with 20 iterations. With our

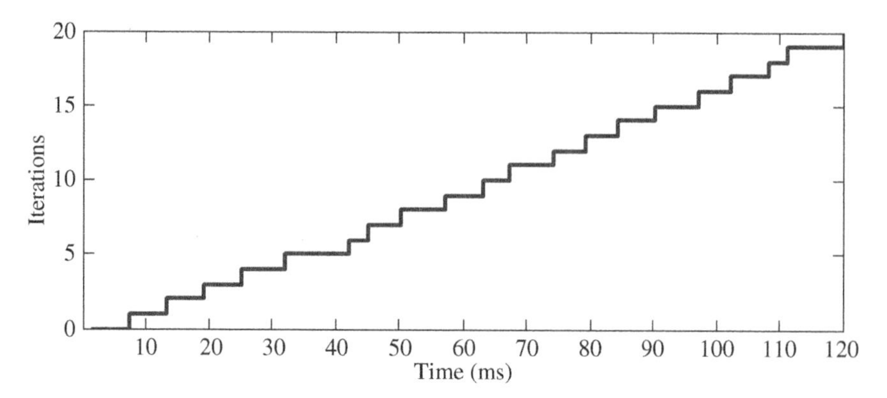

Figure 3.52 Time consumed for 20 iterations with distributed gradient method.

developed platform, each iteration costs a time of 6 ms on an average. During the experimentation process, the control setting is supposed to update five times before being deployed for control action. Consequently, the control update interval should be set larger than 30 ms.

Here, we set the interval for generation reference updates to 0.1 seconds. Test results show that frequent updates can be achieved by applying a proper experimental setup. Yet over frequent updates might degrade dynamic performance and cause instability because the tracking abilities of the generators are constrained by the capability of the physical system. Therefore, the updating interval should be chosen carefully. For our test, we use an error method to decide the update interval, and test results show that the value of 0.1 seconds can balance the control performance and technical feasibility, regardless of the constant or variable loading conditions.

3.3.12.1 Constant Loading Conditions

We assume that the loads are constant during the simulation and the proposed control solution is deployed at $t = 10$ seconds. The generation cost during simulation is illustrated in Figure 3.53. The generation cost starts from 812.6 £/h before the control method is used, and it comes down to 804.9 £/h when the optimization process is finished. Note that the obtained optimized results are the same as those obtained by using other centralized methods.

Figure 3.54 demonstrates the generators' speed deviations during the simulation. Note that when the loading condition is at a given constant value, updating rule (3.56) may bring small frequency deviation because of the following reasons: (i) small change of transmission loss brought by the reallocation of active power generations among the generators. For example, in our case, the transmission losses change 0.9 MW during the optimization, which comprises 0.3% of the total

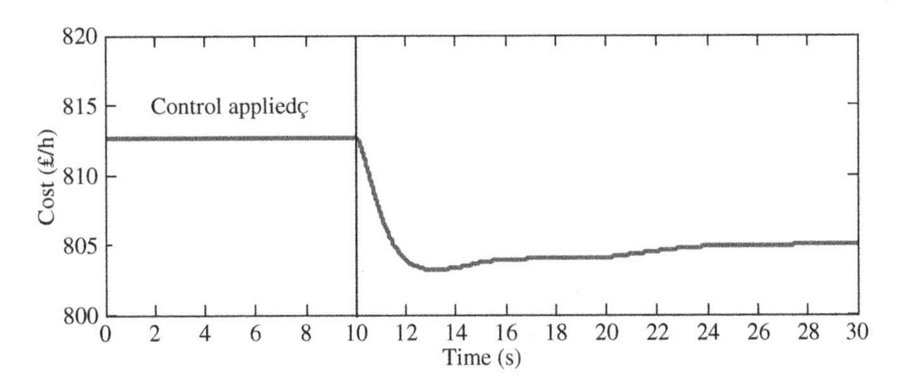

Figure 3.53 Generation cost with proposed control solution.

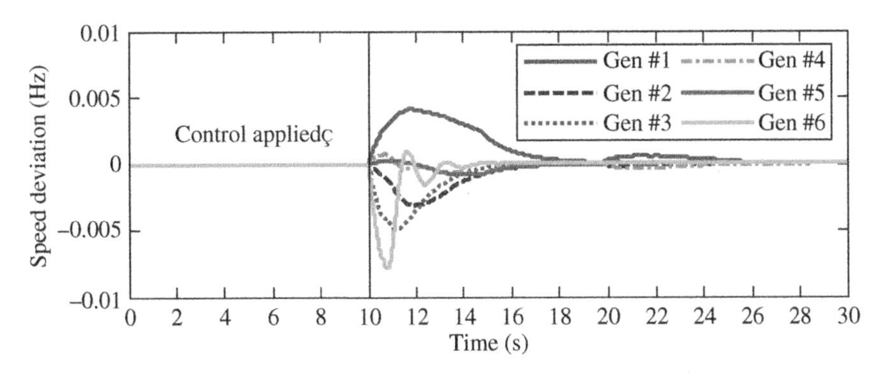

Figure 3.54 Speed deviation of the generators with the distributed solution.

load (283.4 MW). Therefore, the frequency deviation caused by the transmission loss is slight. (ii) Ideally, Eq. (3.56) can make sure that the supply and demand are balanced during the optimization. However, the response speeds of generators are not necessarily the same because the inertia constant, reactance, and other parameters of the generators may not be the same. Therefore, the tracking performance of the updated references varies, which leads to the slight supply-demand imbalance, before reaching the steady state, thus causing frequency deviations.

Note that the frequency deviations caused by the two abovementioned reasons are quite small. Figure 3.54 shows that the maximum frequency deviation reaches only 0.0085 Hz during the optimization. In order to avoid the frequency oscillations, Δf from the nominal value (60 Hz) is fed into the governor control system, which is denoted as $-K_{Pi}\Delta f$ in (3.71). It should be pointed out that the proposed control scheme can also ensure the generators to make timely adjustments against load change during optimization, and this will be demonstrated in the following test case.

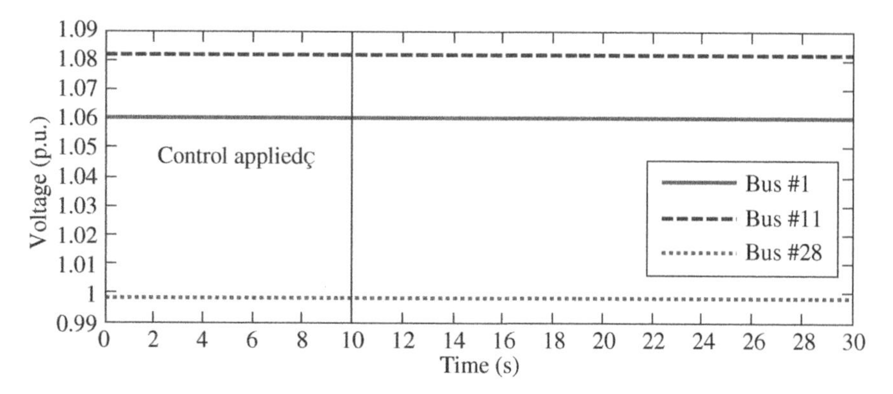

Figure 3.55 Voltages profiles of selected buses with the distributed solution.

Figure 3.55 displays the voltage of the chosen buses, which includes the bus with a generator (bus #1), the highest voltage (bus #11), and the lowest voltage (bus #28). As shown in the figure, the voltages only change slightly during optimization because the reallocation of active power among generators will not influence the voltages too much. Figure 3.53 to Figure 3.55 show that the optimal generation control can be achieved with decent dynamic performance. As the control update interval is 0.1 seconds, the deployed solution can respond in a timely manner under operating condition changes. The perturbation during the control update is also small because of the timely and incremental control update.

The active power outputs of the generators are provided in Figure 3.56. Zoomed-in responses of generation of Gens #5 and #6 are shown in Figure 3.56b. The output of Gen #5 and Gen #6 were clamped to 10 and 12 MW after convergence, which are the generation lower bounds. Therefore, the boundary constraints are not violated by applying the proposed control solution.

3.3.12.2 Variable Loading Conditions

Two load changes are utilized to test the proposed control method in this case. At $t = 10$, the active power loads at bus #12 and bus #21 increase by 20 MW each. At $t = 20$ seconds, these two loads decrease back to their previous values before $t = 10$. Figure 3.57 shows that before the load changes, the optimized generation cost is the same as that of case 1, which is 804.9 £/h. Then, the generation cost increases to 946.8 £/h after the loads increase, which is lower than the cost without optimal control (965.5 £/h). The optimized generation cost drops back to 804.9 £/h when the loads are recovered to their original values .

Figure 3.58 shows that the frequency deviates slightly and finally converges, which indicates that our algorithm is fast and accurate. It is noted that only local frequency is supposed to be measured by the generators. To prevent frequency oscillation, dead band of controllers should be set properly. For example, it is

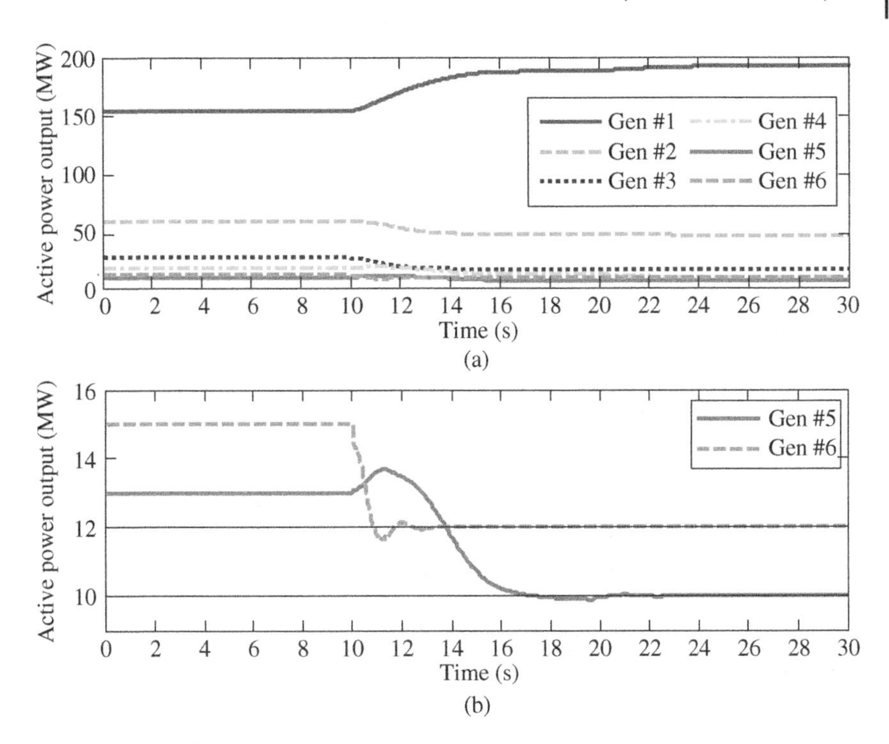

Figure 3.56 Active power outputs of generators.

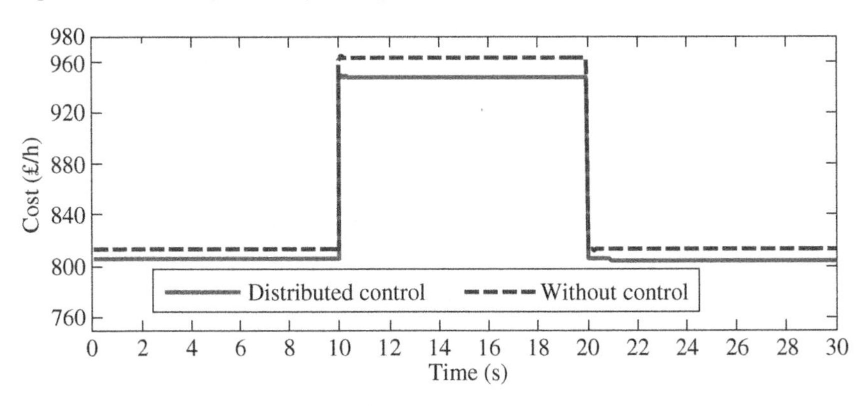

Figure 3.57 Generation cost with proposed control solution.

recommended by NERC that the controller insensitivity of primary frequency control should be ±36 mHz [105]. Moreover, the frequency oscillations can be dampened by frequent IE between the agents. Figure 3.58 demonstrates that there is only a small amount of 0.018 Hz frequency deviation when the change of loads reaches as high as 40 MW.

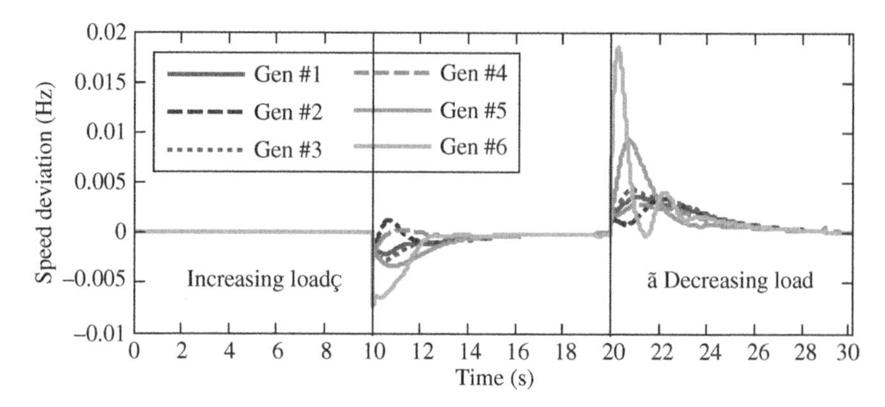

Figure 3.58 Speed deviation of the generators.

Figure 3.59 shows the active power outputs of generators. It can be seen that the outputs of all generators fall within their bounds, indicating the obtained solution is feasible. The output of generators #5 is at the lower bound of 10 MW and the output of #6 is at the lower bound of 12 MW before load increases. Because of the negativity of the weighted gradient sum, these two generators increase their output and finally reaches a level of 14.26 MW. As long as the demands are restored back to their original values, these two generators also decrease their output to their original values, i.e. their lower bounds of 12 MW, which shows good consistency of the optimization algorithm.

For our test, the system frequency can stabilize within 10 seconds, which is quite fast. The proposed algorithm actually realizes optimal generation control online by incorporating the primary control (governor system), secondary control (AGC), and ED.

3.3.12.3 With Communication Channel Loss

The robustness of the algorithm is tested under the condition of loss of a communication channel (e_5 in Figure 3.51), and the test results are shown in Figure 3.60. Following the $N - 1$ rule for communication design ensures that losing any one of the channels will not affect the control performance significantly. Figure 3.60 demonstrates that losing one channel does not degrade the overall control performance dramatically. It is worth pointing out that generally, fewer communication channels does lead to slower converging speed. Therefore, a comprise between converging speed and cost of building the infrastructures of a communication network is necessary during the design phase.

The proposed control solution can be used in other scenarios as well, such as the coordination of virtual power plants (VPPs). VPPs aim at managing a large number of micro/DGs and makes them act like conventional power plants [53]. As more DGs are integrated into the VPP, the proposed control scheme can also

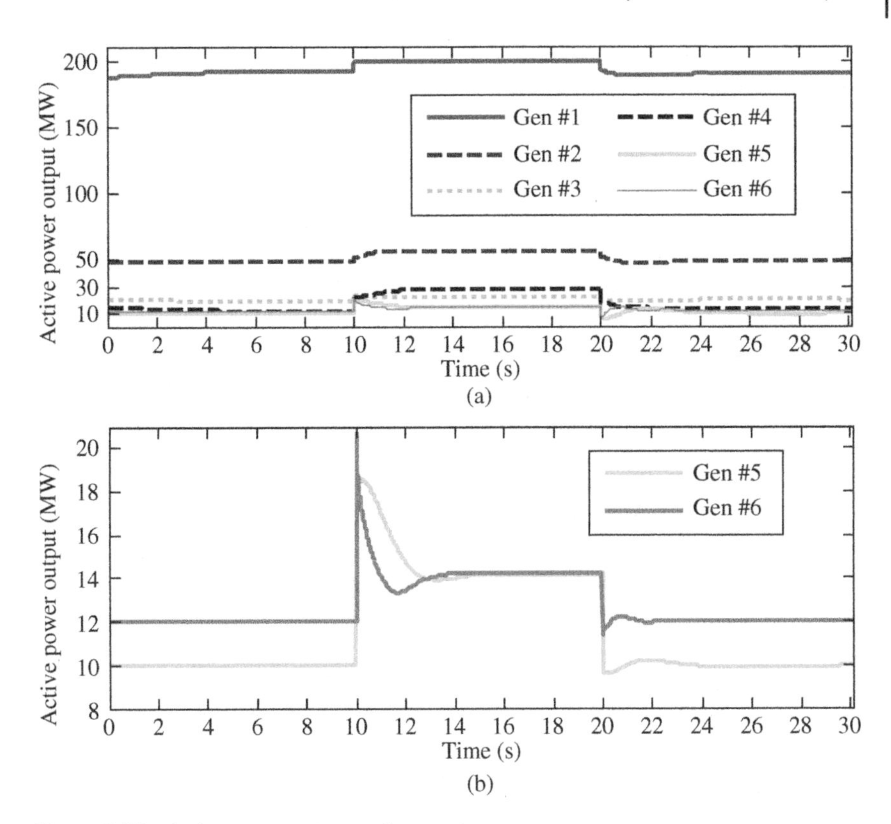

Figure 3.59 Active power outputs of generators.

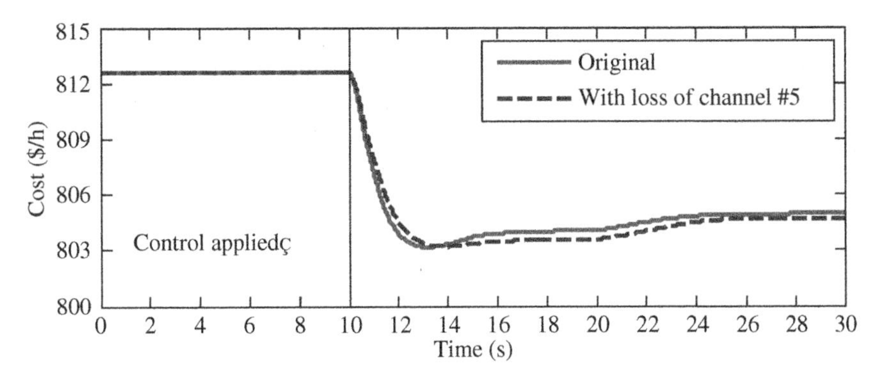

Figure 3.60 Optimization results with one communication channel loss.

realize the real-time optimization by assigning intelligent agents to manage the energy prosumption of VPPs [106].

Another promising application of the proposed control scheme is that we can integrate it with the demand response application to realize the optimization of both energy suppliers and users because demand response is growing more and more active in modern power grids [107]. Demand response is now greatly facilitated by the advanced metering infrastructure (AMI), emerging energy-management controller (EMC) technology, and the concept of aggregator [108]. The aggregator includes energy users that participate in the demand response and actually can be regarded as VPPs to take part in the optimal generation dispatch as its aggregators combine these users into one purchasing unit to trade with the transmission system operator (TSO), retailers, and distributors [109].

The intelligent distributed load management and control system that discussed in [110] can also be realized by integrating demand response with the proposed control algorithm.

It is of significant importance to extend this control method considering the integration of emerging industry activities, e.g. DER, control for VPP, and aggregator. The proposed distributed method only requires low complexity computation and communication and obtains a more viable solution [54]. It can also be integrated with the emerging just-in-time corrective action control such as special protection schemes (SPSs), remedial action schemes (RASs) to render fast control, and optimization solutions for the power system under either normal or abnormal operating conditions [111].

3.3.13 Conclusion and Discussion

In this chapter, we proposed a real-time distributed control solution method to solve the online optimal generation control problem. The solution method is designed based on the constrained gradient algorithm, which can handle both equality and inequality constraints. The proposed solution is implemented based on MAS architecture with the design of the topology of the communication network following the $N-1$ rule. The real-time simulation shows that the proposed control solution is able to realize generation cost reduction in an optimal way while yielding decent dynamic performances under a variety of operating conditions.

3.A Appendix

The governor droop K_{Pi} is 20 for each generator. The inertia of speed governor and turbine is 0.1 and 0.5 seconds, respectively.

The parameters of generators:

- Generator #1: $P_B = 200\,\text{MW}, H = 4.2\,\text{Seconds}, x_d = 1.81\,\text{p.u.}, x_q = 1.76\,\text{p.u.}, x'_d = 0.32\,\text{p.u.}$, $x'_q = 0.35$ p.u., $T'_{d0} = 6.51$ Seconds, $T'_{q0} = 0.34$ Seconds.
- Generator #2: $P_B = 80$ MW, $H = 3.6$ Seconds, $x_d = 2.95$ p.u., $x_q = 2.37$ p.u., $x'_d = 0.69$ p.u., $x'_q = 0.69$ p.u., $T'_{d0} = 6.02$ Seconds, $T'_{q0} = 0.37$ Seconds.
- Generator #3: $P_B = 50$ MW, $H = 3.5$ Seconds, $x_d = 2.95$ p.u., $x_q = 2.92$ p.u., $x'_d = 0.49$ p.u., $x'_q = 0.52$ p.u., $T'_{d0} = 5.66$ Seconds, $T'_{q0} = 0.41$ Seconds.
- Generator #4: $P_B = 35$ MW, $H = 3.5$ Seconds, $x_d = 2.33$ p.u., $x_q = 2.25$ p.u., $x'_d = 0.57$ p.u., $x'_q = 0.62$ p.u., $T'_{d0} = 4.79$ Seconds, $T'_{q0} = 1.96$ Seconds.
- Generator #5: $P_B = 30$ MW, $H = 5.3$ Seconds, $x_d = 6.71$ p.u., $x_q = 6.72$ p.u., $x'_d = 1.34$ p.u., $x'_q = 1.32$ p.u., $T'_{d0} = 5.46$ Seconds, $T'_{q0} = 1.32$ Seconds.
- Generator #6: $P_B = 40$ MW, $H = 4.9$ Seconds, $x_d = 3.62$ p.u., $x_q = 3.67$ p.u., $x'_d = 0.49$ p.u., $x'_q = 0.46$ p.u., $T'_{d0} = 5.69$ Seconds, $T'_{q0} = 1.50$ Seconds.

References

1 Abu-Sharkh, S., Arnold, R.J., Kohler, J. et al. (2006). Can microgrids make a major contribution to UK energy supply? *Renewable and Sustainable Energy Reviews* 10 (2): 78–127.

2 Olivares, D.E., Mehrizi-Sani, A., Etemadi, A.H. et al. (2014). Trends in microgrid control. *IEEE Transactions on Smart Grid* 5 (4): 1905–1919.

3 Femia, N., Petrone, G., Spagnuolo, G. and Vitelli, M. (2005). Optimization of perturb and observe maximum power point tracking method. *IEEE Transactions on Power Electronics* 20 (4): 963–973.

4 Koutroulis, E. and Kalaitzakis, K. (2006). Design of a maximum power tracking system for wind-energyconversion applications. *IEEE Transactions on Industrial Electronics* 53 (2): 486–494.

5 Qu, L. and Qiao, W. (2011). Constant power control of DFIG wind turbines with supercapacitor energy storage. *IEEE Transactions on Industry Applications* 47 (1): 359–367.

6 Rodriguez-Amenedo, J.L., Arnalte, S., and Burgos, J.C. (2002). Automatic generation control of a wind farm with variable speed wind turbines. *IEEE Power Engineering Review* 22 (5): 65–65.

7 de Almeida, R.G., Castronuovo, E.D., and Lopes, J.A.P. (2006). Optimum generation control in wind parks when carrying out system operator requests. *IEEE Transactions on Power Systems* 21 (2): 718–725.

8 Chang-Chien, L., Hung, C., and Yin, Y. (2008). Dynamic reserve allocation for system contingency by DFIG wind farms. *IEEE Transactions on Power Systems* 23 (2): 729–736. https://doi.org/10.1109/TPWRS.2008.920071.

9 Senjyu, T., Datta, M., Yona, A., Sekine, H. and Funabashi, T. (2007). A coordinated control method for leveling output power fluctuations of multiple PV systems. *2007 7th International Conference on Power Electronics*, pp. 445–450.

10 Xin, H., Qu, Z., Seuss, J. and Maknouninejad, A. (2011). A self-organizing strategy for power flow control of photovoltaic generators in a distribution network. *IEEE Transactions on Power Systems* 26 (3): 1462–1473.

11 Dimeas, A. and Hatziargyriou, N. (2004). A multiagent system for microgrids. *IEEE Power Engineering Society General Meeting, 2004*, Volume 1, pp. 55–58.

12 Colson, C.M. and Nehrir, M.H. (2011). Algorithms for distributed decision-making for multi-agent microgrid power management. *2011 IEEE Power and Energy Society General Meeting*, pp. 1–8.

13 Kulasekera, A.L., Gopura, R.A.R.C., Hemapala, K.T.M.U. and Perera, N. (2011). A review on multi-agent systems in microgrid applications. *ISGT2011-India*, pp. 173–177.

14 Zhao, B., Guo, C.X., and Cao, Y.J. (2005). A multiagent-based particle swarm optimization approach for optimal reactive power dispatch. *IEEE Transactions on Power Systems* 20 (2): 1070–1078.

15 Xu, Y. and Liu, W. (2011). Novel multiagent based load restoration algorithm for microgrids. *IEEE Transactions on Smart Grid* 2 (1): 152–161.

16 Xu, Y., Liu, W., and Gong, J. (2011). Stable multi-agent-based load shedding algorithm for power systems. *IEEE Transactions on Power Systems* 26 (4): 2006–2014.

17 Zhang, W., Xu, Y., Liu, W., Ferrese, F. and Liu, L. (2013). Fully distributed coordination of multiple DFIGs in a microgrid for load sharing. *IEEE Transactions on Smart Grid* 4 (2): 806–815.

18 Xiao, L. and Boyd, S. (2006). Optimal scaling of a gradient method for distributed resource allocation. *Journal of Optimization Theory and Applications* 129 (3): 469–488.

19 Nedic, A. and Ozdaglar, A. (2009). Distributed subgradient methods for multi-agent optimization. *IEEE Transactions on Automatic Control* 54 (1): 48–61.

20 Lobel, I. and Ozdaglar, A. (2011). Distributed subgradient methods for convex optimization over random networks. *IEEE Transactions on Automatic Control* 56 (6): 1291–1306.

21 O'Sullivan, J.W. and O'Malley, M.J. (1996). Identification and validation of dynamic global load model parameters for use in power system frequency simulations. *IEEE Transactions on Power Systems* 11 (2): 851–857.

22 Chandorkar, M.C., Divan, D.M., and Adapa, R. (1993). Control of parallel connected inverters in standalone AC supply systems. *IEEE Transactions on Industry Applications* 29 (1): 136–143.

23 Chandrasena, R.P.S., Arulampalam, A., Ekanayake, J.B. and Abeyratne, S.G. (2007). Grid side converter controller of DFIG for wind power generation. *2007 International Conference on Industrial and Information Systems*, pp. 141–146.

24 Chaudhuri, N.R. and Chaudhuri, B. (2013). Adaptive droop control for effective power sharing in multi-terminal DC (MTDC) grids. *IEEE Transactions on Power Systems* 28 (1): 21–29.

25 Chen, C.H. (2008). Economic dispatch using simplified personal best oriented particle swarm optimizer. *2008 3rd International Conference on Electric Utility Deregulation and Restructuring and Power Technologies*, pp. 572–576.

26 Chiang, S.J. and Chang, J.M. (2001). Parallel control of the UPS inverters with frequency-dependent droop scheme. *2001 IEEE 32nd Annual Power Electronics Specialists Conference (IEEE Cat. No.01CH37230)*, Volume 2, pp. 957–961.

27 Katiraei, F. and Iravani, M.R. (2006). Power management strategies for a microgrid with multiple distributed generation units. *IEEE Transactions on Power Systems* 21 (4): 1821–1831.

28 He, J. and Li, Y.W. (2011). Analysis, design, and implementation of virtual impedance for power electronics interfaced distributed generation. *IEEE Transactions on Industry Applications* 47 (6): 2525–2538.

29 Yao, W., Chen, M., Matas, J., Guerrero, J.M. and Qian, Z.M. (2011). Design and analysis of the droop control method for parallel inverters considering the impact of the complex impedance on the power sharing. *IEEE Transactions on Industrial Electronics* 58 (2): 576–588.

30 Guerrero, J.M., Vasquez, J.C., Matas, J., De Vicuña, L.G. and Castilla, M. (2011). Hierarchical control of droop-controlled ac and dc microgrids--a general approach toward standardization. *IEEE Transactions on Industrial Electronics* 58 (1): 158–172.

31 Delghavi, M.B. and Yazdani, A. (2011). An adaptive feedforward compensation for stability enhancement in droop-controlled inverter-based microgrids. *IEEE Transactions on Power Delivery* 26 (3): 1764–1773.

32 Mohamed, Y.A.I. and El-Saadany, E.F. (2008). Adaptive decentralized droop controller to preserve power sharing stability of paralleled inverters in distributed generation microgrids. *IEEE Transactions on Power Electronics* 23 (6): 2806–2816.

33 Bidram, A. and Davoudi, A. (2012). Hierarchical structure of microgrids control system. *IEEE Transactions on Smart Grid* 3 (4): 1963–1976.

34 Olfati-Saber, R., Fax, J.A., and Murray, R.M. (2007). Consensus and cooperation in networked multi-agent systems. *Proceedings of the IEEE* 95 (1): 215–233.

35 Qu, Z. (2009). Cooperative control of dynamical systems: applications to autonomous vehicles. Springer Science & Business Media.

36 Foley, A.M., Leahy, P.G., and McKeogh, E.J. (2010). Wind power forecasting amp; prediction methods. *2010 9th International Conference on Environment and Electrical Engineering*, pp. 61–64.

37 Foley, A.M., Leahy, P.G., and McKeogh, E.J. (2011). Algorithms for distributed decision-making for multiagent microgrid power management. *2011 IEEE Power and Energy Society General Meeting*, pp. 1–8.

38 Lydia, M. and Kumar, S.S. (2010). A comprehensive overview on wind power forecasting. *2010 Conference Proceedings IPEC*, pp. 268–273.

39 Esram, T. and Chapman, P.L. (2007). Comparison of photovoltaic array maximum power point tracking techniques. *IEEE Transactions on Energy Conversion* 22 (2): 439–449.

40 Bahgat, A.B.G., Helwa, N.H., Ahamd, G.E. and El Shenawy, E.T. (2004). Estimation of the maximum power and normal operating power of a photovoltaic module by neural networks. *Renewable Energy* 29: 443–457.

41 Ying, H.H. (1998). Grid integration of wind energy conversion systems: Siegfried Heier. Chichester: Wiley. ISBN: 0-471-97143-X. (2000). *Renewable Energy* 21: 607–608.

42 Luna, A., Lima, F.D.A., Santos, D. et al. Simplified modeling of a DFIG for transient studies in wind power applications. *IEEE Transactions on Industrial Electronics* 58 (1): 9–20.

43 Abad, G., Lopez, J., Rodriguez, M., Marroyo, L. and Iwanski, G. (2011). *Doubly Fed Induction Machine: Modeling and Control for Wind Energy Generation*, vol. 86, 619–625. ISBN: 9781118104965. doi: 10.1002/9781118104965.

44 Tremblay, E., Chandra, A., Lagace, P.J. and Gagnon, R. (2006). Study of grid-side converter control for grid-connected DFIG wind turbines under unbalanced load condition. *2006 IEEE International Symposium on Industrial Electronics*, Volume 2, pp. 1619–1624.

45 Datta, M., Senjyu, T., Yona, A., Funabashi, T. and Kim, C.H. (2009). A coordinated control method for leveling PV output power fluctuations of PV-diesel hybrid systems connected to isolated power utility. *IEEE Transactions on Energy Conversion* 24 (1): 153–162.

46 Mohamed, F.A. and Koivo, H.N. (2007). Online management of microgrid with battery storage using multiobjective optimization. *2007 International Conference on Power Engineering, Energy and Electrical Drives*, pp. 231–236.

47 IEEE Committee Report (1973). Dynamic models for steam and hydro turbines in power system studies. *IEEE Transactions on Power Apparatus and Systems* PAS-92 (6): 1904–1915.

48 Hajagos, L.M. and Basler, M.J. (2005). Changes to IEEE 421.5 recommended practice for excitation system models for power system stability studies. *IEEE Power Engineering Society General Meeting,* 2005, pp. 334–336.

49 Ilic, M.D., Xie, L., and Joo, J. (2011). Efficient coordination of wind power and price-responsive demand--Part I: Theoretical foundations. *IEEE Transactions on Power Systems* 26 (4): 1875–1884.

50 Ipakchi, A. and Albuyeh, F. (2009). Grid of the future. *IEEE Power and Energy Magazine* 7 (2): 52–62.

51 Irisarri, G., Kimball, L.M., Clements, K.A., Bagchi, A. and Davis, P.W. (1998). Economic dispatch with network and ramping constraints via interior point methods. *IEEE Transactions on Power Systems* 13 (1): 236–242.

52 Jaleeli, N., VanSlyck, L.S., Ewart, D.N., Fink, L.H. and Hoffmann, A.G. (1992). Understanding automatic generation control. *IEEE Transactions on Power Systems* 7 (3): 1106–1122.

53 Jiang, L. and Low, S. (2011). "Multi-period optimal energy procurement and demand response in smart grid with uncertain supply. *2011 50th IEEE Conference on Decision and Control and European Control Conference*, pp. 4348–4353.

54 Kar, S. and Hug, G. (2012). Distributed robust economic dispatch in power systems: a consensus + innovations approach. *2012 IEEE Power and Energy Society General Meeting, 2012 IEEE*. IEEE, pp. 1–8.

55 Kirby, B. and Hirst, E. (1998). Generator response to intrahour load fluctuations. *IEEE Transactions on Power Systems* 13 (4): 1373–1378.

56 Ding, Y.M., Hong, S.H., and Li, X.H. (2014). A demand response energy management scheme for industrial facilities in smart grid. *IEEE Transactions on Industrial Informatics* 10 (4): 2257–2269.

57 Binetti, G., Davoudi, A., Naso, D., Turchiano, B. and Lewis, F.L. (2014). A distributed auction-based algorithm for the nonconvex economic dispatch problem. *IEEE Transactions on Industrial Informatics* 10 (2): 1124–1132.

58 Dhar, R.N. and Mukherjee, P.K. (1973). Reduced-gradient method for economic dispatch. *Proceedings of the Institution of Electrical Engineers* 120 (5): 608–610.

59 Lin, C.E. and Viviani, G.L. (1984). Hierarchical economic dispatch for piecewise quadratic cost functions. *IEEE Transactions on Power Apparatus and Systems* PAS-103 (6): 1170–1175.

60 Lin, C.E., Chen, S.T., and Huang, C.-L. (1992). A direct Newton-Raphson economic dispatch. *IEEE Transactions on Power Systems* 7 (3): 1149–1154.

61 Sinha, N., Chakrabarti, R., and Chattopadhyay, P.K. (2003). Evolutionary programming techniques for economic load dispatch. *IEEE Transactions on Evolutionary Computation* 7 (1): 83–94.

62 Damousis, I.G., Bakirtzis, A.G., and Dokopoulos, P.S. (2003). Network-constrained economic dispatch using real-coded genetic algorithm. *IEEE Transactions on Power Systems* 18 (1): 198–205.

63 Selvakumar, A.I. and Thanushkodi, K. (2007). A new particle swarm optimization solution to nonconvex economic dispatch problems. *IEEE Transactions on Power Systems* 22 (1): 42–51.

64 Yao, F., Dong, Z.Y., Meng, K. et al. (2012). Quantum-inspired particle swarm optimization for power system operations considering wind power uncertainty and carbon tax in Australia. *IEEE Transactions on Industrial Informatics* 8 (4): 880–888.

65 Sum-im, T. (2004). Economic dispatch by ant colony search algorithm. *2004 IEEE Conference on Cybernetics and Intelligent Systems*, Volume 1. IEEE, pp. 416–421.

66 Zhang, W., Liu, W., Wang, X., Liu, L. and Ferrese, F. (2015). Online optimal generation control based on constrained distributed gradient algorithm. *IEEE Transactions on Power Systems* 30 (1): 35–45.

67 Nutkani, I.U., Loh, P.C., and Blaabjerg, F. (2014). Cost-based droop scheme with lower generation costs for microgrids. *IET Power Electronics* 7 (5): 1171–1180.

68 Mudumbai, R., Dasgupta, S., and Cho, B.B. (2012). Distributed control for optimal economic dispatch of a network of heterogeneous power generators. *IEEE Transactions on Power Systems* 27 (4): 1750–1760.

69 Mokhtari, G., Nourbakhsh, G., and Ghosh, A. (2013). Smart coordination of energy storage units (ESUs) for voltage and loading management in distribution networks. *IEEE Transactions on Power Systems* 28 (4): 4812–4820.

70 Zhang, Z. and Chow, M.-Y. (2012). Convergence analysis of the incremental cost consensus algorithm under different communication network topologies in a smart grid. *IEEE Transactions on Power Systems* 27 (4): 1761–1768.

71 Loia, V. and Vaccaro, A. (2014). Decentralized economic dispatch in smart grids by self-organizing dynamic agents. *IEEE Transactions on Systems, Man, and Cybernetics: Systems* 44 (4): 397–408.

72 Yang, S., Tan, S., and Xu, J.-X. (2013). Consensus based approach for economic dispatch problem in a smart grid. *IEEE Transactions on Power Systems* 28 (4): 4416–4426.

73 Li, S., Oikonomou, G., Tryfonas, T., Chen, T.M. and Da Xu, L. (2014). A distributed consensus algorithm for decision making in service-oriented internet of things. *IEEE Transactions on Industrial Informatics* 10 (2): 1461–1468.

74 Rahbari-Asr, N., Zhang, Z., and Chow, M.-Y. (2013). Consensus-based distributed energy management with real-time pricing. Power and Energy Society General Meeting (PES), 2013 IEEE. IEEE, pp. 1-5.

75 Rahbari-Asr, N. and Chow, M.-Y. (2014). Cooperative distributed demand management for community charging of PHEV/PEVs based on KKT conditions and consensus networks. *IEEE Transactions on Industrial Informatics* 10 (3): 1907–1916.

76 Vaccaro, A., Loia, V., Formato, G., Wall, P. and Terzija, V. (2015). A self-organizing architecture for decentralized smart microgrids synchronization, control, and monitoring. *IEEE Transactions on Industrial Informatics* 11 (1): 289–298.

77 Bertsekas, D.P. (1977). Monotone mappings with application in dynamic programming. *SIAM Journal on Control and Optimization* 15 (3): 438–464.

78 Bertsekas, D.P. and Shreve, S. (2004). *Stochastic Optimal Control: The Discrete-Time Case.* Belmont, USA: Anthena Scientific

79 Binetti, G., Davoudi, A., Lewis, F.L., Naso, D. and Turchiano, B. (2014). Distributed consensus-based economic dispatch with transmission losses. *IEEE Transactions on Power Systems* 29 (4): 1711–1720.

80 Wood, A.J. and Wollenberg, B.F. (2012). *Power Generation, Operation, and Control.* Wiley.

81 Bertsekas, D. (1982). Distributed dynamic programming. *IEEE Transactions on Automatic Control* 27 (3): 610–616.

82 Christie, R. (1993). *Power Systems Test Case Archive: 30 Bus Power Flow Test Case.* Online resources. Retrieved 2020: https://labs.ece.uw.edu/pstca/pf30/pg_tca30bus.htm

83 Kuo, D.-H. and Bose, A. (1995). A generation rescheduling method to increase the dynamic security of power systems. *IEEE Transactions on Power Systems* 10 (1): 68–76.

84 Berna-Koes, M., Nourbakhsh, I., and Sycara, K. (2004). Communication efficiency in multi-agent systems. *2004 IEEE International Conference on Robotics and Automation, 2004. Proceedings. ICRA'04,* Volume 3. IEEE, pp. 2129–2134.

85 Chowdhury, B.H. and Rahman, S. (1990). A review of recent advances in economic dispatch. *IEEE Transactions on Power Systems* 5 (4): 1248–1259.

86 Wood, A.J. and Wollenberg, B.F. (1996). *Power Generation, Operation, and Control, 32-33.* New York: Wiley.

87 Lin, C.E., Chen, S.T., and Huang, C. (1992). A direct Newton-Raphson economic dispatch. *IEEE Transactions on Power Systems* 7 (3): 1149–1154.

88 Mohammadi, A., Varahram, M.H., and Kheirizad, I. (2006). Online solving of economic dispatch problem using neural network approach and comparing it with classical method. *2006 International Conference on Emerging Technologies,* pp. 581–586.

89 Bakirtzis, A., Petridis, V., and Kazarlis, S. (1994). Genetic algorithm solution to the economic dispatch problem. *IEE Proceedings of Generation, Transmission and Distribution* 141 (4): 377–382.

90 Gaing, Z. (2003). Particle swarm optimization to solving the economic dispatch considering the generator constraints. *IEEE Transactions on Power Systems* 18 (3): 1187–1195.

91 Park, J., Jeong, Y. W., Shin, J. R., & Lee, K. Y. (2010). An improved particle swarm optimization for nonconvex economic dispatch problems. *IEEE Transactions on Power Systems* 25 (1): 156–166.

92 Lian, J., Marinovici, L., Kalsi, K., Du, P., & Elizondo, M. (2012). Distributed hierarchical control of multi-area power systems with improved primary frequency regulation. *2012 IEEE 51st IEEE Conference on Decision and Control (CDC)*, pp. 444–449.

93 Zhang, Y., Gatsis, N., and Giannakis, G.B. (2013). Robust energy management for microgrids with high-penetration renewables. *IEEE Trans. Sustainable Energy* 4 (4): 944–953.

94 Qi, W., Liu, J., and Christofides, P.D. (2011). A distributed control framework for smart grid development: energy/water system optimal operation and electric grid integration. *Journal of Process Control* 21 (10): 1504–1516.

95 Mudumbai, R., Dasgupta, S., and Cho, B. (2010). Distributed control for optimal economic dispatch of power generators. *Proceedings of the 29th Chinese Control Conference*, pp. 4943–4947.

96 Zhang, Z., Ying, X., and Chow, M.-Y. (2011). Decentralizing the economic dispatch problem using a two-level incremental cost consensus algorithm in a smart grid environment. *2011 North American Power Symposium*, pp. 1–7.

97 Zhang, Z. and Chow, M. (2011). Incremental cost consensus algorithm in a smart grid environment. *2011 IEEE Power and Energy Society General Meeting*, pp. 1–6.

98 Srivastava, K. and Nedic, A. (2011). Distributed asynchronous constrained stochastic optimization. *IEEE Journal on Selected Topics in Signal Processing* 5 (4): 772–790.

99 Zhu, M. and Martinez, S. (2012). On distributed convex optimization under inequality and equality constraints. *IEEE Transactions on Automatic Control* 57 (1): 151–164.

100 Xiao, L., Boyd, S., and Kim, S.-J. (2007). Distributed average consensus with least-mean-square deviation. *Journal of Parallel and Distributed Computing* 67 (1): 33–46.

101 Bondy, J.A. (1976). *Graph Theory with Applications*. Oxford: Elsevier Science Ltd.

102 Kundur, P. (1994). *Power System Stability and Control*. New York: McGraw-Hill.

103 Bellifemine, F., Poggi, A., and Rimassa, G. (2001). Developing multi-agent systems with JADE. In: *Intelligent Agents VII Agent Theories Architectures* *and Languages* (ed. C. Castelfranchi and Y. Lesp'erance), 89–103. Berlin, Heidelberg: Springer-Verlag.

104 Alsac, O. and Stott, B. (1974). Optimal load flow with steady-state security. *IEEE Transactions on Power Apparatus and Systems* PAS-93 (3): 745–751.

105 Rebours, Y.G., Kirschen, D.S., Trotignon, M. and Rossignol A Survey of Frequency and Voltage Control Ancillary Services--Part I: Technical features. *IEEE Transactions on Power Systems* 22 (1): 350–357.

106 Wong, V.W.S. (2012). Advanced demand side management for the future smart grid using mechanism design. *IEEE Transactions on Smart Grid* 3 (3): 1170–1180.

107 Papavasiliou, A., Hindi, H., and Greene, D. (2010). Market-based control mechanisms for electric power demand response. 49th IEEE Conference on Decision and Control (CDC), pp. 1891-1898.

108 Yang, P., Chavali, P., and Nehorai, A. (2012). Parallel autonomous optimization of demand response with renewable distributed generators. *2012 IEEE Third International Conference on Smart Grid Communications (SmartGridComm)*, pp. 55–60.

109 Nguyen, D.T., Negnevitsky, M., and de Groot, M. (2011). Pool-based demand response exchange--concept and modeling. *IEEE Transactions on Power Systems* 26 (3): 1677–1685.

110 Zhang, W., Zhou, S., and Lu, Y. (2012). Distributed intelligent load management and control system. *2012 IEEE Power and Energy Society General Meeting*, pp. 1–8.

111 Tomsovic, K., Bakken, D.E., Venkatasubramanian, V., and Bose, A. (2005). Designing the next generation of real-time control, communication, and computations for large power systems. *Proceedings of the IEEE* 93 (5): 965–979.

4

Distributed Reactive Power Control

As for the power system, long-distance reactive power transmission is not possible. Thus, the voltage-related reactive control should be naturally implemented in a distributed manner. In this chapter, we will discuss two types of the reactive power control methods. The first method is based on the well-known artificial intelligence algorithm, Q-learning algorithm. The second method is based on the distributed sub-gradient algorithm. The distributed implementation of control solutions based on a multi-agent system (MAS) framework for these two methods is also discussed in this chapter.

4.1 Q-Learning-Based Reactive Power Control

A reinforcement learning (RL) method based on a fully distributed multi-agent framework for reactive power dispatch is proposed in this section. In this method, two agents exchange information with each other only when their own buses are electrically coupled. The global rewards of RL come from a consensus-based global information collection algorithm. In addition, this algorithm is demonstrated to be reliable and efficient. In order to reach the goal of minimizing the active power loss and satisfy operational constraints at the same time, a distributed Q-learning algorithm is implemented. The proposed method is able to learn from scratch without an accurate system model. The results from the simulation of different sizes of power systems show that the method has great computational efficiency and gains nearly optimal solutions. Also, the results show that appropriate prior knowledge can highly accelerate the learning algorithm and reduce the existence of unnecessary disturbances. This method is a good potential candidate for online implementation.

Distributed Energy Management of Electrical Power Systems, First Edition.
Yinliang Xu, Wei Zhang, Wenxin Liu, and Wen Yu.

Published 2021 by John Wiley & Sons, Inc.

4.1.1 Introduction

To decrease real power loss and improve the voltage profile of the power system without violating certain operational constraints, optimal reactive power dispatch (ORPD) is widely used. ORPD can be realized by appropriately changing control circumstances such as generator bus voltage reference, transformer tap setting, capacitor bank switching, etc. Although various kinds of traditional optimization algorithms have been proposed to deal with ORPD problem, such as linear programming [1], Newton method [2], and interior point (IP) method [3, 4], there exist some shortcomings of these methods including sensitivity to initialization and mathematical characteristics of objective functions, known as convexity, differentiability, and continuity.

In recent years, intelligent optimization algorithms were applied to deal with the weaknesses of the above traditional algorithms. These algorithms mimic certain physical phenomena to reach ideal results. In this way, these algorithms have the advantages of computational efficiency [5] and have the ability to avoid the calculation of derivatives. In [6, 7], algorithms based on ant colony and particle swarm optimization were proposed for ORPD to reach the minimal real power loss with constraints. These kinds of algorithms not only need global information but also need to be implemented offline in a centralized way.

In order to collect global operating conditions and deal with the huge amount of data, it is necessary for centralized control schemes to have complicated communication networks and a powerful central control system. Therefore, centralized schemes appear expensive to implement and sensitive to failures of a single point, while distributed schemes appear more flexible, low cost, and reliable to implement [8, 9]. Therefore, they are regarded as an ideal choice for next-generation power systems.

MAS, one of the most popular distributed control solutions, has the advantages of surviving single-point failures [10, 11] and decentralized data processing. Thus, MAS can realize the efficient distribution of tasks, which finally accelerates the operation and decision-making process [12]. In the area of artificial intelligence, agent-based technology has been regarded as a promising paradigm for conceptualizing, designing,and implementing software systems. In recent years, many scholars have applied MAS-based methods to deal with the problem of power system reconfiguration and restoration [8, 9, 11]. However, there still remain some problems with the existing algorithms. One of the most important shortcomings is that most methods were only validated through simulations rather than rigorous analysis so that the convergence and stability of these algorithms still remain under-discussion.

Known as one of the most complex systems around the world, the power system holds incredible complexity, so that it is very difficult to propose a direct model-based design for a power system. Thus, a desirable algorithm should avoid the analysis of such complex models. RL, which iteratively learns the optimal act from experiences in an unknown system, meets the above requirement. During the learning process, the action of the system is updated based on a reward signal which judges the performance of the system. Whereas in distributed RL, the calculation and distribution of global reward signals, which are important for autonomous agents' learning process, still remain a difficult problem.

The implementation of an ideal ORPD solution should have some characteristics, i.e. distributed, adaptive, optimal, online, and suitable. "Adaptive" represents the algorithm that has the ability to adapt to various kinds of operating conditions. "Optimal" represents that the result should gain optimal real-power loss and voltage profile. In order to achieve the online implementation, the algorithm should also have the advantage of computational efficiency. Algorithms holding these characteristics are more likely to keep up with the developing tendency of power systems [8–12].

To meet the needs of power systems and deal with the problems of the existing solutions, in this work, a multi-agent system-based reinforcement learning (MASRL) is proposed. To reach the objective of minimizing real power loss without violating certain operating constraints, an integration of two algorithms is realized, which is called the MASRL algorithm. This new algorithm combines average-consensus-based algorithms and distributed RL algorithms.

Because of the fully distributed MAS framework, the proposed solution not only holds properties of centralized solutions but also holds properties of distributed solutions. Further demonstration of the effectiveness of the proposed solution comes from simulation studies.

4.1.2 Background

Through a discovery algorithm based on average consensus, the fully distributed solution achieves not only the calculation of global information but also the distribution of global information. RL is used to deal with distributed optimization. Some introductions about these two algorithms are presented below.

4.1.3 Algorithm Used to Collect Global Information

There are two reasons due to which discovering global information becomes a challenging problem. The first one is that only local communications are useful. The other one is that the power network is extremely complex. To deal with these problems, an average-consensus theorem-based global information collecting algorithm is designed. In the average-consensus theorem, important information is guaranteed to be shared in a distributed way by local information of each agent [13]. Moreover, consensus algorithm has attracted great interests of many scholars in different areas, and a survey is accessible in [14]. Considering the global information collection algorithm, the process of updating information for agent i can be presented as

$$x_{i,m+1} = x_{i,m} + \sum_{j=1}^{n} a_{i,j}(x_{j,m} - x_{i,m}) \tag{4.1}$$

where $x_{i,m}$ and $x_{j,m}$ are the local information collected by agents i and j at iteration m, while $x_{i,m+1}$ have the same meaning at iteration $m+1$. $a_{i,j}$ is the coefficient of information communicated between agents i and j. Moreover, n is the number of agents in the information collecting process. As shown in Eq. (4.2), by rigorous stability analysis, it can be observed that only when the coefficients a_{ij}s meet certain constraints [9, 15], will all x_is converge to the same value based on rigorous stability analysis.

$$x_{i,\infty} = \frac{1}{n}\sum_{i=1}^{n} x_{i,0}, \qquad i = 1 \sim n \tag{4.2}$$

Based on the consensus theory, the whole information exchange process is modeled as a linear system based on discrete time, as shown in Eq. (4.3).

$$X_{m+1} = DX_m \tag{4.3}$$

Respectively, X_m and X_{m+1} are the vectors of collected information at the mth and $(m+1)$th iterations, and D is a sparse iteration matrix. The time needed for convergence is determined by Eq. (4.4) [13].

$$\tau = \frac{1}{\log_E(|\lambda_2|)} \tag{4.4}$$

In the above equation, $E(0 < E < 1)$ is a measurement of tolerance for error and is predefined. $|\lambda_2|$ is the second largest eigenvalue of D.

From Eq. (4.4), it can be observed that the time needed for convergence decreases asymptotically by a factor of E. While another factor $|\lambda_2|$ determines the converging speed. Therefore, $|\lambda_2|$ can be regarded as the evaluation of the speed of an information discovery algorithm. References [16, 17] show that the convergence for any systems can be guaranteed through the information discovery algorithm as long as the discovered global information can be expressed as a summation of local information.

4.1.4 Reinforcement Learning

RL is a subarea of machine learning. It mainly focuses on what an agent is supposed to do in an unknown environment so that it can reach maximal cumulative reward [18, 19]. RL learns how to take action in an optimal way based on its experience in the unknown environment [20]. Major advantages of RL include [21, 22]:

(i) Only need less accurate or even no environment model. This is a critical advantage because, in very complex systems, the environment is too difficult to be abstracted as a model.
(ii) The ability to learn from scratch and the online learning process.
(iii) A relatively clear and simple evaluation of the quality of a solution. While other methods need complicated mathematical operations such as gradient or inversion of a matrix, it only needs a clear reward function for evaluation. This property largely reduces the computational complexity and relieves some mathematical property restrictions imposed on the objective function.
(iv) It can avoid local optimum by following the probabilistic transition rules. In recent years, RL algorithms are applied in various kinds of domains, including traffic light control, robot games, resource management, etc. A survey of RL application can be found in [23]. However, the distributed implementation of the RL algorithm still remains a difficult problem. The major difficulty is the calculation of global reward. This paper solves this problem through an average-consensus-based global information discovery algorithm.

4.1.5 MAS-Based RL Algorithm for ORPD

In ORPD problems, there are three control variables. The first one is the generator bus voltage magnitude, the second one is the capacitor bank switching, and the third one is the transformer tap setting. The former two variables can directly change the injected reactive power, and the last one can make a difference to the Y-bus matrix of the power system [15], which will indirectly make a difference to power flow as

$$P_{Gi} - P_{Li} - V_i - \sum_{j=1}^{n} V_j Y_{i,j} \cos(\delta_i - \delta_j - \theta_{i,j}) = 0 \tag{4.5}$$

$$Q_{Gi} + Q_{Ci} - Q_{Li} - V_i \sum_{j=1}^{n} V_j Y_{i,j} \sin(\delta_i - \delta_j - \theta_{i,j}) = 0 \tag{4.6}$$

In the above equations, n is the number of buses, P_{Gi}, $Q - Gi$ are the active and reactive power generation at bus i, P_{Li}, Q_{Li} are the active and reactive power load at bus i, Q_{Ci} is the capacitor bank reactive power compensation at bus i, $Y_{i,j}$, $\theta_{i,j}$ are the Y-bus admittance matrix elements, V_i and V_j are the voltage magnitudes at bus i and j, and δ_i and δ_j are the voltage phase angles at bus i and j, respectively.

The definition of the global reward function of distributed RL-based optimization is related to the objective and constrains of ORPD. Once the global reward is defined, a distributed Q-learning algorithm can be applied to search the optimal solution for ORPD.

4.1.6 RL Reward Function Definition

Before implementing RL, the reward function needs to be defined for the evaluation of solution performance. Based on Eq. (4.1), the signal of global reward can be calculated as a combination of local signals, as shown in Eq. (4.7).

$$r^k = \frac{1}{n}\sum_{i=1}^{n} r_{i,0}^k \tag{4.7}$$

where $r_{i,0}^k$ is the local reward based on local signals. Because ORPD aims to minimize real power loss and satisfy certain constraints at the same time, the definition of signals for the local reward needs to include two parts to address these two factors, as shown in Eq. (4.8).

$$r_{i,0}^k = \begin{cases} 0, & \text{if constraints are violated} \\ \left(\frac{1}{K+P_{i,\text{loss}}^k}\right), & \text{otherwise} \end{cases} \tag{4.8}$$

As shown in (4.8), there are three constraints that need to be checked. The first one is the magnitude of the local bus voltage, the second one is local injected reactive power, and the third one is current through connected transmission lines. If anyone of them is violated, it can be observed that the local reward signal becomes zero. K is a constant used to decrease the sensitivity of $r_{i,0}^k$ according to $P_{i,\text{loss}}^k$, while $P_{i,\text{loss}}^k$ is the local active power loss calculated based on Eq. (4.9). In this chapter, K is set to be 5% of the total load divided by the number of transmission lines in the system in the process of normal operating conditions.

$$P_{i,\text{loss}}^k = \sum_{j=1}^{N_i} g_i j^k \left[(V_i^k)^2 + (V_j^k)^2 - 2V_i^k V_j^k \cos(\delta_i^k - \delta_j^k)\right] \tag{4.9}$$

where N_i is the set of indexes of buses corresponding to bus i and g_{ij}^k is the real part of Y_{ij}^k.

In Eq. (4.9), the local active power loss can be calculated by sharing the information of voltage magnitude and phase angle between connected node agents (NAs). It is also possible to directly calculate the loss by the measured active power at the transmitting and receiving ends of the transmission line. After the calculation of local active power losses, the total loss of active power can be shown as:

$$P^k_{\text{Loss}} = \frac{1}{2}\sum_{i=1}^{n} P^k_{i,\text{loss}} \tag{4.10}$$

Because the definition of global reward is the summation of local rewards, the global reward signal can be discovered by the above information collection algorithm. The signal of global reward can be regarded as the evaluation of the performance of a candidate solution, and even the total loss of active power defined as Eq. (4.10) is indirectly calculated. Larger global reward indicates a better current solution.

4.1.7 Distributed Q-Learning for ORPD

After discovering the global rewards through the NAs, implementing the distributed RL algorithm through RL agents becomes possible. It can be observed that the number of RL agents and the number of reactive control devices are equal.

In this chapter, several types of reactive control devices are considered, including capacitor banks, tap changing transformers, and PV generators.

Distributed Q-learning is applied in this chapter for optimization. Q-learning, first introduced by Watkins in 1989 [24], aims to learn an action-value function that represents the ideal utility of a series of actions taken in a given state. These actions follow a fixed rule. Q-leaning is able to evaluate the ideal utility of actions without a clear environment model.

Because each RL agent independently chooses its own action at every learning step in the ORPD problem, the whole problem can be regarded as a single-stage Q-learning problem. Single stage indicates that the agents directly implement the transition from one control action to another. For every possible action, agents have a Q value that represents an evaluation of the usefulness of this action. Then, based on the reward received for the action, these values will be updated in every step [25]. Equation (4.11) shows the traditional centralized Q-learning algorithm.

$$Q^{k+1}(a_1^{k+1}, \cdot, a_i^{k+1}, \cdot, a_n^{k+1}) \leftarrow (1-\alpha)Q^k(a_1^k, \cdot, a_i^k, \cdot, a_n^k) + ar^k \tag{4.11}$$

where a_i is the agent's chosen action, r_k is the global reward of step kth, α is the learning rate while $0 < \alpha < 1$, and $Q^k(.)$ is the Q-value for the candidate solution.

$\prod_{i=1}^{n} n_{ai}$ represents the size of the Q-table, where n_{ai} is the number of possible actions taken by agent i and n is the number of RL agents. When the number of agents and accessible actions increases, the size of the Q-table will geometrically increase.

A distributed Q-learning algorithm proposed in [25, 26] is implemented in this chapter. In this algorithm, $\sum_{i=1}^{n} n_{ai}$ is the size of the Q-table, which requires less memory compared to the one in centralized Q-learning. The rule used for updating the distributed Q-Learning is shown in Eq. (4.12).

$$Q^{k+1}(a_i^{k+1}) \leftarrow (1-\alpha)Q^k(a_i^k) + ar^k \tag{4.12}$$

The balance between exploration and exploitation in the process of RL implementation is very important. Too much exploitation causes failure in finding the global optimal solution, while pure exploration costs too much time degrading the Q-learning algorithm's performance. In this chapter, the transition from current random action to the later optimal action with the highest Q value is defined based on metropolis criterion [27] shown in Eq. (4.13)

$$a_i^k = \begin{cases} a_{bi}^k, & \exp\left(\frac{Q(a_{ri}^k)-Q(a_{bi}^k)}{T^k}\right) < \varepsilon \\ a_{ri}^k, & \text{otherwise} \end{cases} \tag{4.13}$$

where a_i^k is agent is control action and $\varepsilon(0 < \varepsilon < 1)$ is set to be relatively small. T adjusts the balance between exploration and exploitation. As exploitation surpasses exploration, the value of T decreases until reaching some lower limits. T can be initialized as a large value, decreasing at each step based on $T^{k+1} = p \times T^k$ with $0 < p < 1$. Table 4.1 describes the flowing chart of the proposed MASRL algorithm.

4.1.8 MASRL Implementation for ORPD

The algorithms proposed previously are applied to the node and RL agent, respectively. The number of NAs and the number of buses are equal, while the number of RL agents equals to the number of reactive power control devices. Figure 4.1 gives an illustration of the function modules of each kind of agents and operation in MASRL. It is shown in Figure 4.1 that local measurement of NA provides local

Table 4.1 MASRL for optimal reactive power dispatch.

1. Initialize ε, T^0, $Q_i(.) = 0$ with $i = 1,2\ldots N$
2. Repeat until the optimal Q-tables have been obtained
    2.1 Each agent generates an action a_{ri}^k based on the learned experience
    2.2 Each agent selects its best action a_{bi}^k based on the current Q-table
    2.3 Choose $a_i^k = a_{bi}^k$, if $\exp[(Q(a_{ri}^k) - Q(a_{bi}^k))/T^k] < \varepsilon$, otherwise $a_i^k = a_{ri}^k$
    2.4 Evaluate the overall performance of distributed actions using global reward r^k obtained through global information discovery
    2.5 Each agent updates its Q-table independently
    2.6 Update T^{k+1} according to the temperature-dropping criterion
3. Each agent chooses greedy-optimal action (a_i^*), which maximizes the accumulated reward

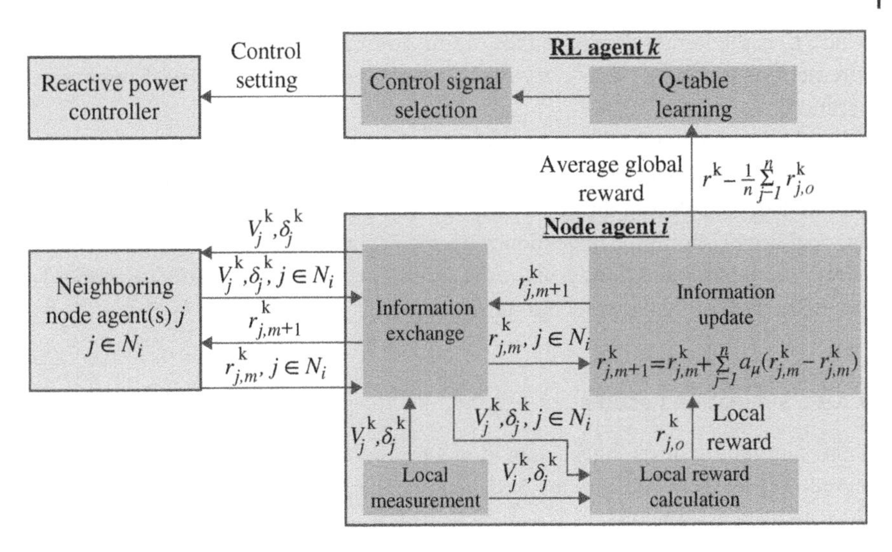

Figure 4.1 Operation of MASRL for ORPD.

voltage and phase angle to both information exchange module and local reward calculation module. Then, the local voltage and reward are exchanged between the information exchange module and its neighboring agents. Also, necessary information for the global reward discovery is provided by the information exchange module.

When a consensus of the global reward is reached, RL agents will use it to update their local Q-table. The updating process will continue until the table does not change anymore. In that case, the learning process will be considered to be convergent. In addition, the current Q-table will decide the optimal action for the associated generator voltage regulating, transformer tap changing, or capacitor bank. Only when the system operating condition changes, will the Q-tables be updated again, such as under load fluctuation.

There are four steps in each iteration of optimization: implementation of the updated control signal according to local Q-table, the measurements of local signals for calculation of local reward, finding global reward, and the updating process of the local Q-table. Because of the fact that measurements can only be available after the convergence of system response and the convergence of the global reward collecting process, most time is used for measurement of signal and discovery of global reward. As a result, estimation of the optimization process can be shown as (4.14)

$$\sum_{i=1}^{K_i}(T_{mi} + T_{di}) \tag{4.14}$$

where K_i is the iteration required for optimization and T_{mi} is time spent for measurement at iteration i while T_{di} is the time used for finding global reward at iteration i.

However, there are three reasons that lead the above estimation to become inaccurate: first, successful information delivery at the designated time is hard to be guaranteed, considering the confusion of autonomous agents; second, because of the operating conditions' dependence on K_i, T_{mi}, and T_{di}, estimating them accurately is very hard; and finally, real-time simulation also relates to the implementing methods of both software and hardware.

4.1.9 Simulation Results

Some simulation tests have been done to the proposed MASRL algorithm, including the Ward–Hale 6-bus and system, the IEEE 30-bus system, and the IEEE 162-bus system [28].

4.1.10 Ward–Hale 6-Bus System

In Figure 4.2, there are six buses, seven transmission lines, and five reactive power control variables ($|V_2|$, T_{3-4}, T_{5-6}, Q_{c4}, Q_{c6}) in the ward–Hale power system, where $|V_2|$ is the terminal voltage reference of the PV bus, T_{3-4} is the tap setting for changers between buses 3 and 4, the same as T_{5-6}, and Q_{c4} is the reactive power demand for the capacitor banks of buses 4, the same as Q_{c6}. Each RL agent associates with a reactive power controller. The tap changing transformers do not require additional NAs. Through monitoring information exchanging between node neighboring agents $b3$–$b4$ along with agents $b5$–$b6$, the global reward signal can be obtained, respectively. In this way, it becomes possible to decrease the number of

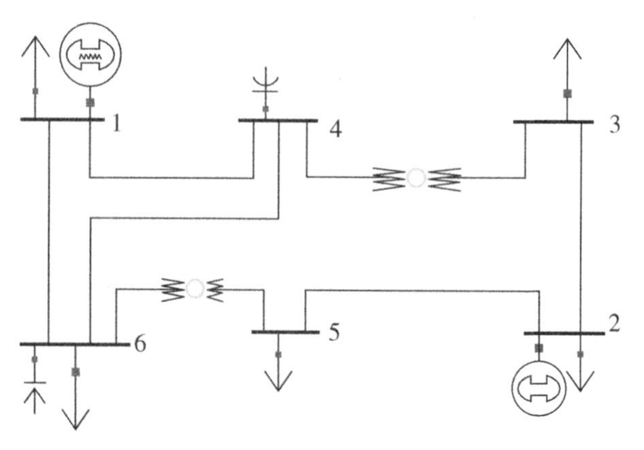

Figure 4.2 The Ward–Hale 6-bus system.

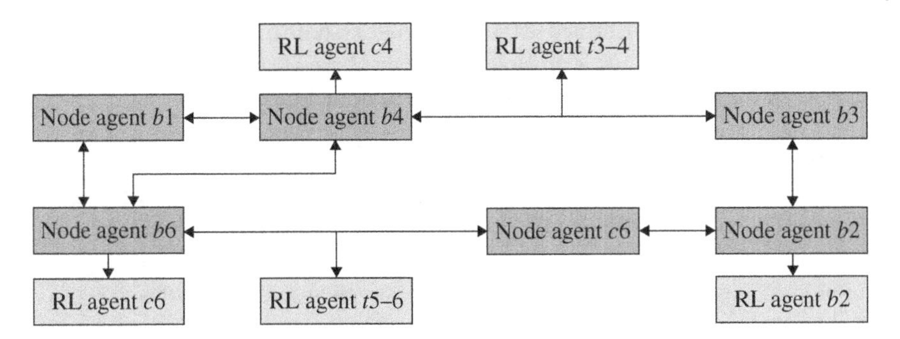

Figure 4.3 Implementation of MASRL for the Ward–Hale system.

NAs, which further reduces the implementation cost and increases the communication efficiency. Thus, as shown in Figure 4.3, there are only six nodes and five RL agents in the system, which can be regarded as a demonstration of the "fully distributed" characteristic of the designed solution.

The allowable range of voltage of the PV generator is (0.95, 1.05) with a step size of 0.005. Changing by a step size of 0.01, the tap changing transformer is set between 0.9 and 1.1. In addition, the capacitor banks give reactive power generations between (0, 0.30) with a step size of 0.01. Therefore, $213 \times 312 = 8\,899\,821$ is the number of total possible solutions.

Considering the goal of minimizing real power loss and satisfying the operational constraints at the same time, there are two kinds of constraints included in simulation studies. The first one is that the range of bus voltages are set among 0.95–1.05. The other one is that the range of the reactive power productions of two generators should be limited between −0.2 and 0.5.

To accomplish the information updating process, seven branches and seven independent coefficients need to be determined. Table 4.2 shows how these coefficients are determined based on different algorithms. From (4.4), it can be observed that there exists a negative correlation between the magnitude of the second largest eigenvalue ($\|\lambda\|$) and the converging speed. Thus, through the comparison of $\|\lambda\|$s of the iteration matrices, the fastest converging speed belongs to the Mean Metropolis method [13].

Table 4.2 Independent coefficients for information exchange.

	a_14	a_16	a_23	a_25	a_34	a_46	a_56	$\|\lambda_2\|$
Uniform	1/6	1/6	1/6	1/6	1/6	1/6	1/6	0.83
Metropolis	1/4	1/4	1/3	1/3	1/4	1/4	1/4	0.74
Mean Metropolis	1/3	1/3	2/5	2/5	1/3	2/7	1/3	0.66

There are two tests performed to the Ward–Hale system: the first is learning from scratch while the other one is previous knowledge-based learning. The results and analysis of these simulations are presented in later sections.

4.1.10.1 Learning from Scratch

Setting the five loads to rated values, the learning process is presented in Figures 4.4 and 4.5. The process of learning from scratch needs about 2500 iterations to reach convergence. From the fact that the number of investigated solution (2500) is far less than the number of possible solutions (8 899 821), it is obvious that the algorithm is computationally efficient.

It can be observed that at some times, the converged value of the global reward can be reached before the final convergence. Although, in this case, the algorithm already gets an ideal solution. The agents do not know whether better solutions can be reached or not. If the time is limited, the algorithm can just end when a relatively good solution is found before convergence.

Because a current solution must first be deployed before evaluation, sometimes there are voltage violations, as shown in Figure 4.5, and these bad candidate solutions will create undesirable disturbances. To deal with this problem, there are two

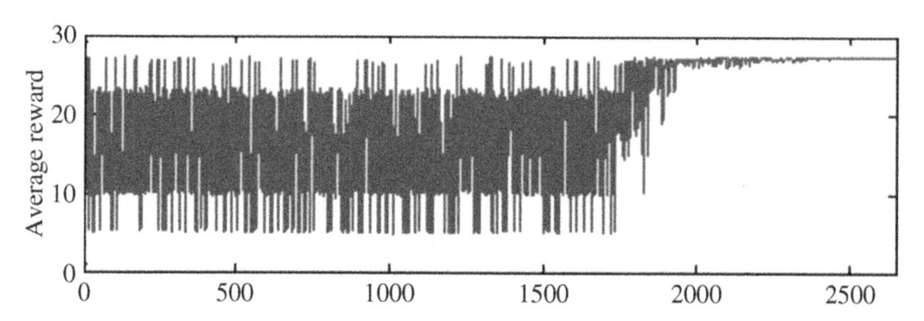

Figure 4.4 Updating process of average reward of the Ward-Hale system (learning from scratch).

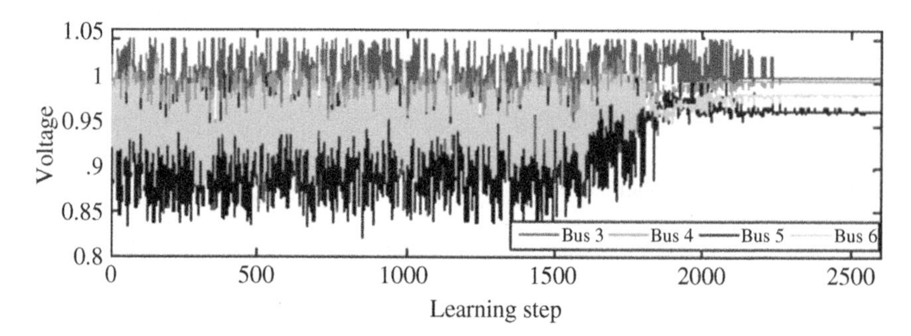

Figure 4.5 Updating process of voltage magnitudes of the Ward–Hale system (learning from scratch).

Table 4.3 Normalized Q-Tables of the five RL agents.

Q-Table for −V2−		*Q-Tables for T_{3-4} and T_{5-6}*			*Q-Tables for Q_{c4} and Q_{c6}*		
Control	*−V2−*	*Control*	T_{3-4}	T_{5-6}	*Control*	Q_{c4}	Q_{c6}
0.95	24.6	*0.9*	51.9	65.4	*0*	19.4	37.6
0.955	23.3	*0.91*	55.8	67.3	*0.01*	21.3	34.3
0.96	22.8	*0.92*	57.8	70.3	*0.02*	23.7	42.1
0.965	31.8	*0.93*	67.3	72.5	*0.03*	33.7	42.5
0.97	45.7	*0.94*	65.2	75.4	*0.04*	48.1	43.8
0.975	42.1	*0.95*	67.5	80.2	*0.05*	45.8	46.9
0.98	46	*0.96*	70.3	85.3	*0.06*	53	48.7
0.985	45.3	***0.97***	68.3	**100**	*0.07*	52.8	45.7
0.99	65.2	*0.98*	79.5	85.2	*0.08*	45	28.8
0.995	70.1	*0.99*	82.4	83.5	*0.09*	34.6	36.7
1	69.7	*1*	85.4	80.6	*0.1*	49.4	38.6
1.005	57.2	*1.01*	89.3	83.9	*0.11*	31.5	51.8
1.01	54.8	***1.02***	**100**	82.8	*0.12*	64.1	44.9
1.015	63.1	*1.03*	80.4	70.5	*0.13*	53.6	35.7
1.02	62.6	*1.04*	68.9	72.6	*0.14*	48.1	50.8
1.025	65.2	*1.05*	62.3	78.2	*0.15*	66	58.4
1.03	68.5	*1.06*	64.1	63.9	*0.16*	69	37.5
1.035	76.2	*1.07*	53.3	60.2	*0.17*	66.9	59.5
1.04	75.2	*1.08*	49.2	55.3	*0.18*	67.9	62.5
1.045	80.7	*1.09*	47.2	49.5	*0.19*	64.4	77.6
1.05	**100**	*1.1*	40.1	44.7	*0.2*	69.9	70.8
					0.21	67.2	74.8
					0.22	64.3	74.4
					0.23	64.5	75
					0.24	70.8	82.5
					0.25	73.8	83.7
					0.26	79.8	87.4
					0.27	82.3	80.8
					0.28	85.6	83
					0.29	87.5	85
					0.3	**100**	**100**

ways: the first one is pre-evaluating a current solution using some analysis based on simple models or heuristic rules. Depending on the result of pre-evaluation, a solution will be accepted or discarded. The second is collecting some initial knowledge using offline learning. In this method, prior knowledge-based learning can significantly reduce the learning time and lower the possibility of undesirable disturbances.

Table 4.3 shows the converged Q-tables of the five RL agents. Normalization has been done to the Q-values in each column of Table 4.3 (with a maximum value of 100). Choosing the actions that gain maximum value to be the optimal solution corresponding to the simulation, the optimal solution in Table 4.3 is 1.05, 1.02, 0.97, 0.30, and 0.30 for $||V_2||$, T_{3-4}, T_{5-6}, Q_{c4}, and Q_{c6} respectively. In the solution, 0.0899 is the value of active power loss.

An examination of the optimality of the solution from MASRL has been done. The best solution given by a centralized discrete particle swarm optimization (DPSO)-based method [29] is 1.05, 1.03, 0.98, 0.30, and 0.30, while the corresponding active power loss is 0.0899. Through a comparison of results from DPSO and MASRL, it can be concluded that MASRL can reach almost the same solutions as the DPSO. Although large-scale systems may yield multiple solutions with comparable performance, online applications only need to find a nearly global best solution.

4.1.10.2 Experience-Based Learning

The goal of the test is to check the impact of prior knowledge to the performance of the MASRL, randomly setting the loading levels to be either 15% larger or smaller than rated values and initializing the Q-tables based on former test. Figures 4.6–4.8 show the simulation results.

In Figure 4.6, compared to the previous test, the speed of convergence of this test is much faster. It only takes around 80 iterations while the previous test costs

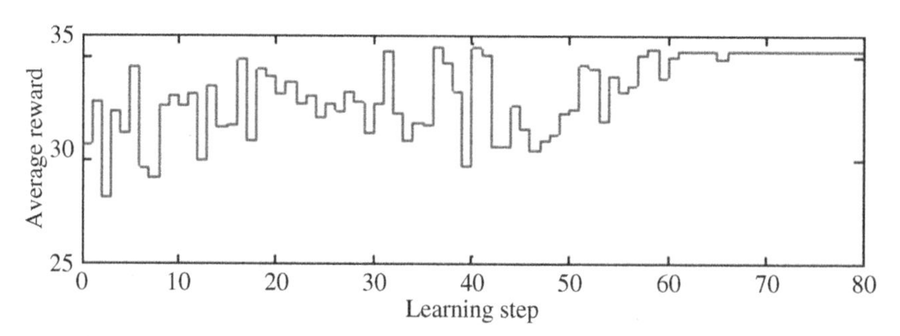

Figure 4.6 Updating process of average reward of the Ward–Hale system (learning based on prior knowledge).

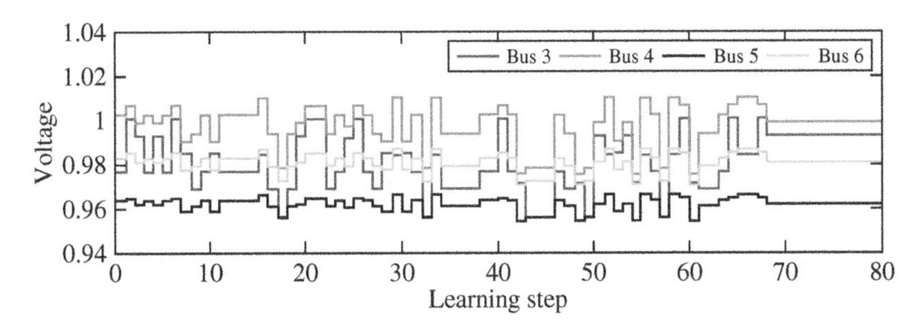

Figure 4.7 Updating process of steady-state voltage magnitudes of the Ward–Hale system (learning based on prior knowledge).

2500 iterations. Although the difference of loading level exists, prior knowledge is able to significantly accelerate the learning process. In addition, it is reasonable to make the prediction that if operating points remain the same as the one tested, the algorithm will converge even faster.

In Figure 4.7, it can be observed that the steady-state voltage of non-generator buses always stays within the range of 0.95–1.05. Besides, investigating the transient states is also necessary. Figure 4.8 shows the dynamic responses of bus voltage in the learning process.

In Figure 4.8, it can be observed that the system converges fast and the voltages remain within the limit. It can be concluded that during the process of continuous learning, undesirable disturbances and the violation of constraints all decrease.

Because of the fact that reward signal calculation can only be measured after the system becomes stable, Figure 4.8 can be viewed as an estimation of the speed of the MASRL algorithm. It only takes less than half a second for the bus voltage to converge.

Because of the fact that T_{di} is very smaller compared to T_{mi}, the overall time for the learning process is estimated as $0.5 \times 80 = 40$ seconds, and this estimation is

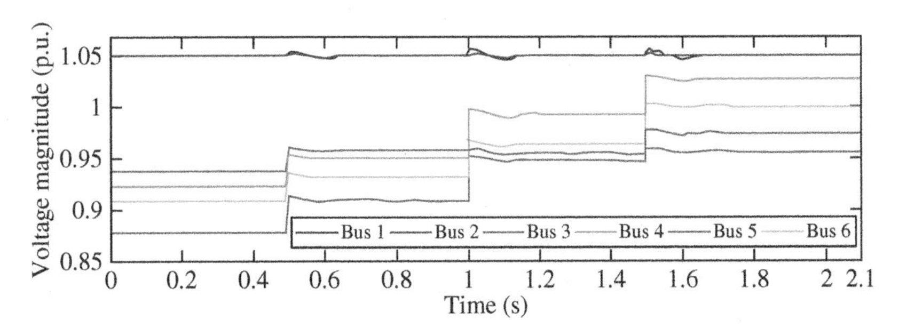

Figure 4.8 Dynamic voltage response during learning of the Ward–Hale system (learning based on prior knowledge).

very conservative. Most of the times, there is no need to wait for half a second to decide the performance of a potential solution.

4.1.10.3 IEEE 30-Bus System

In order to test the performance of the designed algorithm under the condition of a larger searching space, this paper implements the MASRL algorithm to the IEEE 30-bus system. As shown in Table 4.4, there are eight control variables, one half for tap changing transformer and the other half for capacitor banks.

In addition, every control variable for the tap changing transformers is set to be 1 of 21 possible values within [0.9, 1.1], while each control variable for the capacitor banks is set within [0, 20]. Therefore, the overall number of possible solutions is $21^8 = 3.7823 \times 1010$. It is very challenging to find a good solution from all these solutions.

As mentioned before, the MASRL algorithm goes through two different operating conditions. In the first one, the whole Q-tables are set to zero and the loads are set to rated values. In the second, initial Q-tables come from the converged Q-tables and loads are randomly generated, and Figures 4.9–4.12 show the simulation results.

In Figures 4.4 and 4.9, it can be observed that it takes 4000 iterations for the learning process to reach convergence, and the speed of the algorithm seems independent of the size of searching space, which is an ideal characteristic.

Thus, the algorithm is computationally efficient under the condition of a huge searching space. Figures 4.10 and 4.11 show the fact that not only the learning process can be largely accelerated by prior knowledge but also the possibility of constraint violations can be reduced by prior knowledge. In Figure 4.12, once a candidate solution is deployed, the system's response reaches stability within around 1 second. Similarly, if previous experience is available, it only takes about 140 seconds for the optimization process to converge.

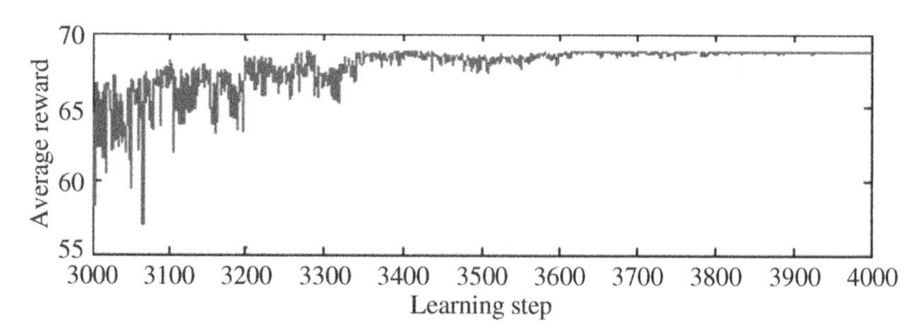

Figure 4.9 Updating process of the average reward of the IEEE 30-bus system (learning from scratch).

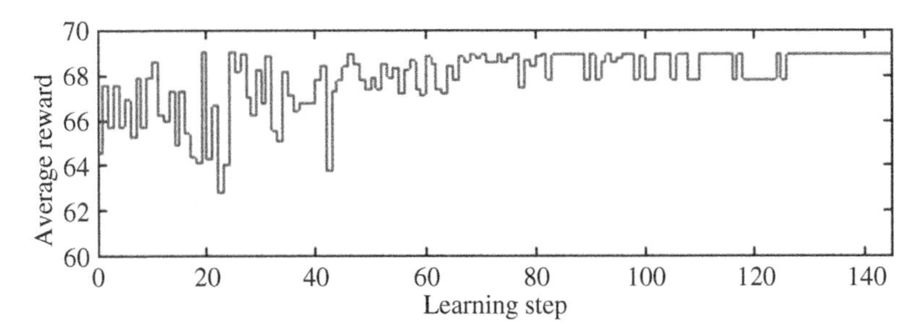

Figure 4.10 Updating process of average reward of the IEEE 30-bus system (learning based on prior knowledge).

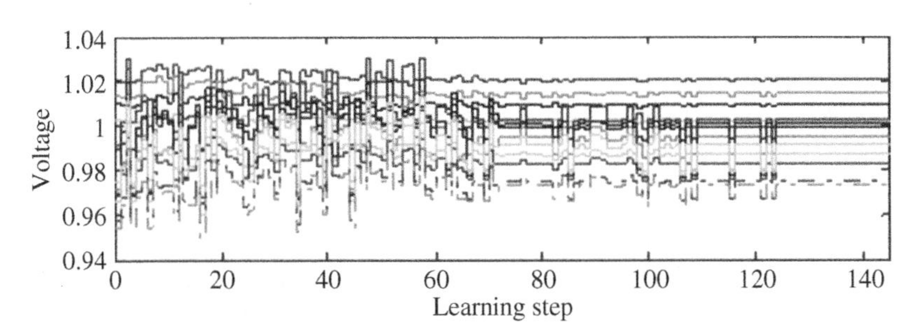

Figure 4.11 Updating process of steady-state voltage magnitudes of the IEEE 30-bus system (learning based on prior knowledge).

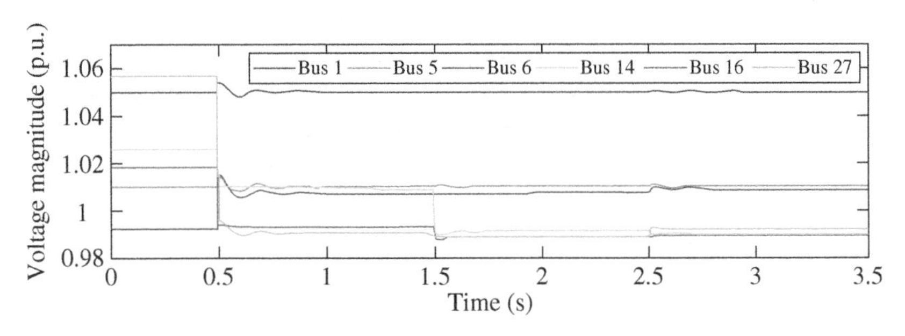

Figure 4.12 Dynamic voltage response during learning of the IEEE 30-bus system (learning based on prior knowledge).

As shown in Table 4.4, there are some evaluations about the optimality of the results from the MASRL algorithm. Compared to the solutions from centralized DPSO and IP algorithm, the solution obtained from MASRL is as good as the solution from DPSO and appear better than the solution from IP algorithm.

Table 4.4 Comparison of solutions obtained using MASRL, DPSO, and IP (IEEE 30-bus system).

Variable	MASRL	DPSO	IP
$T_{4\text{–}12}$	1.05	1.05	0.932
$T_{6\text{–}9}$	1.05	1.06	0.978
$T_{6\text{–}10}$	0.9	0.9	0.969
$T_{28\text{–}27}$	1	0.99	0.968
Q_{c7}	0.18	0.15	0.16
Q_{c14}	0.04	0.04	0.02
Q_{c16}	0.1	0.07	0.1
Q_{c30}	0.04	0.04	0.03
Ploss (MW)	17.94	17.93	18.15

4.1.10.4 IEEE 162-Bus System

In this test, the application scenario is a large-scale power system. The MASRL algorithm need to deal with 20 control variables: 10 for reactive power dispatch, 6 for tap changing transformer, and 4 for capacitor banks. For the tap changing transformers, variables are set to be 1 of the 21 possible values in the range of [0.90, 1.10]. For capacitor banks, the range is [0, 20]. As a result, the total number of possible solution is around $2110 = 1.6680 * 10^{13}$. As shown in Figures 4.13 and 4.14, MASRL is tested in the conditions with and without prior knowledge, respectively. The same as before, the results show that the prior knowledge largely accelerates the learning process. Table 4.5 shows that the performance of MASRL seems better than the performance of IP and almost the same as that of DPSO

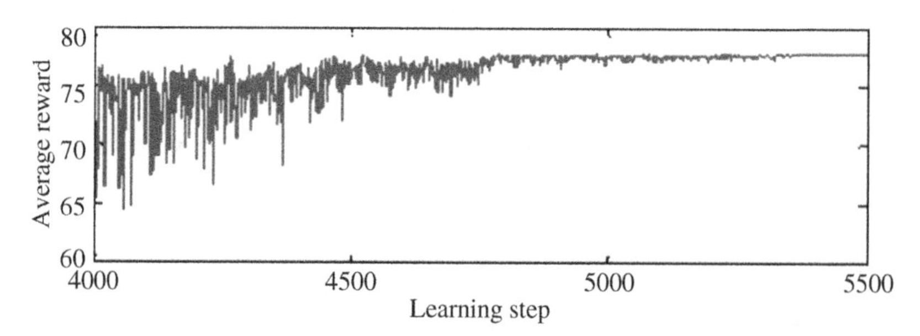

Figure 4.13 Updating process of average reward of the IEEE 162-bus system (learning from scratch).

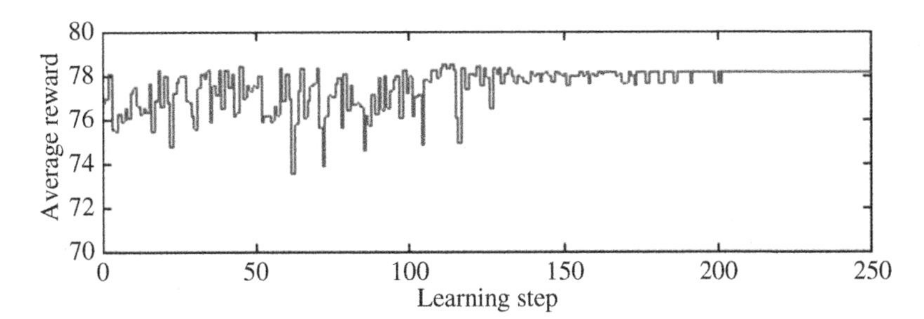

Figure 4.14 Updating process of average reward of the IEEE 162-bus system (learning based on prior knowledge).

Table 4.5 Comparison of solutions obtained using MASRL, DPSO, and IP (IEEE 162-Bus system).

Variable	MASRL	DPSO	IP
T^{22-39}	1.04	1.06	1.108
T^{52-118}	1.09	1.09	1.043
T^{75-130}	1	1	1.025
T^{94-109}	1.04	1.1	1.025
T^{120-14}	1	1	0.975
$T^{144-141}$	1.1	1.08	1.025
Q^{c27}	0.17	0.11	0.16
Q^{c52}	0.03	0.2	0.05
Q^{c98}	0.15	0.04	0.07
Q^{c147}	0.13	0.11	0.14
Ploss (MW)	162.82	162.78	162.92

4.1.11 Conclusion

This paper proposes a novel, fully distributed MASRL solution for ORPD. MASRL solution can not only avoid some problems of centralized algorithms but also gain a high-quality result, which is comparable or even better than that of centralized algorithms. These advantages come from the integration of two algorithms: one is an information discovery algorithm based on consensus and the other one is a distributed Q-learning algorithm.

The responsibility of the first algorithm is calculating the global rewards based on local communications. It is proved that this information discovery process is stable for any size and any configuration of systems. Moreover, this algorithm appears robust in different operating conditions.

For the second algorithm, there are many ideal advantages. First, this distributed Q-learning algorithm needs less memory and appears more computationally efficient compared to other algorithms. Second, although sometimes not converging to global optimum, this algorithm is able to provide nearly optimal solutions. Third, this algorithm can work with a relatively less accurate system model, which avoids model impreciseness. Finally, the efficiency of the algorithm can be further improved through continuous online learning.

In order to decide which algorithm to choose (MASRL or centralized optimization algorithms), one should balance the factors of cost, speed, reliability, and flexibility. Future work should focus on implementing the proposed solution to hardware agents and using real-time simulation to test the MAS.

4.2 Sub-gradient-Based Reactive Power Control

The development and utilization of renewable energy in the energy system show strong vitality in realizing the sustainable development of human society and nature. In the current context, an increasing number of devices of reactive power control cooperate into the power distribution system, which calls for upgrading the existing reactive power control solutions. This paper puts forward a fully distributed MAS in terms of the optimal reactive power control (ORPC) solution to ameliorate voltage profiles of the power grids while working with a higher energy efficiency under various operating conditions. In the new scenario, only data measured locally or get from the adjacent buses are required by the reactive controller so as to update its control settings. Based on the moderate assumptions, this paper derives the updated rules of the sub-gradient algorithm. The results show that the stable-state performance of the proposed method is comparative to that of the centralized optimization scheme. Because the structure of communication topology is simple and the data amount addressed in the process is cut down, this solution shows an advantage of timely response to changes in operating conditions. The validity of the scheme is verified by simulation studies of power systems weight disparate scales.

4.2.1 Introduction

Research studies on ORPC are front-burner issues in the electric system. The major purpose of ORPC is composed of two different parts, reducing system active losses and ameliorating voltage profiles, respectively [30]. Control variables in the ORPC system are involved with voltage references of generator buses, reactive power generation of var sources, the tap ratios of transformers, and other OPRC variables. The main constraints correspond to the limits of the generator and load bus voltage, the tap ratios of transformers, the reactive power sources, and the balance between supply and demand in the system [31].

Presently, the methods adopted by reactive power control are a three-level control pattern [32]. The primary reactive power control, as local control, takes charge of controlling the voltage of a certain bus, mostly the generator bus. The reactive power injection over a regional voltage zone is regulated by the secondary reactive power control via a centralized control method to keep the voltages of the pilot buses within normal ranges. Both the primary and secondary controls here give up the reactive power injection optimization, in which the target is to minimize power losses and improve voltage profiles. Thus, tertiary ORPC is configured to realize the reactive power generation/distribution over the considered power grid.

Scholars have developed variable optimization methods and techniques to address the ORPC problem, such as gradient method [33], linear programming [34, 35], IP method [3, 4], and computational intelligence-based algorithms [36, 37]. The performances of the power generation grids can be improved to some extent by applying these optimization approaches. Nonetheless, because of the current centralized deployment philosophy, most reactive power control solutions cannot realize optimal online control because of high-density data to be processed and the latency intrinsic in complicated communication systems.

OPRC research studies are now facing new challenges as we are thriving to build a smarter grid. Massive distributed generation (DG), equipped in smart grids, makes the OPRC problem more complicated to solve. The integration of these DGs demands more reactive power control equipment for voltage support, subject to the frequent change of the operating conditions, which requires updating existing OPRC solutions with the aim of increasing responsiveness.

Because response speed can be improved through computing technology, researchers have developed the MAS architecture to solve the problem of ORPC [38, 39]. Extant MAS-based methods developed for reactive power control endeavor to split the primary optimal problems and then solve them by applying distributed computing techniques [40–43]. Nevertheless, because the majority of these methods still need specific agents to coordinate the operations of all other agents, these methods are not fully distributed. Therefore, these methods are not resilient and still suffer a single-point failure. Furthermore, constructing a control center to manage significant amounts of reactive power control devices that are geographically distributed is quite costly. In addition, considering the intermittent nature of renewable resources, the conventional snapshot-based optimal algorithms' lack of adaptability restrain their deployment for online applications, which require frequent update control settings.

Xu *et al.* proposed a distributed solution method for ORPD in their paper [44]. This method is designed based on a global information discovery algorithm and a distributed Q-learning algorithm. This learning-based solution circumvents the dilemma to analyze complicated power system models. Additionally, it is easy to implement and is adaptive to the change in operating conditions. However, it has shortcomings. The first one lies in undesirable disturbances that are caused by the learning process required for the Q-learning algorithm. The second one is that the optimization process requires too many steps to converge, especially when learning from scratch. To surmount these limitations, model-based analysis can be applied to reduce the possibility of random control updates and increase the rate of convergence.

The above analysis shows that for smart grids, an ideal OPRC solution ought to be distributed, highly efficient in computation, and fit for online optimization. In this section, we propose an MAS-based distributed solution for ORPC. The solution, which is based on a distributed sub-gradient algorithm, is appropriate for distributed computing. In addition, different from traditional solutions, the proposed solution does not require a specialized agent to coordinate other agents. Therefore, the solution can greatly reduce the chance of single-point failure, which is prone to happen for the centralized or semi-distributed solutions mentioned above. Moreover, the solution proposed here is flexible and scalable, suitable for power systems, and has various sizes and tautologies. As shown later, the proposed solution is robust to a specific type of failure. With a timely response due to the reduced communication delays and updates of control action, the MAS-based solution can make the online optimization of smart grids into realization.

Because of the proposed fully distributed solution, there is no need to coordinate with certain agents. Therefore, the probability of single-point failure can be dramatically reduced in either an integrated or the above mentioned semi-distributed agent-based solutions. Furthermore, the presented solution can be implemented in different topologies flexibly. Subsequently, it is demonstrated that the solution is robust against certain failures. The MAS can realize online optimization of the intelligent grid for control action updates and the reduction in communication latency.

In this section, we will first introduce the problem formulation. Then, we will briefly introduce the sub-gradient algorithm used for optimization. Following that, we provide the derivation of equations for distributed sub-gradient calculations. After discussing the issues related to the implementation, we present the simulation results for different sizes of power systems with the proposed solution. Concluding remarks are provided at the end of this section.

4.2.2 Problem Formulation

In this section, it is assumed that reactive power control devices such as capacitors and load tap transformers have been equipped or deployed. The OPRC proposed is designed to minimize actual power losses and improve the voltage profiles of the grid by means of redistributing the reactive power.

The objective function of minimizing the active power loss is expressed as [45]:

$$\min P_L = \sum_{j=1}^{n}\sum_{k=1}^{n} V_j V_k G_{jk} \cos(\delta_j - \delta_k) \tag{4.15}$$

where V_j is the voltage magnitude at bus j, δ_j is the phase angle at bus j, n is the number of buses, and G_{jk} is the conductance component of Y-bus element Y_{jk}.

On the purpose of improving voltage profiles of the system, the corresponding objective function is defined as

$$D_v = \sum_{i=1}^{n} (V_i - V_i^*)^2 \tag{4.16}$$

where V_i and V_i^* are the actual and required bus voltage magnitudes, respectively, for bus i.

To achieve a good balance between obtaining an expected voltage pattern and minimizing the active power losses [46], the function is defined as a weight sum of P_L and D_v:

$$f = W_1 P_L + W_2 D_v \tag{4.17}$$

where W_1 and W_2 refer to the weight coefficients for P_L and D_v, respectively, which can be determined by the operator of the system.

In this chapter, the control variables we considered for ORPC include capacitor bank switching, voltage references of the generator buses, and transformer tap settings, and these variables are denoted with the following vector:

$$\begin{aligned} \mathbf{u} &= [\mathbf{Q}_C, \mathbf{V}_G, \mathbf{T}_{LTC}]^T \\ &= [Q_{c1}, Q_{c2} \cdots Q_{cl}, V_{g1}, V_{g2} \cdots V_{gp}, t_{t1}, t_{t2} \cdots t_{tm}]^T \end{aligned} \tag{4.18}$$

where l, p, and m separately refer to the number of capacitor banks, generators, and transformers contributing to ORPC; Q_{ci} is the injected reactive power of *ith* capacitor bank; V_{gi} is the bus voltage reference of generator i; and t_{ti} is the tap setting of transformer i.

The control variables should be bounded as [47]:

$$\begin{cases} \underline{Q}_{ci} \le Q_{ci} \le \overline{Q}_{ci} & 1 \le i \le l \\ \underline{V}_{gi} \le V_{gi} \le \overline{V}_{gi} & 1 \le i \le p \\ \underline{T}_{ti} \le T_{ti} \le \overline{T}_{ti} & 1 \le i \le m \end{cases} \tag{4.19}$$

where "_" and "−;" represent the lower and upper bounds of the control variables, respectively.

Additionally, supply and demand balance of the system should be maintained, which can be formulated as constraints given in Eqs. (4.20) and (4.21) [45].

$$P_{Gj} - P_{Lj} - V_j \sum_{k=1}^{n} V_k (G_{jk} \cos \delta_{jk} + B_{jk} \sin \delta_{jk}) = 0 \tag{4.20}$$

$$Q_{Gj} + Q_{Cj} - Q_{Lj} - V_j \sum_{k=1}^{n} V_k (G_{jk} \sin \delta_{jk} - B_{jk} \cos \delta_{jk}) = 0 \tag{4.21}$$

Here, P_{Gj}, Q_{Gj} are the active and reactive power generations at bus j, P_{Lj}, Q_{Lj} are the active and reactive power load at bus j, Q_{Cj} refers to the injected reactive power of the capacitor bank at bus j, $G_{jk} + jB_{jk} = Y_{jk}$ is the element of the admittance matrix (Y-bus), corresponding to the branch connecting bus j and bus k, and $\delta_{jk} = \delta_j - \delta_k$ is defined as the voltage phase angle difference between these two buses.

According to the above mentioned analysis, ORPC can be expressed as a constrained optimization problem in which the objective function is given by Eq. (4.17) and the constraints follow Eqs. (4.19)–(4.21).

The convexity of the active power loss function is demonstrated over a wide range of operating conditions in [48]. Thus, in this section, we assume that the objective function given in Eq. (4.15) is convex with respect to the control variables. Besides, the convexity of D_v defined as Eq. (4.16) is obvious. Consequently, we can draw a conclusion that the additive function f defined in Eq. (4.17) is convex, and the overall ORPC problem is, thus, a constrained convex optimization problem.

The commonly used IP methods for convex optimization are centralized. Under the context of accommodating increasing renewable energy, current control philosophy is moving toward shifting from the conventional centralized paradigm into a distributed control architecture [49]. Consequently, a distributed control solution is much preferred to realize ORPC for the power system with large amounts of reactive power control devices. Because the sub-gradient algorithm can be easily implemented in a distributed manner, we apply it to solve the ORPC problem here. Compared with the IP method, the sub-gradient-based solution method takes less memory and computation and is suitable to solve large-scale problems with decomposition techniques [50, 51].

4.2.3 Distributed Sub-gradient Algorithm

For convenience, we first give a brief introduction to the distributed sub-gradient algorithm.

For a convex function $f(\mathbf{u}) : \mathbf{R}^n \to \mathbf{R}$, the sub-gradient algorithm can be expressed as [52, 53]:

$$\mathbf{u}^{k+1} = \mathbf{u}^k - \alpha_k \nabla f^k \tag{4.22}$$

where ∇f^k is the sub-gradient of f at u^k. The details on optimality, convergence speed, and error analysis can be found in [53].

Notably, the above algorithm does not require that "∇f^k" to align with the steepest descent direction.

Rewrite Eq. (4.22) into a distributed form:

$$u_i^{k+1} = u_i^k - \alpha_k \nabla_i f^k \tag{4.23}$$

where u_i^k refers to the ith component of $\mathbf{u}^{(k)}$. Note that the update of u_i^{k+1} is only decided by its preceding state u_i^k and sub-gradient $\nabla_i f^k$. If it is possible to determine the sub-gradient in a distributed way, the algorithm will become a distributed algorithm.

For ORPC problem solving, it is not easy to compute the sub-gradients in a distributed way because of the following reasons. The first reason lies in that P_L is a function of state variables ($\mathbf{V}$ and $\boldsymbol{\delta}$) rather than control variables ($\mathbf{u}$) because it is difficult to express P_L w.r.t. the control variables, which makes it difficult to evaluate the performances of control combinations. Secondly, the power system is actually a complicated and coupled nonlinear system, wherein the change in one control variable will result in the changes of quite a lot of state variables. Relatively accurate computation of sub-gradients requires to carryout sensitivity analysis [26, 54]. This technique needs to calculate the inverse of the global information-based matrix and thus is not suitable for distributed control implementation for future smart grids.

As mentioned earlier, there is no need to find the steepest descent direction for the sub-gradient algorithm. Therefore, mild assumptions and approximations can be utilized to develop the equations for sub-gradient calculation.

The method introduced in [49] numerically approximates the sub-gradient:

$$\nabla_i f(\mathbf{u}^k) = \frac{f(\mathbf{u}^k) - f(\mathbf{u}^{k-1})}{u_i^k - u_i^{k-1}} \tag{4.24}$$

According to the formula Eq. (4.24), both $f(\mathbf{u}^k)$ and its previous value, $f(\mathbf{u}^{k-1})$ are required for the calculation of the sub-gradient. The author in [49] finds the global information, $f(\mathbf{u}^k)$ using the so-called *coupled oscillator* first appeared in [55, 56]. One problem with Eq. (4.24) is that it is difficult to choose the appropriate time interval between two consecutive control updates, u^k and u^{k-1}. For one thing, shorter time interval cannot guarantee the convergence of the *coupled oscillator* algorithm. For another, the larger time interval may result in inaccuracies for sub-gradient calculations, especially under the circumstances of severe operating conditions changes. Researchers in [49] set the time interval as 1.2 minutes. The entire optimization process may take more than 20 minutes to converge. Here, in this chapter, in order to improve the convergence speed as well as the accuracy for sub-gradient calculation, we propose to calculate the sub-gradients based on the current state of the power system based on the mild assumptions.

4.2.4 Sub-gradient Distribution Calculation

In this section, three assumptions are utilized for simplifying the design of distributed sub-gradients algorithms as follows:

Hypothesis 1: *The change in reactive power at bus i* (Q_i) *only leads to a change in the voltage magnitude at bus i* (V_i).
Hypothesis 2: *The change in transformer tap setting* (t_i) *only leads to the change in power system loss* (P_L).
Hypothesis 3: *The change in active power injection at bus i* (P_i) *only results in the change of voltage phase angle at bus i* (δ_i).

Noticeably, there is no possibility of online analysis on the basis of elaborated power system models for the complication of the power system. Hence, it is necessary to apply suitable assumptions on online optimization for the sake of simplification. During algorithm design, these assumptions exploit the P–δ and Q–V characteristics that are extensively used in studies of power systems such as the fast decoupled load flow and decoupled active and reactive power controls. A simple algorithm appropriate for online distributed computing can be obtained through these assumptions.

The equations of distributed sub-gradient calculation can be derived as follows. The sub-gradient vector $\nabla f(\mathbf{u})$ can be represented as:

$$\nabla f(\mathbf{u}) = \left[\frac{\partial f}{\partial Q_{c1}} \cdots, \frac{\partial f}{\partial Q_{cl}}, \frac{\partial f}{\partial V_{g1}} \cdots, \frac{\partial f}{\partial V_{gp}}, \frac{\partial f}{\partial t_{t1}} \cdots, \frac{\partial f}{\partial t_{tm}} \right]^T \tag{4.25}$$

As shown in Eq. (4.27), the sub-gradients corresponding to three different control variables are defined, namely, *reactive power sub-gradient* ($\partial f / \partial Q_{ci}$) for capacitor banks, *voltage subgradient* ($\partial f / \partial V_{gi}$) for generators, and *transformer subgradient* ($\partial f / \partial t_{ti}$) for transformers, and the calculation of them will be introduced in the following context.

4.2.4.1 Calculation of $\partial f / \partial Q_{ci}$ for Capacitor Banks

The reactive power sub-gradient can be developed as:

$$\frac{\partial f}{\partial Q_{ci}} = W_1 \frac{\partial P_L}{\partial Q_{ci}} + W_2 \frac{\partial D_v}{\partial Q_{ci}} \tag{4.26}$$

Because W_1 and W_2 are constants, we only need to calculate $\frac{\partial P_L}{\partial Q_{ci}}$ and $\frac{\partial D_v}{\partial Q_{ci}}$.

For the first term in Eq. (4.26), the active power loss PL can be expressed as:

$$P_L = \sum_{j=1}^{n}\sum_{k=1}^{n} V_j V_k G_{jk} \cos(\delta_j - \delta_k) \tag{4.27}$$

Provided the installation of a capacitor bank at bus i, $\frac{\partial P_L}{\partial Q_{ci}} = \frac{\partial P_L}{\partial Q_i}$, and $\frac{\partial D_v}{\partial Q_{ci}} = \frac{\partial D_v}{\partial Q_i}$.

According to the assumption 1, the derivative of P_L w.r.t. Q_i can be expressed as:

$$\frac{\partial P_L}{\partial Q_i} = \sum_{j=1}^{n}\sum_{k=1}^{n} G_{jk} \cos(\delta_j - \delta_k)\frac{\partial(V_j V_k)}{\partial Q_i} \tag{4.28}$$

For Eq. (4.28), the following results can be obtained with different combinations of j, k, and i.

$$\frac{\partial(V_j V_k)}{\partial Q_i} = \begin{cases} 2V_i\frac{\partial V_i}{\partial Q_i} & j = i, k = i \\ V_k\frac{\partial V_i}{\partial Q_i} & j = i, k \neq i \\ V_j\frac{\partial V_i}{\partial Q_i} & j \neq i, k = i \\ 0 & \text{otherwise} \end{cases} \tag{4.29}$$

Rewrite Eq. (4.28) as Eq. (4.32) using Eq. (4.31)

$$\begin{aligned} \frac{\partial P_L}{\partial Q_i} &= \sum_{j=1,j\neq i}^{n} G_{ji} \cos(\delta_j - \delta_i)V_j\frac{\partial V_i}{\partial Q_i} + \sum_{k=1,k\neq i}^{n} G_{ik} \cos(\delta_i - \delta_k)V_k\frac{\partial V_i}{\partial Q_i} + 2G_{ii}V_i\frac{\partial V_i}{\partial Q_i} \\ &= \sum_{j=1,j\neq i}^{n} G_{ji} \cos(\delta_i - \delta_j)V_j\frac{\partial V_i}{\partial Q_i} + G_{ii}V_i\frac{\partial V_i}{\partial Q_i} \\ &\quad + \sum_{j=1,j\neq i}^{n} G_{ij} \cos(\delta_i - \delta_j)V_j\frac{\partial V_i}{\partial Q_i} + G_{ii}V_i\frac{\partial V_i}{\partial Q_i} = 2\frac{\partial V_i}{\partial Q_i}\sum_{j=1}^{n} G_{ij}V_j \cos(\delta_i - \delta_j) \end{aligned} \tag{4.30}$$

The reactive power injection at bus i can be expressed as:

$$Q_i = V_i\sum_{j=1}^{n} V_j(G_{ij} \sin\delta_{ij} - B_{ij} \cos\delta_{ij}) \tag{4.31}$$

Based on assumption 1, the derivative of Eq. (4.31) w.r.t Q_i can be denoted as:

$$\frac{\partial V_i}{\partial Q_i} = \frac{1}{-V_iB_{ii} + \sum_{j=1}^{n} V_j(G_{ij} \sin\delta_{ij} - B_{ij} \cos\delta_{ij})} \tag{4.32}$$

The first term in Eq. (4.26) can be computed using Eqs. (4.32) and (4.34).

According to assumption 1, the second term of the reactive power sub-gradient can be expressed as:

$$\frac{\partial D_v}{\partial Q_i} = 2(V_i - V_i^*)\frac{\partial V_i}{\partial Q_i} \tag{4.33}$$

Now, the second term of Eq. (4.26) can be computed using Eqs. (4.32) and (4.33).

4.2.4.2 Calculation of $\partial f/\partial V_{gi}$ for a Generator

The voltage sub-gradient can be represented as:

$$\frac{\partial f}{\partial V_{gi}} = W_1 \frac{\partial P_L}{\partial V_{gi}} + W_2 \frac{\partial D_v}{\partial V_{gi}} \tag{4.34}$$

For a generator connected at bus i, $\frac{\partial P_L}{\partial V_{gi}} = \frac{\partial P_L}{\partial V_i}$, and $\frac{\partial D_v}{\partial V_{gi}} = \frac{\partial D_v}{\partial V_i}$.

Changes at the terminal voltage of the generator lead to the changes of the reactive power injection at the same bus. Thus, the first term of the voltage sub-gradient given in Eq. (4.34) can be expressed as:

$$\frac{\partial P_L}{\partial V_{gi}} = \frac{\partial P_L}{\partial Q_i} \frac{\partial Q_i}{\partial V_i} \tag{4.35}$$

Based on equation Eq. (4.30), the first term of the right side in Eq. (4.35) can be calculated. Similarly, based on Eq. (4.32), the second term of the right side of Eq. (4.35) can be calculated. Therefore, utilizing Eqs. (4.30), Eq. (4.32), and Eq. (4.35), the first term of voltage sub-gradient can be calculated.

The second term of the voltage sub-gradient is expressed as:

$$\frac{\partial D_v}{\partial V_i} = 2(V_i - V_i^*) \tag{4.36}$$

4.2.4.3 Calculation of $\partial f/\partial t_{ti}$ for a Transformer

By applying assumption 2, the transformer sub-gradient can be expressed as:

$$\frac{\partial f}{\partial t_{ti}} = W_1 \frac{\partial P_L}{\partial t_{ti}} + W_2 \frac{\partial D_v}{\partial t_{ti}} = W_1 \frac{\partial P_L}{\partial t_{ti}} \tag{4.37}$$

For convenience, in the following context, t_{jk} is used to denote the transformer tap setting, with j and k denoting indexes of buses that a transformer t_{ti} is connected to.

For the transformer shown in Figure 4.15, the sub-gradient with respect of transformer tap is acquired based on [57]:

$$\frac{\partial P_L}{\partial t_{jk}} = \frac{\partial P_L}{\partial P_j}\left(-\frac{\partial P_{jk}}{\partial t_{jk}}\right) + \frac{\partial P_L}{\partial Q_j}\left(-\frac{\partial Q_{jk}}{\partial t_{jk}}\right) + \frac{\partial P_L}{\partial P_k}\left(-\frac{\partial P_{kj}}{\partial t_{jk}}\right) + \frac{\partial P_L}{\partial Q_k}\left(-\frac{\partial Q_{kj}}{\partial t_{jk}}\right) \tag{4.38}$$

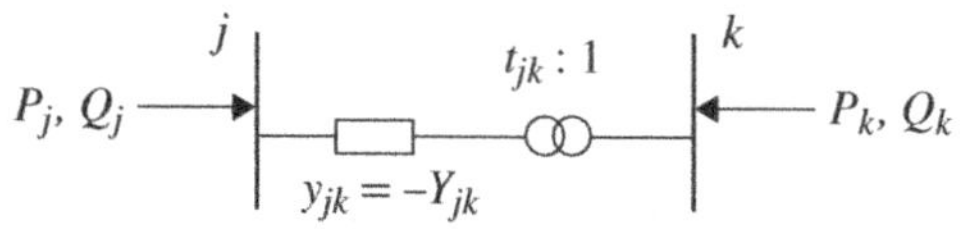

j, k, $t_{jk}:1$, P_j, Q_j, P_k, Q_k, $y_{jk} = -Y_{jk}$

Figure 4.15 Transformer representation.

In light of Eq. (4.27), $\partial P_L/\partial P_j$ can be written as (27) by applying assumption 3.

$$\frac{\partial P_L}{\partial P_j} = \sum_{\alpha=1}^{n}\sum_{\beta=1}^{n} V_\alpha V_\beta G_{\alpha\beta} \frac{\partial \cos(\delta_\alpha - \delta_\beta)}{\partial P_j} \tag{4.39}$$

For different combinations of α, β, and j, we can obtain:

$$\frac{\partial \cos(\delta_\alpha - \delta_\beta)}{\partial P_j} = \begin{cases} -\sin(\delta_j - \delta_\beta)\frac{\partial \delta_j}{\partial P_j}, & \alpha = j, \beta \neq j \\ \sin(\delta_{\alpha j} - \delta_j)\frac{\partial \delta_j}{\partial P_j}, & \alpha \neq= j, \beta = j \\ 0, & \text{otherwise} \end{cases}$$

According to Eqs. (4.40), Eq. (4.39) can be rewritten as follows:

$$\begin{aligned} \frac{\partial \mathrm{P_L}}{\partial \mathrm{P_j}} &= V_j \sum_{\alpha=1}^{n} V_\alpha G_{\alpha j} \sin(\delta_\alpha - \delta_j)\frac{\partial \delta_j}{\partial P_j} - V_j \sum_{\beta=1}^{n} V_\beta G_{j\beta} \sin(\delta_j - \delta_\alpha)\frac{\partial \delta_j}{\partial P_j} \\ &= -V_j \sum_{\alpha=1}^{n} V_\alpha G_{j\alpha} \sin(\delta_j - \delta_\alpha)\frac{\partial \delta_j}{\partial P_j} - V_j \sum_{\alpha=1}^{n} V_\alpha G_{j\alpha} \sin(\delta_j - \delta_\alpha)\frac{\partial \delta_j}{\partial P_j} \\ &= -2V_j \frac{\partial \delta_j}{\partial P_j} \sum_{\alpha=1}^{n} V_\alpha G_{j\alpha} \sin(\delta_j - \delta_\alpha) \end{aligned} \tag{4.40}$$

The active power injection for bus i can be expressed as:

$$P_j = V_j \sum_{\alpha=1}^{n} V_\alpha (G_{j\alpha} \cos\delta_{j\alpha} + B_{j\alpha} \sin\delta_{j\alpha}) \tag{4.41}$$

Then we can calculate $\partial \delta_j/\partial P_j$ with the help of assumption 1:

$$\frac{\partial \delta_j}{\partial P_j} = \frac{1}{V_j \sum_{\alpha=1,\alpha\neq j}^{n} V_\alpha (B_{j\alpha} \cos\delta_{j\alpha} - G_{j\alpha} \sin\delta_{j\alpha})} \tag{4.42}$$

$\partial P_j/\partial P_k$ in Eq. (4.38) can be calculated using Eq. (4.41), and $\partial P_L/\partial Q_k$ there can then be calculated using Eq. (4.30).

The power flow of a transformer branch can be calculated as follows:

$$\begin{aligned} P_{jk} &= V_j^2 g_{jk} + t_{jk} V_j V_k(-G_{jk} \cos\delta_{jk} - B_{jk} \sin\delta_{jk}) \\ Q_{jk} &= -V_j^2 b_{jk} + t_{jk} V_j V_k(-G_{jk} \sin\delta_{jk} + B_{jk} \cos\delta_{jk}) \\ P_{kj} &= t_{jk}^2 V_k^2 g_{jk} + t_{jk} V_j V_k(-G_{jk} \cos\delta_{jk} + B_{jk} \sin\delta_{jk}) \\ Q_{kj} &= -t_{jk}^2 V_k^2 b_{jk} + t_{jk} V_j V_k(-G_{jk} \sin\delta_{jk} + B_{jk} \cos\delta_{jk}) \end{aligned} \tag{4.43}$$

Here, $\partial P_{jk}\partial t_{jk}$, $\partial Q_{jk}\partial t_{jk}$, $\partial P_{kj}\partial t_{jk}$, and $\partial Q_{kj}\partial t_{jk}$ above can be obtained as follows:

$$\frac{\partial P_{jk}}{\partial t_{jk}} = V_j V_k(-G_{jk} \cos\delta_{jk} - B_{jk} \sin\delta_{jk}) \tag{4.44}$$

$$\frac{\partial Q_{jk}}{\partial t_{jk}} = V_j V_k(-G_{jk} \sin \delta_{jk} + B_{jk} \cos \delta_{jk}) \tag{4.45}$$

$$\frac{\partial P_{kj}}{\partial t_{jk}} = 2t_{jk}G_{jk}V_k + V_j V_k(-G_{jk} \cos \delta_{jk} + B_{jk} \sin \delta_{jk}) \tag{4.46}$$

$$\frac{\partial Q_{kj}}{\partial t_{jk}} = -2t_{jk}V_k B_{jk} + V_j V_k(G_{jk} \sin \delta_{jk} + B_{jk} \cos \delta_{jk}) \tag{4.47}$$

We are now able to calculate all the partial derivatives in Eq. (4.38).

Note that the method proposed here only needs the sub-gradients of the objective function related to the control variables, with no need of computing the objective function. In this way, the proposed control solution can adjust the control settings based on the directions of the sub-gradients. The active power loss here is expressed as the formula of bus admittance (**Y-bus**), and magnitudes (**V**) and the phase angle of the bus voltages δ. From the equations for sub-gradient calculation, we can see that the value of the active power loss is not required to calculate the sub-gradients. This indicates that updating local control settings do not require any calculation or measurement of global active power loss. Furthermore, we can obtain all the information required for sub-gradient computation locally or only from the adjacent buses, which then leads to the distributed calculation of sub-gradients.

It should be noted that there are three assumptions used for simplifying the calculation of sub-gradients and then realizing distributed calculations. The simulation results presented later demonstrate that the assumptions made for calculation are reasonable and will not affect the optimality of the proposed control strategy.

4.2.5 Realization of Mas-Based Solution

The control diagram of the ORPC system is shown in Figure 4.16. Each bus is assigned with a NA for acquiring local measurement data and exchanging information with its neighboring NAs. In this chapter, the communication networks of the MAS system are designed such that two NAs of the MAS system can communicate with each other only when their corresponding buses are electrically connected. This kind of practice can facilitate the usage of power line communication, which can then reduce the cost of control implementation. Additionally, this kind of communication topology design can easily realize the information exchanges of voltage amplitudes and phase angles between adjacent buses, i.e. neighboring agents. Yet, the communication topology can be devised to be different from that of the power network. The features of distributed control solutions based on MAS can be found in [16, 17].

As illustrated in Figure 4.16, each reactive power control device is installed with an agent called reactive power control agent (RPCA). Given that the RPCA takes

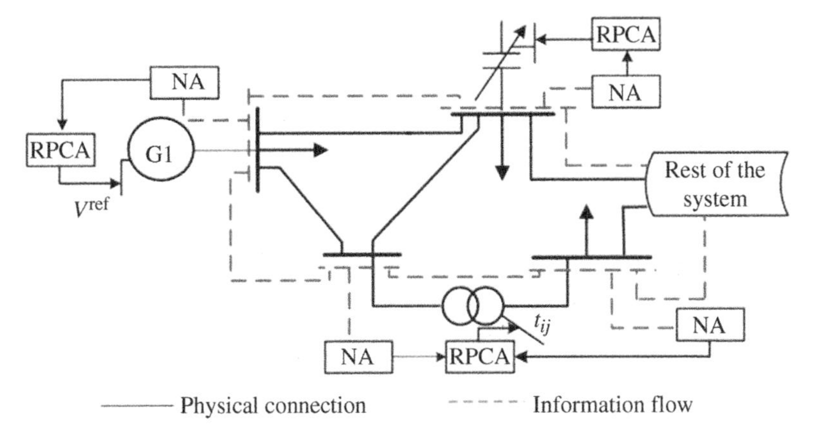

Figure 4.16 The control chart of the distributed OPRC system.

responsibility for generator voltage control or capacitor bank switching, it can receive the information of the local NA. If an RPCA is responsible for transformer tap changing, it is designed to receive information from both NAs of the buses that this transformer is connected to. After obtaining all the necessary information, RPCA will update the control settings by applying the distributed sub-gradient algorithm.

The NAs and RPCAs, controllers similar to the ones described in [44], contains three function modules, namely, data measurement, data processing, and communication. Note that the functions of an agent can be designed case to case, depending on the consideration of different developers and requirements of the system integration. Based on the definition in [58], a basic agent can be a software or hardware entity. In this chapter, an agent is defined as a functional module that combines physical controllers and computing elements. According to our design, the agent can adapt to changes in operating conditions by adjusting the control settings of the related local reactive controller. The following section will discuss issues about measurement and generator control.

4.2.5.1 Computation of Voltage Phase Angle Difference

All three types of calculations need voltage amplitudes and voltage phase angle differences of two adjacent buses. Adjacent buses are connected to the bus physically through power lines. Because each bus is installed with a NA, it is easy to measure the voltage amplitude locally and share it through the distributed communication. However, for most of the existing systems, voltage phase angle measurements are not available if phasor measurement units (PMUs) are not deployed. The reason lies in that a common synchronous frame is required to acquire the measurements of bus voltage phase angles of the system.

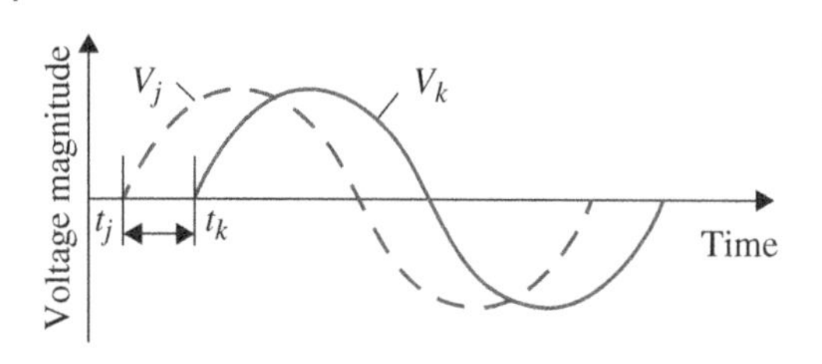

Figure 4.17 Voltage phase angle difference measured by NAs.

Note that for our proposed method, the calculation of the sub-gradients only requires the voltage phase angle difference of two adjacent buses rather than the absolute value of the voltage phase angles of these two buses. According to the design of the NA, it can easily measure the voltage phase angle difference between two adjacent buses. Figure 4.17 shows the principle of calculating the voltage phase angle difference of two adjacent buses by using the corresponding NAs of them. In Figure 4.17, voltage waves V_j and V_k are recorded by NA_j and NA_k, respectively, both of them being recorded with time stamps by using a common clock.

The zero-crossing instants of V_j and V_k are denoted as t_j and t_k. Then, the voltage phase angle difference between V_j and V_k can be calculated as follows:

$$\delta_{jk} = 2\pi f_M (t_k - t_j) \tag{4.48}$$

where f_M is the frequency of the power system measured by NAs.

During this process, the instant of voltage zero-crossing of a bus is recorded by its corresponding agent and transmitted to its neighboring agents for calculating phase angle difference. Therefore, the NAs are only required to share information locally rather than getting involved in the coordination of the central controller, which significantly reduces the amount of communication data as well as the time consumed for communication.

4.2.5.2 Generation Control for ORPC

For the generator that participates in ORPC, an automatic voltage control system is deployed for this generator's voltage regulation. The reallocation of reactive power will lead to a change of active power loss of the power system and thus change the active power demand of the overall system. In order to keep an efficient balance between supply and demand of active power, the active power/frequency control system is adopted for the generator generally based on the droop control or PI control. The active power/frequency control systems are typically realized on the basis of droop control or PI control. The generation control scheme for ORPC is illustrated in Figure 4.18. As shown in the control signal updating rule and the formulae for sub-gradient calculation, an RPCA generates the local control signal based on the measured voltage amplitudes and voltage phase angle difference from local and neighboring agents and bus admittance. Because instants of

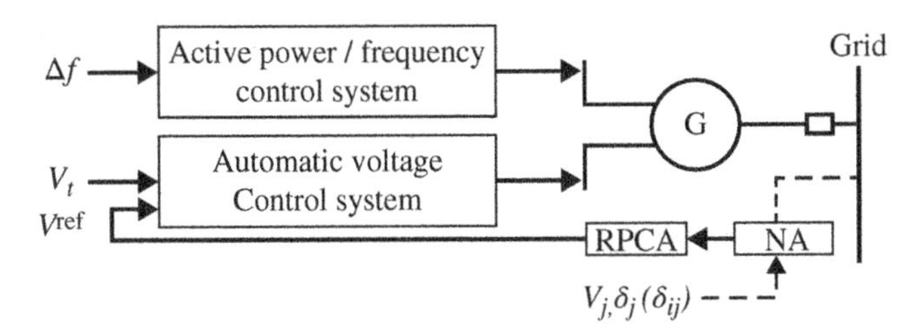

Figure 4.18 OPRC generation control scheme.

zero-crossing will be utilized to compute the voltage phase angle difference of two connected buses, it is necessary to synchronize them using a common clock signal (for example, a GPS signal). Once all the required information has been obtained through local measurement and communication, the RCPAs will update and then deploy local control signals consecutively. This process will be repeated until the optimal solution is found, or the change of the operating conditions triggers a new round of optimization.

4.2.6 Simulation and Tests

We evaluate the presented control scheme with the 6-bus Ward-Hale system, the IEEE 30-bus system, and the 1062-bus system in this part.

4.2.6.1 Test of the 6-Bus Ward–Hale System

There are six reactive power controllers in this 6-bus system [31] and their parameters are shown in Table 4.6. Based on the system, we devise two test scenarios. The first test scenario only considers active power loss ($W_1 = 2$ and $W_2 = 0$) while both the active power loss and voltage profiles ($W_1 = 2$ and $W_2 = 1/6$) are considered in the second scenario. For comparison of the performance between the centralized control solution and the distributed control solution, we implement these two different solutions as in Figure 4.19. For the distributed control solution,

Table 4.6 Summary for reactive power controller.

Control variables	Lower bound	Upper bound
Q_{c4}	0	0.15
Q_{c6}	0	0.30
V_{g2}	1.1	1.15
T_{4-3}	0.9	1.1
T_{6-5}	0.9	1.1

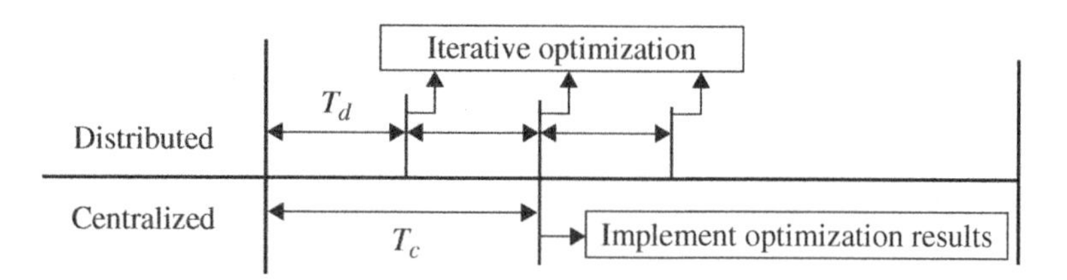

Figure 4.19 Realization of centralized and distributed control solutions.

the control settings are updated every two seconds ($T_d = 2s$) until convergence. Because this algorithm is quite simple for calculation and control update, two seconds is enough for this distributed solution to finish one round of update. In order to evaluate the optimality of the obtained solution, we also utilize the centralized algorithm to acquire the global optimal solution. Because the centralized solution often requires more time for data acquisition and processing, we set the update interval for the controlled settings to four seconds ($T_c = 4$ seconds). Besides, these two control solutions are both deployed at t = 4 seconds with a simulation duration of 25 seconds.

To implement the proposed distributed sub-gradient algorithm, each agent only needs to process a small amount of data in every step of updating. Therefore, using inexpensive hardware, such as an Android device based on a cortex-A8 processor, can easily implement this algorithm. In general, data acquisition and processing can be completed in one cycle (16.67 ms for 60 Hz system). Our experiences with MAS hardware implementation shows that information exchange between two agents takes less than 5 ms with the actual control being completed within a fraction of a millisecond. Therefore, a step of control update can be completed in less than 0.1 seconds, regardless of the size or configuration of the power system. This is due to the simplicity of the applied distributed algorithm, which consumed quite a little time for communication and computation. In fact, we can set the time interval between control updates to a very small value, even less than one seconds. For our implementation here, this value is set to a range of 2–10 seconds, which can be achieved without difficulty.

Test 1 ($W_1 = 2$ and $W_2 = 0$)

For this scenario $W_2 = 0$, the control objective only aims at minimizing the active power loss of the system. The simulation results of this scenario are provided in Figures 4.20–4.25.

As can be seen from Figure 4.20, the converged values of the objective function of the distributed and centralized solutions are the same, and they are remarkably lower than the case without applying optimization.

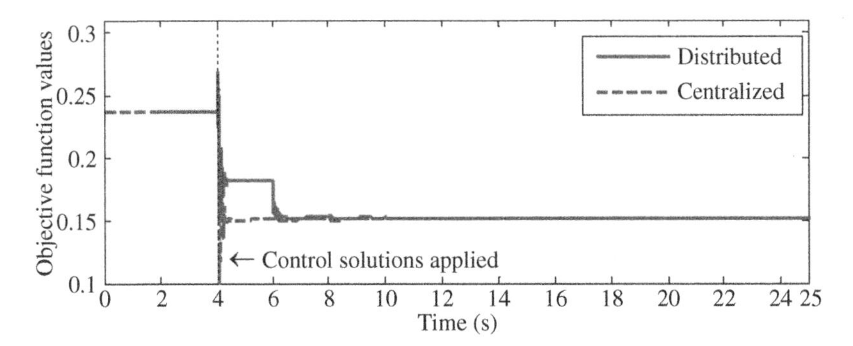

Figure 4.20 The evolution of objective function with different control solutions.

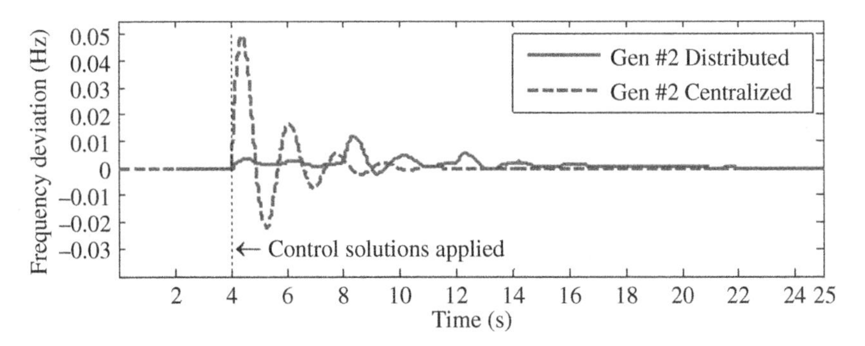

Figure 4.21 Velocity misalignment with different control solutions.

Figure 4.21, shows the normalized speed deviations for generator #2 (the generator with maximum speed deviation) with different control schemes. For the distributed control, the maximum speed deviation is 0.012 Hz, which is much smaller than that of the centralized scheme (0.050 Hz).

In the designed distributed control solution, we update the control settings in an incremental way, enforcing a relatively small change for control variables. However, for the centralized control solution, the control settings are only updated after global optima have been obtained, which enforces large change for the control settings of control devices. Consequently, the overall response of the system under the proposed distributed control is smoother and thus much preferred.

Figure 4.22 depicts the voltage profiles with the proposed distributed solution. As shown in figure, in the absence of ORPC, the lowest bus voltage value is only 0.857 p.u. (at bus #4), while after implementing the proposed solution, the lowest bus voltage increases to 0.942 p.u. (at bus #3). This shows that although the optimization of this scenario only considers the active power loss, the voltage profiles can still be improved to some extent.

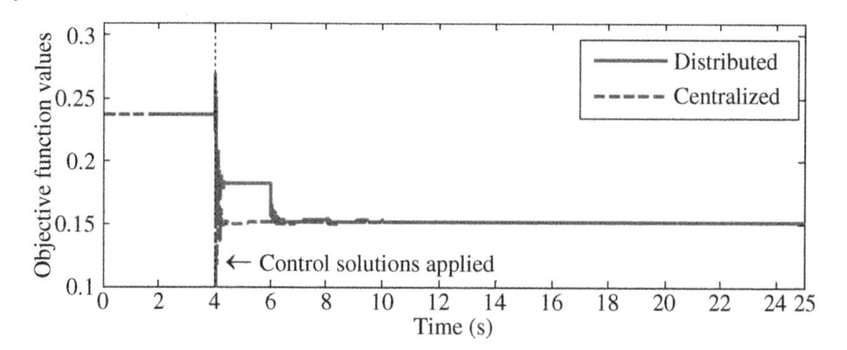

Figure 4.22 Voltage curve in distributed control solutions.

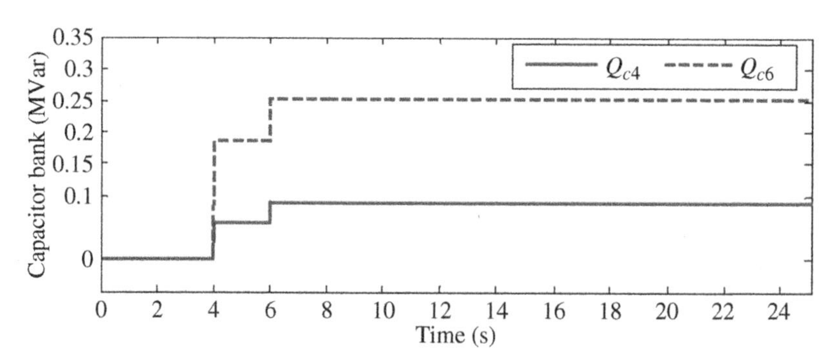

Figure 4.23 The control sequences of capacitor banks solutions.

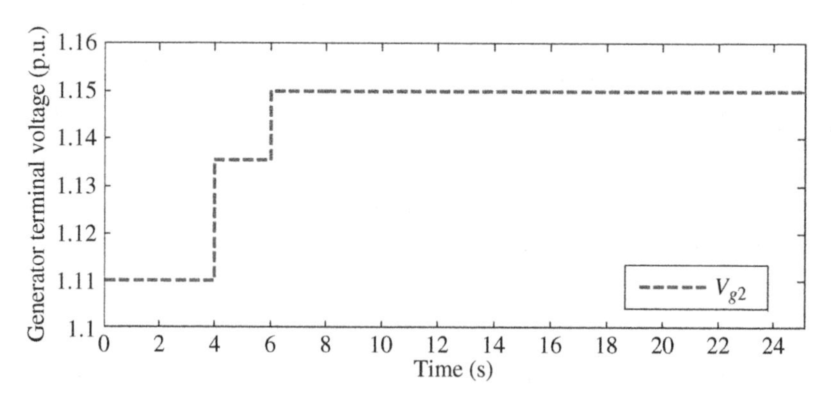

Figure 4.24 The control sequences of generator voltage.

The updating of control settings for generator terminal voltages, capacitor banks, and tap changing transformers are shown in Figures 4.23–4.25. Note that the generator terminal voltage V_{g2} reaches its upper bound at six seconds after the convergence for this scenario, as shown in Figure 4.24.

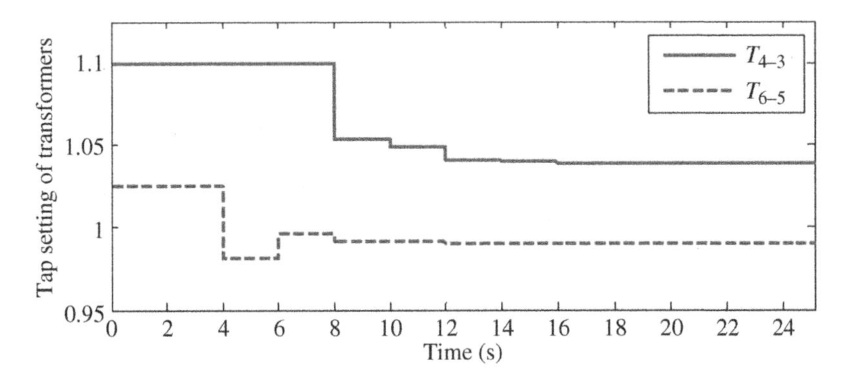

Figure 4.25 The control sequences of transformer tap setting.

Test 2 ($W_1 = 2$ and $W_2 = 1/6$)

In the second test scenario, the preferred voltage amplitudes for the six buses are set as 1.04, 1.15, 1.0, 1.0, 1.0, and 1.0 p.u., respectively (Figures 4.26 and 4.27). The optimization results are illustrated in Figure 4.26. Because the voltage profiles are considered in this scenario, the values of the objective function of this scenario are much higher than the values of test 1.

Figure 4.27 shows the evolution of the voltage profiles during optimization. Compared with test 1, the voltage profiles here are closer to the desired voltage levels. Note that there may be cases that the voltage of the specified bus deviates from the predefined value. The reason lies in that the optimization of this scenario is a compromise between minimizing active power loss and keeping a satisfying voltage level, which cannot ensure the voltage of a specific bus to be exactly the same as the predetermined value.

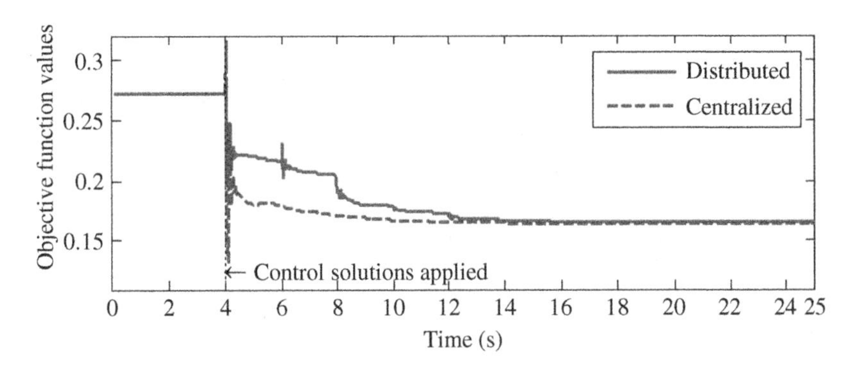

Figure 4.26 The evolution of the objection function value.

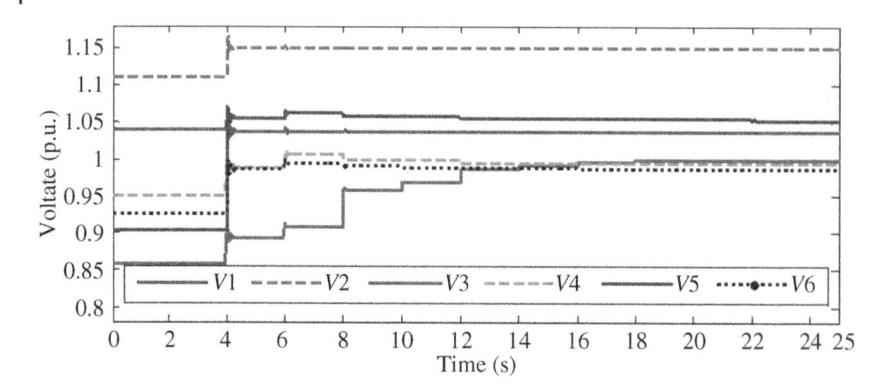

Figure 4.27 Voltage profiles with distributed control solution.

4.2.6.2 Test of IEEE 30-Bus System

The parameters of the IEEE 30-bus system used for the test can be found in [59]. The system comprises 48 branches, 6 generator buses, and 22 load buses, with 5 generator buses being the PV buses and 1 generator bus being the swing bus. The ORPC control variables we used for tests contain nine capacitor bank switches, five generator (PV) bus voltage control settings, and four transformer tap settings.

Figure 4.28 shows the control implementation of the system of IEEE 30-bus. Here, we also test two scenarios, in the first scenario considering the normal operating conditions and in the second scenario considering abnormal operating conditions such as data loss and controller failure. The comparison with the centralized method is also provided.

In these two test scenarios, the expected voltage magnitudes for all load buses are set to 1.0 p.u., and desired values for generators buses are set to 1.05 for bus #1, at 1.02 for bus #2, at 1.01 for bus #5, at 1.01 for bus #8, at 1.05 for bus #11, and at 1.05 for bus #13 p.u., respectively.

Test Under Normal Operating Conditions

Both distributed and centralized solutions are deployed according to Figure 4.19. Because the system is relatively large, we set T_d to 4 seconds and T_c to 10 seconds. In addition, in order to evaluate the performance of the proposed solution under the system operating conditions change, two changes of operating conditions are simulated. At $t = 35$ seconds, the active power load at bus 12 is increased by 5 MW, while the reactive power load being increased by 5 Mvar simultaneously, and the active power load at bus 21 is also increased by 5 MW at the same instant. At $t = 45$ seconds, the previously increased load is decreased back to the original value before before $t = 35$ seconds. In this test, we set the weight factors W_1 and W_2 severally as 2 and 1/10. Both solutions are deployed at $t = 10$ seconds with a simulation duration of 60 seconds.

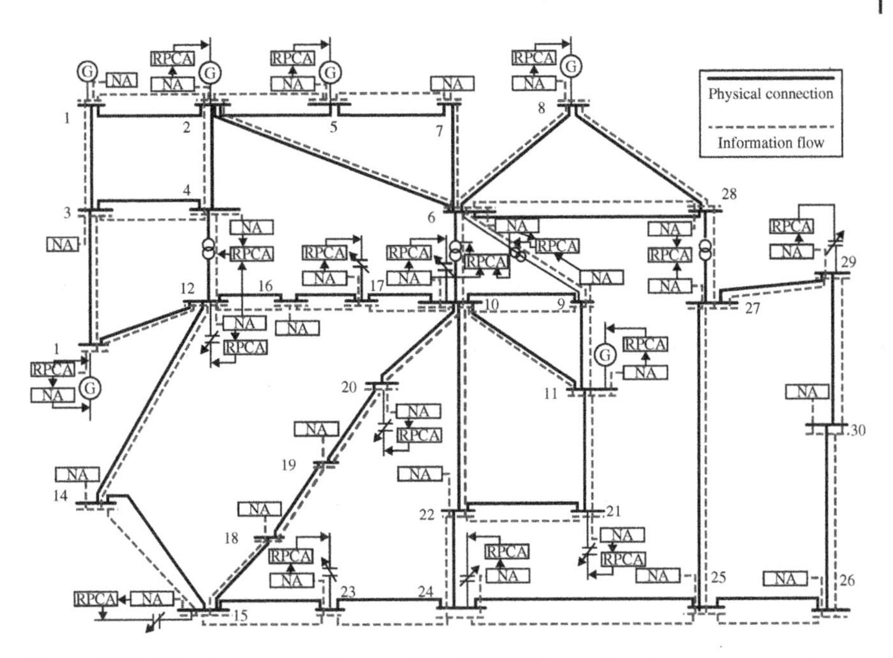

Figure 4.28 The deployment diagram of the IEEE 30-bus test system.

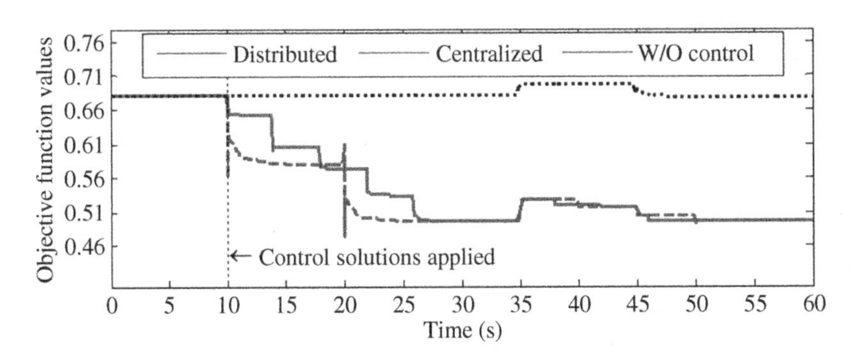

Figure 4.29 Evolution of the objective function.

The changes in the objective function values with different control solutions are shown in Figure 4.29. At the steady state, the proposed control solutions achieve the same optimal performance as the solution for centralized control, whereas it can respond in a timely manner to reach new optimal operating points under sudden changes (from 35 to 50 seconds).

Here, we set the update interval for updating control settings of the centralized solution to 10 seconds. However, in practice, this interval is generally much greater

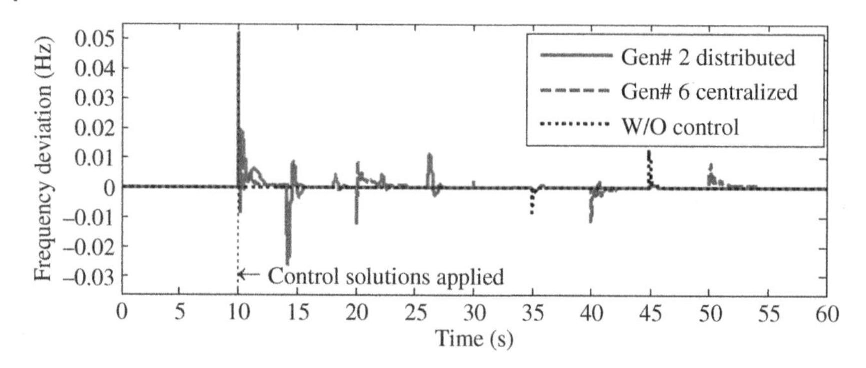

Figure 4.30 Speed deviations with different control solutions.

because the centralized controller needs to collect a large amount of data via a complex communication network and also to process them. Because of the simplicity of distributed computing technology and communication networks, distributed solutions, however, can respond in a timely manner. Consequently, when the system operating conditions change continuously and unexpectedly, the distributed solution can provide a better response. The speed deviations of generators with different control schemes are shown in Figure 4.30. It can be seen that the maximum speed deviation for the distributed solution is only 0.0265 Hz (Gen#2), while the centralized solution results in the maximum deviation of 0.0515 Hz at Gen#6. Accordingly, the dynamic performance of the distributed control solution is more preferred.

Figures 4.31–4.36 show the capacitor banks, generator terminal voltage, and tap changer control settings. As shown in these figures, the rate of convergence of the optimization process is relatively fast, and the convergence speed does not increase dramatically as the size of the system increases. Therefore, the model-based distributed ORPC algorithm proposed in this chapter can respond faster than that in [44].

We also compare the presented control scheme with the multi-agent-based particle swarm optimization (MASPSO) method discussed in [42]. According to [42], the agent size of MASPSO is set to $6 * 6$. The MASPSO control updating interval is fixed at 4 seconds, which is the same as the proposed method. Based on our test results, each iteration of MASPSO takes approximately 0.8 seconds. With the control updating interval being set to 4 seconds, around five iterations can be compeleted for MASPSO scheme.By taking the underlying delays and uncertainties into consideration, we conduct three iterations of optimization within 4 seconds. As can be seen in Figure 4.34, the MASPSO converges at $t = 50$ seconds (after approximate 30 iterations).

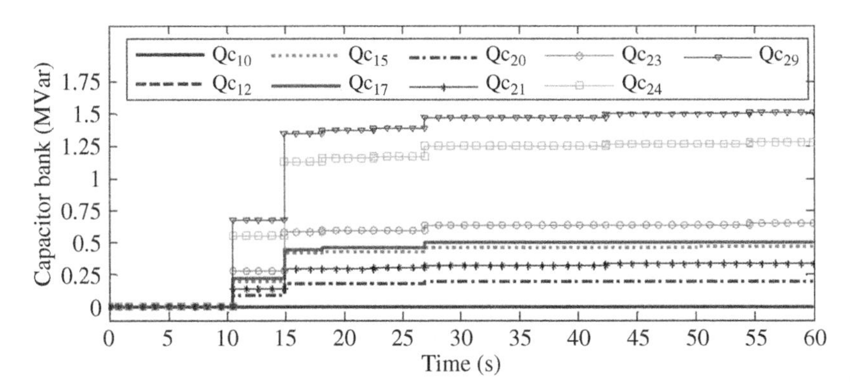

Figure 4.31 Control sequences of capacitor banks.

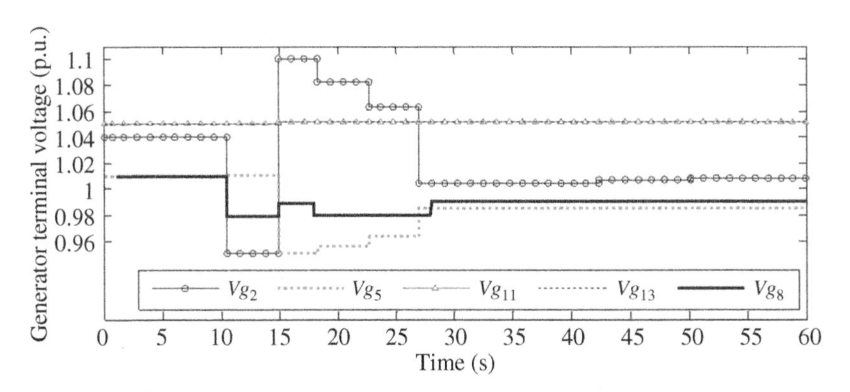

Figure 4.32 Control sequences of generator terminal voltages.

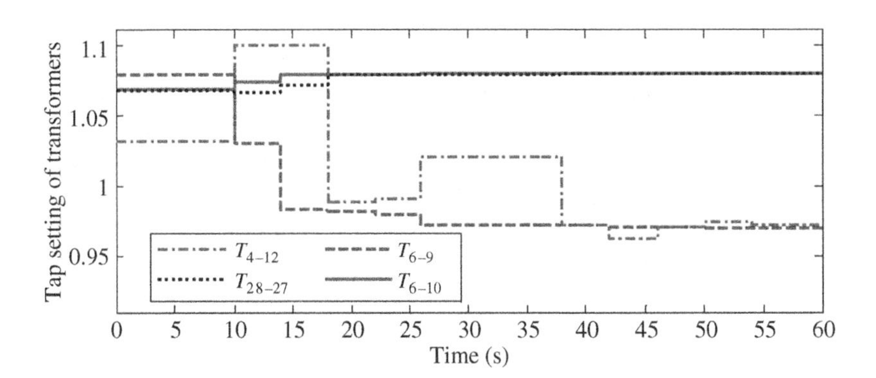

Figure 4.33 Control sequences of transfer tap settings.

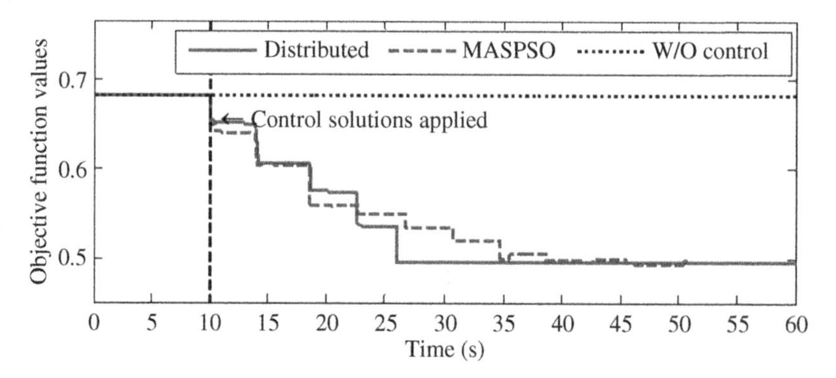

Figure 4.34 Comparison with MASPSO.

For each iteration of the MASPSO method, the corresponding agent (particle) needs to conduct power flow studies, calculate the fitness function, spread its information to the whole environment, and then update the control references using the predefined operators. Considering a large number of agents are involved in operations for each iteration, with each agent being responsible for processing a significant amount of data, the time consumed for each iteration of MASPSO will be much longer than that of the proposed method. The proposed method also converges in a much faster manner because its optimization is directed by model-based analyses, which is much better than the random search-based PSO technique.

Test with Data Loss and Controller Failure

In this part, two scenarios are simulated wherein the first scenario is the case with data loss and the second scenario being the case with controller failure. Figure 4.35 shows the simulation results under data loss. The data loss occurs at $t = 20$ seconds because of the malfunction of the communication between the agents associated with bus 27 and bus 29. Accordingly, the RPCA at branches 27–28 (transformer) and bus #29 (capacitor) are excluded for further control updates. At $t = 25$ seconds, the compromised communication is restored, and the influenced agent recovers their normal operations.

As shown in the simulation results, the data loss due to the malfunction of communication slows down the overall converging speed. If the communication can be recovered afterward, the overall optimization process will resume, and this case will not degrade the optimality of the final obtained solution. Yet, the data loss does slow the convergence speed slightly.

It should be pointed out that large communication latency also leads to the slowdown of the converging speed. As mentioned earlier, it takes less than 22 ms to

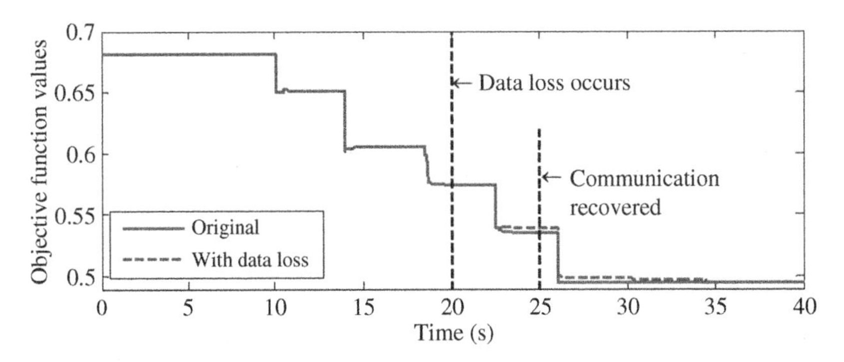

Figure 4.35 Optimization process under data leakage.

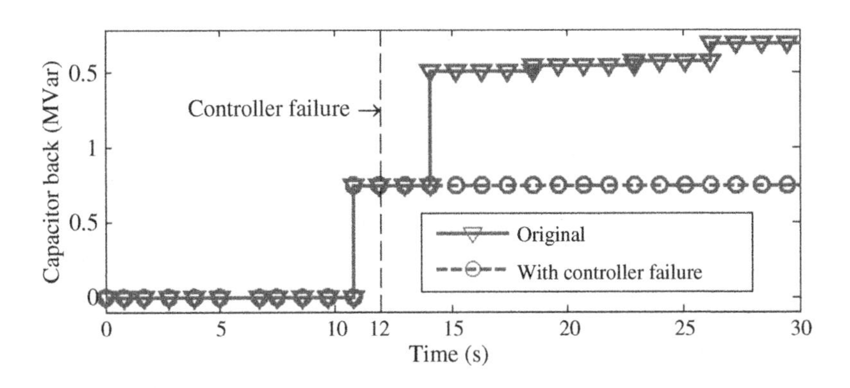

Figure 4.36 Control settings of capacitor bank $Q_{C_{29}}$ during optimization.

complete one step of the control update in the test setup. As 22 ms is the maximum time that different iterations might require for our experimental setup, short communication latency can be tolerated. Because the interval of control update is set to four seconds, the communication latency less than four seconds has no effect on the performance of the proposed control algorithm. However, if the communication latency is greater than the predefined control updating interval, i.e. four seconds in this case, its impact on the optimization process can thence be treated the same as the data loss.

For the control failure scenario, we assume that the agent associated with the capacitor bank at bus #29 failed at 12 seconds. Figures 4.36 and 4.37 show the

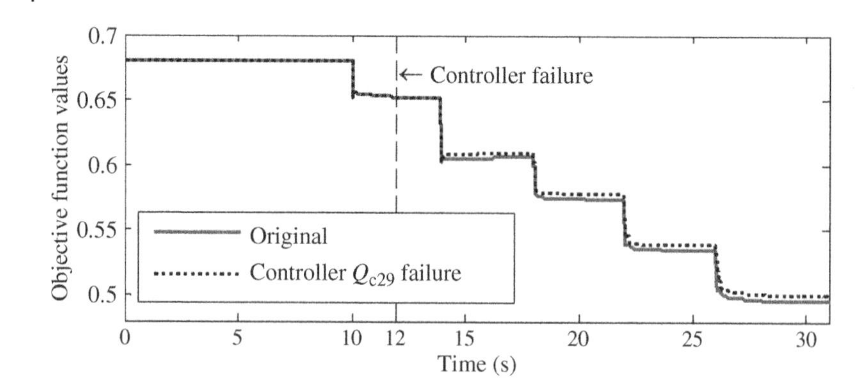

Figure 4.37 Optimization proceedings with and without agent loss.

corresponding test results with this controller failure. As shown in Figure 4.36, after the failure of this control agent, then the output of the capacitor bank is held fixed with the capacitor at this bus being excluded for further operation. Because the rest of the system is still working after the occurrence of the failure, the remaining agents continue to proceed with the optimization process. As can be seen from Figure 4.37, the converged value of the objective function is slightly higher than that without the controller failure. Thus, the controller failure does degrade the optimality of the obtained solution.

Considering that the proposed control solution is fully distributed, the control scheme does not require a central controller and a complex communication network. Under the circumstances of the loss of agents or communication links between agents, the remaining working agents can still interact with each other to realize ORPC for the system. As demonstrated in Figure 4.37, the loss of the agent may inevitably degrade the optimality of the obtained solution, and the loss of the communication links may reduce the convergence speed of the optimization process. Nevertheless, because of the merits of autonomously operating agents, the distributed solution exhibits strong resistance against the abovementioned abnormality. Additionally, it is less susceptible to a single-point failure compared with the traditional centralized solution.

The results of simulation show that it takes six iterations to converge for the five-reactive controller. Ward–Hale system also converges after 6 iterations for the 18-controller IEEE 30-bus system and 23 iterations. Note that the convergence rate of the proposed distributed control solution does not increase significantly as the system size and the number of reaction controllers increase for the studied cases. Yet, future work should dedicate to investigate its performance with large-scale systems.

4.2.7 Conclusion

In this chapter, we investigate a fully distributed MAS-based solution for online ORPC of power systems. This solution can be easily implemented by using MAS with a relatively simple communication network configuration. Based on the mild assumption, this chapter designs a sub-gradient-based supporting algorithm for ORPC. Because the algorithm and underlying communication network are simple, the proposed solution enables online optimal reactive power control of future smart grids.

The proposed fully distributed solution here does not need powerful centralized controller and requires relative few computational resources. Furthermore, considering the characteristics of distributed control, it is less susceptible to single-point failure, hence more reliable.

References

1. Stott, B., Marinho, J.L., and Alsac, O. (1979). Review of linear programming applied to power system rescheduling. *IEEE Conference Proceedings Power Industry Computer Applications Conference, 1979. PICA-79*. IEEE, pp. 142–154.
2. Monticelli, A. and Liu, W.-H.E. (1998). Adaptive movement penalty method for the Newton optimal power flow. *IEEE Transactions on Power Systems* 7 (1): 56–64.
3. Santos, J.R., Lora, A.T., Exposito, A.G. and Ramos, J.L.M. (2003). Finding improved local minima of power system optimization problems by interior-point methods. *IEEE Transactions on Power Systems* 18 (1): 238–244.
4. de Souza, A.C.Z., Honorio, L.M., Torres, G.L., and Lambert-Torres, G. (2004). Increasing the loadability of power systems through optimal-local-control actions. *IEEE Transactions on Power Systems* 19 (1): 188–194.
5. Pham, D. and Karaboga, D. (2000). *Intelligent Optimisation Techniques: Genetic Algorithms, Tabu Search, Simulated Annealing and Neural Networks*. London: Springer London.
6. Vlachogiannis, J.G., Hatziargyriou, N.D., and Lee, K.Y. (2005). Ant colony system-based algorithm for constrained load flow problem. *IEEE Transactions on Power Systems* 20 (3): 1241–1249.
7. Zhao, B., Guo, C.X., and Cao, Y.J. (2004). Improved particle swam optimization algorithm for OPF problems. *IEEE PES Power Systems Conference and Exposition, 2004*. IEEE, pp. 233–238.

8 Solanki, J.M. and Schulz, N.N. (2005). Using intelligent multi-agent systems for shipboard power systems reconfiguration. *Proceedings of the 13th International Conference on Intelligent Systems Application to Power Systems*. IEEE, 3pp.

9 Huang, K., Cartes, D.A., and Srivastava, S.K. (2007). A multiagent-based algorithm for ring-structured shipboard power system reconfiguration. *IEEE Transactions on Systems, Man, and Cybernetics, Part C (Applications and Reviews)* 37 (5): 1016–1021.

10 Wooldridge, M. (2002). *An Introduction to Multiagent Systems*. Wiley.

11 Dimeas, A.L. and Hatziargyriou, N.D. (2005). Operation of a multiagent system for microgrid control. *IEEE Transactions on Power systems* 20 (3): 1447–1455.

12 Logenthiran, T., Srinivasan, D., and Wong, D. (2008). Multi-agent coordination for DER in MicroGrid. *2008 IEEE International Conference on Sustainable Energy Technologies*. IEEE, pp. 77–82.

13 Xiao, F., Wang, L., and Jia, Y. (2008). Fast information sharing in networks of autonomous agents. *2008 American Control Conference*. IEEE, pp. 4388–4393.

14 Olfati-Saber, R., Fax, J.A., and Murray, R.M. (2007). Consensus and cooperation in networked multi-agent systems. *Proceedings of the IEEE* 95 (1): 215–233.

15 Korres, G.N., Katsikas, P.J., and Contaxis, G.C. (2004). Transformer tap setting observability in state estimation. *IEEE Transactions on Power Systems* 19 (2): 699–706.

16 Xu, Y. and Liu, W. (2011). Novel multiagent based load restoration algorithm for microgrids. *IEEE Transactions on Smart Grid* 2 (1): 152–161.

17 Xu, Y., Liu, W., and Gong, J. (2011). Stable multi-agent-based load shedding algorithm for power systems. *IEEE Transactions on Power Systems* 26 (4): 2006–2014.

18 Busoniu, L., Babuska, R., Schutter, B.D., and Ernst, D. (2010). *Reinforcement Learning and Dynamic Programming Using Function Approximators*. CRC Press.

19 Sutton, R.S. and Barto, A.G. ((2018)) *Reinforcement Learning: An Introduction, Adaptive Computation and Machine Learning*. The MIT Press, Cambridge, Massachusetts, US.

20 Vlachogiannis, J.G. and Hatziargyriou, N.D. (2004). Reinforcement learning for reactive power control. *IEEE Transactions on Power Systems* 19 (3): 1317–1325.

21 Antonio Martın H., J. and de Lope Asiaın, J. (2007). A distributed reinforcement learning control architecture for multi-link robots-experimental validation. *ICINCO-ICSO*, pp. 192–197.

22 Fernndez, F. and Parker, L.E. (2001). Learning in large cooperative multi-robot systems. *International Journal of Robotics and Automation* 16 (4): 217–226.

23 Bu, L., Babu, R., De Schutter, B. et al. (2008). A comprehensive survey of multiagent reinforcement learning. *IEEE Transactions on Systems, Man, and Cybernetics, Part C (Applications and Reviews)* 38 (2): 156–172.

24 Watkins, C.J. C.H. (1989). Learning from delayed rewards. PhD thesis. Cambridge: King's College.

25 Matignon, L., Laurent, G.J., and Le Fort-Piat, N. (2007). Hysteretic Q-learning: an algorithm for decentralized reinforcement learning in cooperative multi-agent teams. *2007 IEEE/RSJ International Conference on Intelligent Robots and Systems. IEEE*, pp. 64–69.

26 Menezes, T.V. and Da Suva, L.C.P. (2006). A method for transmission loss allocation based on sensitivity theory. *Power Engineering Society General Meeting, 2006. IEEE*. IEEE, 6pp.

27 Guo, M., Liu, Y., and Malec, J. (2004). A new Q-learning algorithm based on the metropolis criterion. *IEEE Transactions on Systems, Man, and Cybernetics, Part B (Cybernetics)* 34 (5): 2140–2143.

28 Christie, R. (2000). Power systems test case archive. *Electrical Engineering*. University of Washington. https://www2.ee.washington.edu/research/pstca (accessed 21 July 2020).

29 AlRashidi, M.R. and El-Hawary, M.E. (2007). Hybrid particle swarm optimization approach for solving the discrete OPF problem considering the valve loading effects. *IEEE Transactions on Power Systems* 22 (4): 2030–2038.

30 Turitsyn, K., Sulc, P., Backhaus, S., and Chertkov, M. (2011). Options for control of reactive power by distributed photovoltaic generators. *Proceedings of the IEEE* 99 (6): 1063–1073. https://doi.org/10.1109/JPROC.2011.2116750.

31 Sharif, S.S., Taylor, J.H., and Hill, E.F. (1996). On-line optimal reactive power flow by energy loss minimization. *Proceedings of 35th IEEE Conference on Decision and Control*, Volume 4 (December 1996), pp. 3851–3856. https://doi.org/10.1109/CDC.1996.577262.

32 Rebours, Y.G., Kirschen, D.S., Trotignon M., and Rossignol, S. A survey of frequency and voltage control ancillary services – Part I: Technical features. *IEEE Transactions on Power Systems* 22 (1): 350–357. https://doi.org/10.1109/TPWRS.2006.888963.

33 Laughton, M.A. (1970). Power-system load scheduling with security constraints using dual linear programming. English. *Proceedings of the Institution of Electrical Engineers* 117: 2117–2127(10).

34 Zhang, G. (1986) On-line network constrained reactive power control using an incremental reactive current model. *Proceedings of the 2nd International Conference Power Systems Monitoring Control*, pp. 156–161.

35 Wu, F.F., Gross G., Luini, J.F. (1979). A two-stage approach to solving large-scale optimal power flows. *Power Industry Computer Applications Conference, 1979. PICA-79. IEEE Conference Proceedings*. IEEE, pp. 126–136.

36 Yoshida, H., Kawata, K., Fukuyama, Y., Takayama, S. and Nakanishi, Y. A particle swarm optimization for reactive power and voltage control considering voltage security assessment. *IEEE Transactions on Power Systems* 15 (4): 1232–1239.

37 Yan, W., Liu, F., Chung C.Y., and Wong, K.P. (2006). A hybrid genetic algorithm-interior point method for optimal reactive power flow. *IEEE Transactions on Power Systems* 21 (3): 1163–1169.

38 Lin, C.H. and Lin, S.-Y. (2008). Distributed optimal power flow with discrete control variables of large distributed power systems. *IEEE Transactions on Power Systems* 23 (3): 1383–1392.

39 Cheng, X., Zhang, Y., Cao, L., Li, J., Shen, T. and Zhang, S. (2005). A real-time hierarchical and distributed control scheme for reactive power optimization in multi-area power systems. *Transmission and Distribution Conference and Exhibition: Asia and Pacific, 2005 IEEE/PES*. IEEE, pp. 1–6.

40 Leeton, U. and Kulworawanichpong, T. (2012). Multi-agent based optimal power flow solution. *Power and Energy Engineering Conference (APPEEC), 2012 Asia-Pacific*. IEEE, pp. 1–4.

41 Nagata, T., Hatano, R., and Saiki, H. (2009). A multi-agent based distributed reactive power control method . *Power & Energy Society General Meeting, 2009. PES'09. IEEE*. IEEE, pp. 1-7.

42 Zhao, B., Guo, C.X., and Cao, Y.J. (2005). A multiagent-based particle swarm optimization approach for optimal reactive power dispatch. *IEEE Transactions on Power Systems* 20 (2): 1070–1078.

43 Nasri, M., Farhangi, H., Palizban, A. and Moallem, M. (2012). Multi-agent control system for real-time adaptive VVO/CVR in Smart Substation. *Electrical Power and Energy Conference (EPEC), 2012 IEEE*. IEEE, pp. 1-7.

44 Xu, Y., Zhang, W., Liu, W., and Ferrese, F. (2012). Multiagent-based reinforcement learning for optimal reactive power dispatch. *IEEE Transactions on Systems, Man, and Cybernetics, Part C (Applications and Reviews)* 42 (6): 1742–1751.

45 Vaisakh, K. and Rao, P.K. (2008). Optimal reactive power allocation using PSO-DV hybrid algorithm. *India Conference, 2008. INDICON 2008. Annual IEEE*, Volume 1. IEEE, pp. 246–251.

46 Augugliaro, A., Dusonchet, L., Favuzza, S. and Sanseverino, E.R. (2004). Voltage regulation and power losses minimization in automated distribution networks by an evolutionary multiobjective approach. *IEEE Transactions on Power Systems* 19 (3): 1516–1527.

47 Nicholson, H. and Sterling, M.J.H. (1973). Optimum dispatch of active and reactive generation by quadratic programming. *IEEE Transactions on Power Apparatus and Systems* 2: 644–654.

48 de la Torre, S. and Galiana, F.D. (2005). On the convexity of the system loss function. *IEEE Transactions on Power Systems* 20 (4): 2061–2069.

49 Vaccaro, A., Velotto, G., and Zobaa, A.F. (2011). A decentralized and cooperative architecture for optimal voltage regulation in smart grids. *IEEE Transactions on Industrial Electronics* 58 (10): 4593–4602.

50 Lobel, I. and Ozdaglar, A. (2011). Distributed subgradient methods for convex optimization over random networks. *IEEE Transactions on Automatic Control* 56 (6): 1291.

51 Xu, Y., Zhang, W., Liu, W. et al. (2014). Distributed subgradient-based coordination of multiple renewable generators in a micro

52 Bhattacharyya, B. and Goswami, S.K. (2009). Sensitivity based evolutionary algorithms for reactive power dispatch. *International Conference on Power Systems, 2009. ICPS'09. IEEE*, pp. 1–7.

53 Barbarossa, S. (2005). Self-organizing sensor networks with information propagation based on mutual coupling of dynamic systems. *Proceedings of IWWAN* 2005.

54 Barbarossa, S. and Scutari, G. (2007). Decentralized maximum-likelihood estimation for sensor networks composed of nonlinearly coupled dynamical systems. *IEEE Transactions on Signal Processing* 55 (7): 3456–3470.

55 Mamandur, K.R.C. and Chenoweth, R.D. (1981). Optimal control of reactive power flow for improvements in voltage profiles and for real power loss minimization. *IEEE Transactions on Power Apparatus and Systems* 7: 3185–3194.

56 Wooldridge, M. (2009). *An Introduction to Multiagent Systems*. Wiley.

57 Lee, K.Y., Park, Y.M., and Ortiz, J.L. (1985). A united approach to optimal real and reactive power dispatch. *IEEE Transactions on Power Apparatus and Systems* 5: 1147–1153.

5

Distributed Demand-Side Management

Demand management and response are being recognized as important drivers for active user participation in the energy market. It is an adjustment in the power consumption of an electric utility customer to better match the demand for power with the supply. Traditional application of demand response (DR) usually requires a centralized powerful control center and a two-way communication network between the system operators and energy end users. Yet, the increasing user participation in smart grids may limit their applications. In this chapter, we will investigate two methods that can realize the distributed DR or demand management. First, we discuss a distributed solution for incentive-based management (LM) program. The LM problem for this method is formulated as a constrained optimization problem aiming at maximizing the overall utility of users while meeting the requirement for load reduction requested by the system operator and is solved by using a distributed dynamic programming (DDP) algorithm. For the second method, we focus on controlling of plug-in electric vehicles (PEVs) that are connected to a gird to minimize the energy loss of the system. To this end, the charging rates of PEVs are controlled in an optimized way. The optimal control solution is based on a consensus algorithm, which aligns each PEV's interest with the system's benefit. The control strategy can be implemented based on a multi-agent system (MAS) framework, which only requires information exchanges among neighboring agents. Both the methods introduced in this chapter are distributed methods, and the corresponding control solutions also exhibit the merits of other distributed control solutions. Different tests are also designed for the evaluation of these two discussed methods.

Distributed Energy Management of Electrical Power Systems, First Edition.
Yinliang Xu, Wei Zhang, Wenxin Liu, and Wen Yu.

Published 2021 by John Wiley & Sons, Inc.

5.1 Distributed Dynamic Programming-Based Solution for Load Management in Smart Grids

Because of the desire for economic and environmental benefits, a more intelligent, effective, reliable, and flexible power grid is demanded by both energy providers and users [1, 2]. The term "smart grid" has been used to address such a power grid with enhanced functionalities. There are two remarkable advantages of the smart grid. First, more renewable sources could be accommodated in such a smart grid. Second, more active participants from energy end users will be accepted in the smart grid. This active participation from the users is beneficial to promote the efficiency of the power market through reducing power generation cost, providing sufficient reserve margin, and facilitating to keep the stability of the overall system [3].

The load management (LM) program, also known as DR, has attracted much attention as one of the promising options for user participation. It functions to adjust the electricity demand of the end users in response to the change in electricity price over the time horizon or in response to the incentive payments or discount rates that are designed to decrease or increase energy consumption when the system capacity is insufficient or the reliability is compromised [4]. According to the estimation of EPRI, if deployed, DR could reduce the peak demand of the country by 45 000 MW [5]. Meanwhile, according to the suggestions from the Battle group, DR could provide the benefits equivalent to as high as tens of millions of dollars by carrying out simple price mechanisms [6]. A benefitcost analysis had been conducted by the US Federal Energy Regulatory Commission, which indicates that the incorporation of LM within the regional energy market would achieve more than $60 financial benefits [7].

There are mainly two forms of LM programs, one is incentive-based programs (IBP) and the other is price-based programs (PBP) [8]. For IBP, the participants are rewarded with financial benefits or discount rates for their energy bills to reduce their load demand when required by the program initiator, during peak demand, or high electricity price periods. IBP has been operated by many utilities or third-party organizations such as California-based PG&E and Pennsylvania-New Jersey Maryland power market [9, 10]. PBP aims to alleviate the demand peak and demand valley by providing dynamic electric prices to the customers. These rates include critical peak pricing, extreme day pricing, real-time pricing, etc. [11]. PBP has been adopted in the deregulated market by numerous utilities [12].

During the past decades, the ways of implementing LM had been enriched, and potential market values have also been improved. However, reviews on the experience of LM vary [13]. In general, current LM programs are inept to some extent and are not competent for continuous and repeated use of energy by the consumers. There exist many problems with the practical implementation of LM, such as reliability drop resulting from frequent schedule readjustments, interruption of communication, or sudden operating changes during the LM period. In addition, the participants of the LM program may also be concerned about their comfort and business continuity when they carry out their LM schedules. Thus, a more reliable and automated LM solution is highly desired.

However, some opponents believe that owners who have LM capacity were in unrelated business and the grid operator should not count on them. Yet, the practice of large-scale LM has demonstrated its significance in enhancing the system reliability and reducing the operation cost. The integration of intermittent resources, such as the wind and solar, is more than ever, demanding the utilization of LM as a top-tier dispatchable resource [14]. However, integrating multiple LM resources into the system involves handling the various parallel and multichannel communications among the LM components, which make the control system of the current LM program even more complicated. Apparently, with the vertical-based and central-controlled mechanism, the conventional LM solution is unlikely to solve such onerous tasks properly.

There exist intolerable weaknesses in the centralized solutions – for instance, the single-point failure and the inapplicability under particular conditions [15]. For the centralized LM solutions, the control center requires to collect all information from energy users and to process huge amount of data [16]. Such an operation mechanism could not generate effective commands in a very short time, particularly during the peak demand period. Because of the lack of necessary knowledge about the controllability of loads, the exploitation of LM in both residential and industrial sectors [17] has been limited. For the current LM or DR programs that apply the direct or interruptible load control, the equipment of participants is required to be shut down or started by utilities within a relatively short interval of time. For the energy users without remote control functionalities, they are not entitled to the LM program even if they are willing to do so. Thus, these above mentioned limitations of the centralized LM approach cannot fully exploit the potential of LM programs.

Various distributed solutions have been proposed to address the issues confronted by current LM programs. A distributed LM algorithm is developed for solving the PEV charging problem by utilizing a congestion price mechanism [18]. However, such an algorithm requires to acquire the essential unified price information through a centralized manner. A distributed LM strategy utilizing an alternating direction method of multipliers is developed in [19], and this method not only requires all energy users to report their information regarding the load utilization to the system operator but also requires the system operator to send the control signals back to all participants. This mechanism of two-way communication requires a communication system of high bandwidth because the number of energy users that participate in the LM can be quite large. Although the adoption of the so-called aggregators may alleviate the heavy burden of communication, the operation of LM programs also requires to avoid the heavy communication tasks between system operator/aggregator and energy users. To this end, we need to develop a flexible and active LM solution with a distributed framework and communication-efficient mechanism as well.

In this chapter, a distributed LM solution to reduce the peak load in smart grids has been proposed. Here, the LM problem is formulated as an optimization problem that aims to maximize the total utility of all energy users. A DDP method is applied to solve this problem to yield a distributed solution. In the proposed solution, each energy end user is denoted by a load management agent (LMA). An LMA could exchange information with its neighboring LMAs. During the LM period, an LMA first receives information regarding the load settings and incentive of this LM event that is broadcasted by the system operator. Then, the LMAs cooperate with each other in a distributed manner, aiming at maximizing their total utilities while meeting the criteria of load reduction.

The following sections of this chapter will be organized as follows: Section 5.1.1 and Section 5.1.2 will illustrate how the LM problems formulate under the proposed LM system. Section 5.1.3 will propose a DDP algorithm to solve the LM problem. Section 5.1.4 will present case studies based on simulation, while Section 5.1.5 will present the conclusion.

5.1.1 System Description and Problem Formulation

Figure 5.1 presents the proposed LM system. An incentive-based mechanism is adopted here because it can be used in both regulated and deregulated power markets. Each user is assigned an intelligent agent, i.e. LMA, to manage its load. The load of a user can be the load of a single physical device or the virtual aggregated "load" of several physical devices, via the so-called gateway introduced in [20].

When an LM event is triggered during the peak load period, the system operator (utility) will first calculate the anticipated electricity demand for all participants

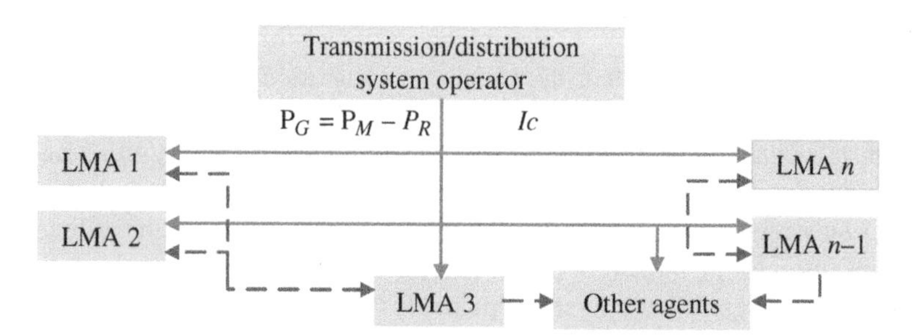

Figure 5.1 Design of the proposed LM system. Source: Based on Zhang et al. [20].

after the LM process is completed. We denote this demand with P_G, and it is calculated according to the requirement for demand reduction, P_R. P_G is calculated as $P_G = P_M - P_R$, with P_M being the current running load. Then, the system operator broadcasts the information of P_G and LM incentive, I_c, to all LMAs. Once this information is received by the LMAs, they will cooperate with others autonomously to realize the common LM objective, without a coordinator or a centralized control center. Each LMA is designed such that it can exchange information with its neighboring agents and update its setting of load usage according to the rules given by the DDP algorithm. The topology of the communication network that supports the information exchange among these distributed LMAs can be designed to be the same as the topology of an electric network because the proposed LM solution requires communication links among neighbors only. Such a communication design leads to a relatively small financial cost by utilizing some particular communication technology, such as power line communication [21]. Nevertheless, other forms of topologies could also be used [16].

By adopting such a structure and mechanism for LM, these LMAs act as a coalition, which aims at maximizing the overall utility while meeting the requirement of load reduction. Consequently, this process can be transferred to solving an optimization problem, which will be introduced in the next section.

5.1.2 Problem Formulation

Assume that there are n participants available for an LM program. Generally, the system operator does not have access to the users' devices because of the lack of right of control. It is only in charge of broadcasting information about P_G and I_C to all participants. As discussed previously, the LMAs cooperate with others autonomously to maximize the utility of their coalition society with the demand reduction required by system operator being also achieved.

Consider that an LM participant can shed a portion of its load. We denote the status of a load with state variable x_i^k as:

$$x_i^k = \begin{cases} 1 \text{ if the } k\text{th load sector is on} \\ 0 \text{ if the } k\text{th load sector is off} \end{cases}$$

where $i = 1, 2, \dots, n, k = 1, 2, \dots, n_i$. Here, n_i refers to the number of load sectors of participant i. Given that the LM could control all the n_i load units of a participant and there will be 2^{n_i} load reduction settings. Hence, a participant could adjust the value of x_i^k to signify the quantity of load shedding.

Let P_{Li}^k denote the kth load sector of participant i before an LM event. Therefore, the utility of participant i during the LM event is then expressed as:

$$U_i = \sum_{k=1}^{n_i} x_i^k W_i^k P_{Li}^k - I_c \sum_{k=1}^{n_i} x_i^k P_{Li}^k \tag{5.1}$$

Here, W_i^k is the weight factor for k^{th} load sector of user i, which is predefined. Usually, such a factor is defined in terms of either the level of priority of the load or the production of unit electricity consumption [22]. Note that the first term of the right hand side of the above equation refers to the benefits of participant i by consuming energy, while the second term refers to the loss of incentives if the corresponding loads are still running during LM.

For the purpose of maximizing the entire utility of all the participants, i.e. the utility of the LM coalition society, we have the following objective function:

$$\max \sum_{i=1}^{n} U_i = \sum_{i=1}^{n} \left(\sum_{k=1}^{n_i} x_i^k W_i^k P_{Li}^k - I_c \sum_{k=1}^{n_i} x_i^k P_{Li}^k \right) \tag{5.2}$$

The constraints to signify the requirement of load reduction given by the system operator is:

$$\sum_{i=1}^{n} \sum_{k=1}^{n_i} x_i^k P_{Li}^k = P_G \tag{5.3}$$

Because $\sum_{i=1}^{n} I_c \sum_{k=1}^{n_i} x_i^k P_{Li}^k = I_c * P_G$, we formulate the LM problem as a constrained optimization problem as:

$$\begin{cases} \max \sum_{i=1}^{n} \sum_{k=1}^{n_i} x_i^k W_i^k P_{Li}^k \\ \text{subject to} \sum_{i=1}^{n} \sum_{k=1}^{n_i} x_i^k P_{Li}^k = P_G \end{cases} \tag{5.4}$$

The constrained optimization problem formulated in (5.2)–(5.4) can be used to model the practical LM problems [23]. Note that such an optimization problem can be considered as a 0-1 knapsack or bin-packing problem, which can be effectively solved by utilizing dynamic programming (DP).

Traditional methods are not capable of yielding an autonomous solution for LM because most of them are deployed in a centralized way. When the number of users participating in the LM program is small, the communication traffic and low latency may be tolerable. Yet, when more and more users with multiple electric devices (load sectors) are enrolled for LM, the communication traffic or low-latency issue becomes a serious issue because the centralized controller needs to collect all the data of the users (devices). Another method regarding the centralized implementation is the control right of the users' devices. Generally, the energy users are not willing to grant the system operator the right of control to access their electric devices because allowing this can lead to unbearable interruptions of their energy usage. Accordingly, an autonomous solution with situation awareness is urged for both the energy users and energy providers.

To date, distributed intelligence is making headway in applications of smart grids. By (i) building a sensor network covers our transmission, distribution, and local premise of energy consumers and (ii) by integrating with communication networks, intelligent electric devices, etc., the distributed control and optimization solution will drive the current power gird to be a more reliable, more efficient, and secure "smart grid" [24]. This motivates us to design a distributed algorithm that can solve the LM problem in a distributed manner, thus rendering an autonomous LM solution.

5.1.3 Distributed Dynamic Programming

5.1.3.1 Abstract Framework of Dynamic Programming (DP)

The framework of DP introduced in [25] will be utilized to elaborate the DDP here. We use S to denote the set of feasible states, with its elements being defined as state variable, $\mathbf{x}$. F refers to the set of extended real-valued functions $J : S \rightarrow [-\infty, +\infty]$ on S. $\forall J_1, J_2 \in F$, the following notation is used for convenience:

$$\begin{cases} J_1 \le J_2 \text{ if } J_1(\mathbf{x}) \le J_2(\mathbf{x}) \ \forall \mathbf{x} \in S \\ J_1 = J_2 \text{ if } J_1(\mathbf{x}) = J_2(\mathbf{x}) \ \forall \mathbf{x} \in S \end{cases} \tag{5.5}$$

Let $H : S \times F \rightarrow [-\infty, +\infty]$ be the mapping that is monotone in the sense that for all $\mathbf{x} \in S$

$$H(\mathbf{x}, J_1) \le H(\mathbf{x}, J_2), \ \forall J_1, J_2 \in F \text{ with } J_1 \le J_2 \tag{5.6}$$

The objective of DP is to find a function $J^* \in F$ such that:

$$J^*(\mathbf{x}) = \inf_{x \in S} H(\mathbf{x}, J^*), \ \forall \mathbf{x} \in S \tag{5.7}$$

We define the mapping $T : F \rightarrow F$ as:

$$T(J)(\mathbf{x}) = \inf_{\mathbf{x} \in S} H(\mathbf{x}, J) \tag{5.8}$$

$T(\ )$ refers to a series of operation or calculation procedures that are collectively defined as the operator to map the objective function to its optimum. This process can be stated as equivalent to finding the fixed point of T within F such that

$$J^* = T(J^*) \tag{5.9}$$

The LM problem in (5.4) aims to find the optimal solution of maximizing the entire participants' utility provided that P_G is greater than the entire generation. Then, the DP process can accordingly be described as follows:

$$\begin{cases} f_k^* = \min \{f_{k-1}^* - x_k W_k P_{Lk}\}, \ f_0^* = 0 \\ \text{subject to } \sum_{i=1} x_i W_i P_{Li} \le P_G \end{cases} \tag{5.10}$$

where x_k is the kth element of $\mathbf{x}$, and $k = 1, 2, \ldots, n$.

We defined H and J as follows:

$$\begin{cases} H(x_k, J^*) = J^*(x_1, ...x_{k-1}) - x_k W_k P_{Lk} & (5.11a) \\ J^*(x_1, x_2, ...x_{k-1}) = f^*_{k-1} & (5.11b) \end{cases}$$

Thus, the mapping T is then defined as:

$$T(J)(x_1, \dots, x_k) = \inf_{x \in S} H(x_k, J^*) \quad (5.12)$$

Based on the definition stated above, the LM problem can obviously be generalized as a DP problem. It should be pointed out that the original utility maximization problem given in (5.4) is transformed to a minimization problem given in (5.11a). Because of the nondecreasing feature of J and the non-negative characteristics of x_k, W_k, and P_{Lk}, H defined above would be monotone.

5.1.3.2 Distributed Solution for Dynamic Programming Problem

For our proposed LM solution, each LM participant (load/user) is designated by an agent for distributed computation. For a system with n agents, there are total n state variables in the state space S. Each agent is in charge of calculating the values of the solution function J^* at x_i. Agent j is defined as the neighboring agent of i ($j \neq i$) if there is the communication link between agent i and agent j.

We use $N(i)$ to represent the set of agent i's neighbors. Note that j is not considered as the neighbor of i if the value of J on x_j does not influence the values if $T(J)$ on x_i. Thus, to calculate $T(J)$ on x_i, agent i only needs to acquire information regarding the values of J on $x_j, j \in N(i)$, and possibly on x_i.

We propose a two-stage process for agents' cooperation to obtain the optimal solution of the LM problem, and the first stage is defined as the information discovery stage, whereas the second stage is defined as the state update stage. Each agent is designed to have two buffers per neighbor J_{ij}, and these two buffers are denoted with J_{ij} and $\mathbf{x}_{ij}$. J_{ij} is used to store the latest estimates of the solution function J^* from agent j, while $\mathbf{x}_{ij}$ is used to store the states corresponding to J_{ij}. Additionally, an agent, say agent i, is designed to have another two buffers to store its own estimates of the solution function J^* and corresponding states, which are denoted as J_{ii} and $\mathbf{x}_{ii}$, respectively. For each iteration stage, agent i will first communicate with its neighboring agents to acquire information of the latest estimates of the solution function J^* and the corresponding state variables during the information discovery stage. Then, this agent calculates its new estimate of the optimal solution (J^*) and states ($\mathbf{x}$) in the state update stage.

We summarize the update regulation for the DDP algorithm as follows:

Stage 1: Information discovery (ID)

$$\begin{cases} J_{ij}[t+1] = J_{jj}[t] \\ \mathbf{x}_{ij}[t+1] = \mathbf{x}_{jj}[t] \end{cases} \tag{5.13}$$

Stage 2: State update (SU)

$$\begin{cases} J_{ii}[t+1] = \inf_{x_i \in S} H(J_{ii}[t], J_{ij}[t+1], x_i) \\ \mathbf{x}_{ii}[t+1] = \arg\{\inf_{\mathbf{x} \in S} H(J_{ii}[t], J_{ij}[t+1], x_i)\} \end{cases} \tag{5.14}$$

According to [26], the converged values of J^* and $\mathbf{x}^*$ can be written as:

$$\begin{cases} \lim_{t\to\infty} J_{ij}[t] = J_{ii}[t] = J^* \\ \lim_{t\to\infty} \mathbf{x}_{ij}[t] = \mathbf{x}_{ii}[t] = \mathbf{x}^* \end{cases} \tag{5.15}$$

The conditions for convergence are as follows:

1. There exists a positive scalar P such that, for every agent, there is at least one information stage for this agent to communicate with its neighboring agents and also at least one SU stage of this agent, for every P steps of iterations;
2. There exist two functions $\underline{J}$ and $\overline{J}$ such that the set of all functions $J \in F$ with $\underline{J} \leq J \leq \overline{J}$ belong to F, and

$$\overline{J} \geq T(\overline{J}), T(\underline{J}) \geq \underline{J} \tag{5.16}$$

and

$$\lim_{t\to\infty} T^t(\overline{J})(\mathbf{x}) = J^*(\mathbf{x}) \tag{5.17a}$$

$$\lim_{t\to\infty} T^t(\underline{J})(\mathbf{x}) = J^*(\mathbf{x}) \tag{5.17b}$$

The first condition here implies that both the stages of ID and SU are mandatory for the convergence. However, there is no other requirement for the timing and sequence of these two stages during the iteration. Note that it is not necessary to carry out SU stage after each ID stage. In other words, the SU stage can be carried out after the execution of several ID stages. Hence, the algorithm can be implemented by using asynchronous communication protocols. The second condition ensures the existence of a fixed point for the LP problem [27]. It can be observed that during the optimization, an agent exchanges data with its neighboring agents at stage 1 only. The shared data includes two parts: (i) header information and (ii) intermediary optimization data. The header information includes information of agent ID and iteration number, which is a 32-bit data. The dimension of the optimization data is determined by the dimension of state variables.

J_{ij} is the scalar number and $\mathbf{x}_{ij}$ is an n-dimensional vector with n representing the number of load sectors. We use double-type data to store the optimization data. Consequently, the size of the data for information exchange is $32 + (n + 1) \times 2$. Note that the volume of data is linearly proportional to the size of the system here.

The number of iterations needed for the agents to reach their optimum can be used to signify the complexity of the proposed DDP algorithm. As mentioned above, the formulated LM problem in (5.4) is actually a 0-1 knapsack problem that can also be translated into the shortest path problem. According to Dijkstra, for centralized implementation of the DP algorithm, the computation complexity is $O(n)$. Here, n denotes the number of nodes (load sectors). For distributed implementation, the complexity is scaled down by a factor of n because the computation effort is shared among agents, as shown in (5.13) and (5.14). Each agent will exchange information with its neighboring agents in stage 1 and then update its state in stage 2. If we use $n^e_{\max}$ to denote an agent's maximum number of neighboring agents, the maximum computation needed for these two stages are bounded by $n^e_{\max}$. Consequently, the computation complexity of the proposed DDP algorithm is $O(n)$ rather than $O(n^2)$.

Theoretically, the solution function satisfying (5.16) exists and can be found. Yet, it is not easy to provide the off-the-shell formula in practice. Nevertheless, the DDP problem could still converge to a fixed point, which is at least a local optimal solution because of the non-convexity of the LP problem.

In this chapter, we develop an index to evaluate the performance of the DDP algorithm:

$$I_p = \frac{f_d^*}{f_g^*} \times \frac{t_g}{t_d} \tag{5.18}$$

f_d^* and f_c^* refer to the values of the objective function obtained by the DDP and global centralized algorithms, respectively. t_d and t_c refer to the corresponding time consumption of these two algorithms. A large value of I_p indicates the high performance of the algorithm. In the simulation part, we will use the index defined here to evaluate the proposed DDP-based solution.

During the period of an LM event, once the agent of a participant acquires the information about total demand and incentive (I_c and P_G), it will first initialize its state variables with feasible load settings. Following that, the agent will communicate with its neighboring agents for exchanging the information regarding the latest states and solution functions. This process follows the procedures of stage 1 given in (5.13). In stage 2, the agent determines its states by utilizing the knowledge of up-to-date information obtained from stage 1, as shown in (5.14). These two stages repeat until the convergence of the algorithm is achieved.

In the process of each iteration, an agent needs to only communicate with its neighboring agents for information exchange and it updates its states locally.

Through the proposed LM solution, the computation efforts are distributed across multiple agents. Practically, a centralized controller and a complicated communication structure will not be required for this implementation. The following section will also demonstrate that our proposed distributed solution is flexible and is adaptive to the operating condition changes.

5.1.4 Numerical Example

Here, we first provide a simple example of three energy users (agents) to demonstrate the proposed distributed solution. In this example, the system operator requires the energy users to reduce the demand by 30 MW for one hour with the incentive being given as \$0.5/kWh. The baseline of the load, in this case, is 90 MW, resulting in a total target load setting of 60 MW for the LM participants aggregated by the aggregator. The load baselines for users No. 1, No. 2, and No. 3 are 20, 30 (10, 20), and 40 MW, and the weights of their load are set to 2, 3, and 4, respectively. User No. 2 has two load levels, i.e. 10 and 20 MW. Figure 5.2 shows the graph of the communication network for LMAs.

As for the LM process, each agent first initializes its load settings for optimization with a feasible value (generally its load baseline will be proper).

We set the maximum number of iterations to 10 for this example. Notice that the buffers $\mathbf{x}_{ii}$ or $\mathbf{x}_{ij}$ used to store the estimated states of agent i are vectors. Table 5.1 provides one of feasible options for agent initialization. Buffer $\mathbf{x}_{11}$ used to store the states agent No. 1 is initialized with a vector [1 (0 0) 0], wherein agent No. 2 and

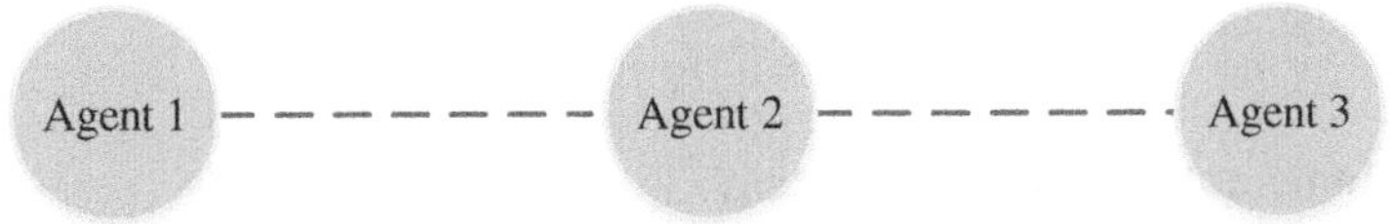

Agent 1 - - - - - - - Agent 2 - - - - - - - Agent 3

Figure 5.2 Topology of the communication network for agents.

Table 5.1 Initialization of agents.

Agent	States ($\mathbf{x}_{ii}\backslash\mathbf{x}_{ij}$)			Utility ($J_{ii}\backslash J_{ij}$)		
1	$\mathbf{x}_{11}$	$\mathbf{x}_{12}$	—	J_{11}	J_{12}	—
	[1, (0 0), 0]	[1, (0 0), 0]	—	40	40	—
2	$\mathbf{x}_{21}$	$\mathbf{x}_{22}$	$\mathbf{x}_{23}$	J_{21}	J_{22}	J_{23}
	[0, (1 1), 0]	[0, (1 1), 0]	[0, (1 1), 0]	90	90	90
3	—	$\mathbf{x}_{32}$	$\mathbf{x}_{33}$	—	J_{32}	J_{33}
	—	[0, (0 0), 1]	[0, (0 0), 1]	—	160	160

No. 3's initial states are set "OFF" because their states remain unknown to agent No. 1 before the start of the optimization process. Here, the state of agent No. 2 is initialized as (0 0) because its corresponding user has two load sectors. The buffer J_{11}, which is used to store the estimated optimal utility of agent No. 1, is initialized with 40, and this value is calculated according to the initial state, $\mathbf{x}_{11}$.

As shown in Figure 5.2, agent No. 1 has only one neighbor, i.e. agent No. 2. The buffers $\mathbf{x}_{12}$ and J_{12} of agent No. 1 are then used to store the latest states and corresponding estimated solution function of agent No. 2. In addition, these two buffers are initialized as $\mathbf{x}_{12} = \mathbf{x}_{11}$ and $J_{12} = J_{11}$. Buffers for agents No. 2 and No. 3 are initialized in the same way.

Figure 5.3a presents the update of agents' utilities in the process of optimization. As can be observed, the values of utility functions for all the agents, J^*, are monotonically increasing because of the feature of the DP algorithm. The converged utility is 220, which is the maximum utility these agents can obtain considering the constraint of the load reduction requirement of 30 MW.

Figure 5.3b,c presents the update of load settings in the process of optimization for agent No. 1 and No. 2. To meet the requirement of minimum load reduction, agent No. 2 had switched OFF its first load sector, with only the second one being switched ON. Note that the converged solutions of all the agents are exactly the same. We can easily verify that the second load sector of user No. 2 and load of user No. 3 should be set ON to achieve the maximum utility of this LM coalition. In this case, the converged solution is also the globally optimal LM solution, and corresponding payment for contributing the load reduction is increased $\$0.5 * 30 * 10^3 = \$15\,000$. Note that the algorithm takes only three iterations to converge in this case.

5.1.5 Implementation of the LM System

Figure 5.4 provides the implementation of a typical LM system with a total of 14 agents. The information, (P_G and I_c), will be broadcasted by the system operator to all LMAs in the system via the communication network. Users can choose to participate in the LM program, or not. If a user chooses not to participate, the corresponding LMA of this user is set to the deactivated mode.

The communication network between the system operator and LMAs is built by utilizing the technique called general packet radio services (GPRS), which is a technique widely used for the data transmission service of mobile phones and remote meter reading. Upon receiving the signal from the system operator, the active LMAs start to search their neighboring agents to comprise the coalition for the LM program.

The communications of neighboring agents for information exchange can be realized by using the off-the-shell wireless communication protocol such as WiFi

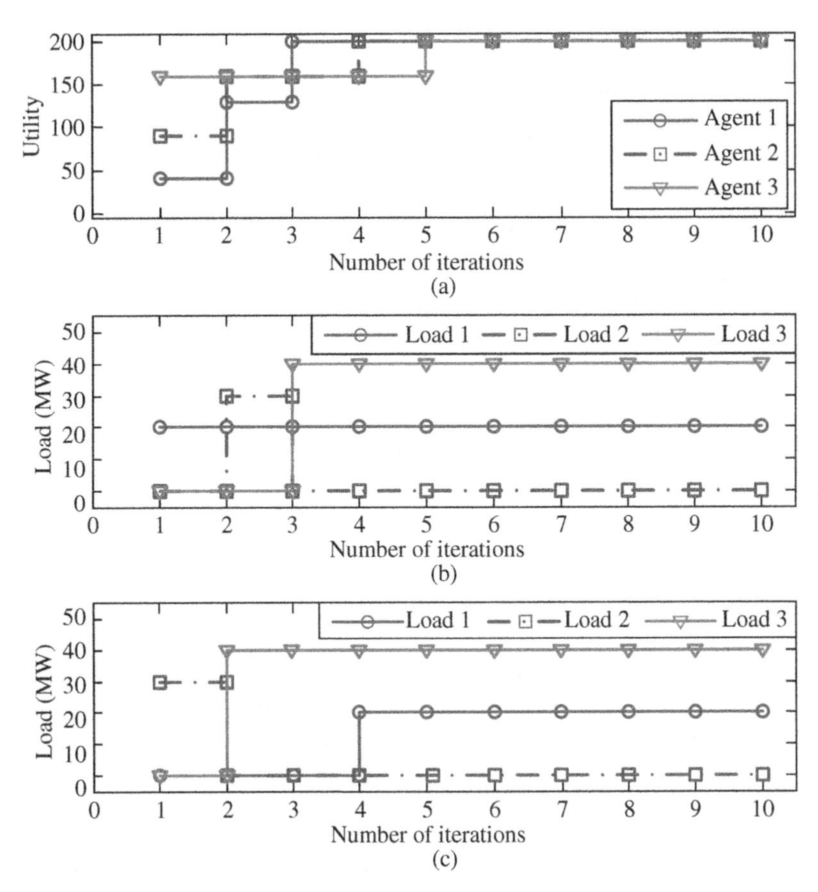

Figure 5.3 LM optimization process with a three-agent system. (a) Profiles of utility for agents during optimization. (b) Update of load settings during optimization for agent No. 1. (c) Update of load settings during optimization for agent No. 2.

and Zigbee or wired communication such as fiber optic or power line communication techniques. The wireless communication techniques generally have relatively short transmission ranges and are fitted for the household or community-level LM programs. The wired communication techniques, on the other hand, have longer ranges of transmission and can be used for industry-level programs. As for the software-level implementation, the JADE (Java Agent Development Framework), which is a software framework developed for MAS applications based on Java language, can be utilized for customized design. This kind of JADE-based system can be distributively implemented and the configuration can be easily controlled via a remote GUI [28, 29].

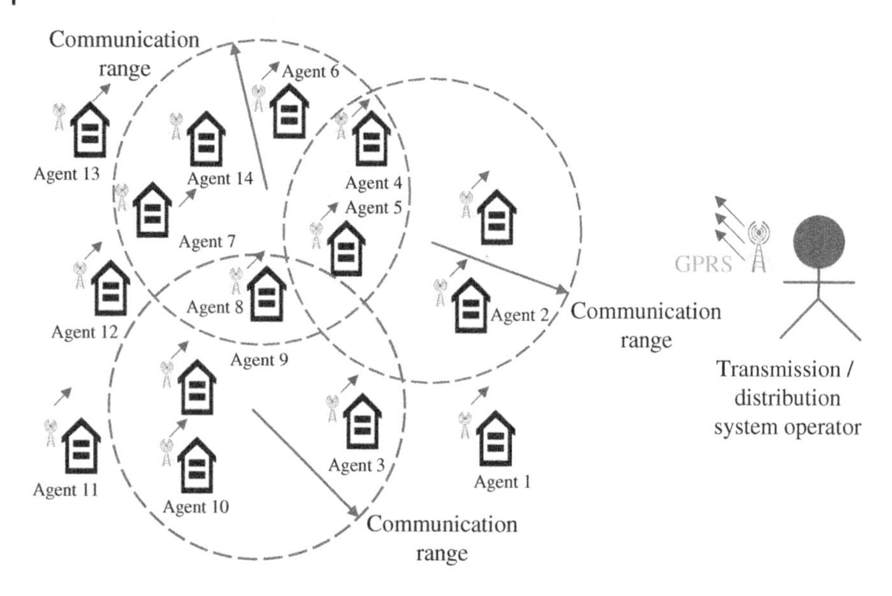

Figure 5.4 Implementation of the proposed LM system.

It should be pointed out that the system operator only broadcasts the signal of LM demand and incentive to energy users. It does not require to collect the information of the current usage of the load sectors or to access the control right of loads. The agent of a specific energy user makes its decisions locally while cooperating with other agents to fulfill the LM objective. In this process, the agent does not need to send information to the system operator or to receive a control signal from the system operator. Consequently, the problem of lack of control right of load for the system operator is avoided.

5.1.6 Simulation Studies

This section will first utilize the IEEE 14-bus system to demonstrate the DDP algorithm. Following this, a larger system with more agents will be used to evaluate the performance of the proposed LM solution.

5.1.6.1 Test with IEEE 14-bus System

The parameters of loads of the IEEE 14-bus system can be found in [30], and we assume that there is a user at each bus. The load reduction requirement given by the system operator is 140 MW, with the incentive being \$0.50/kWh for the qualified users. Table 5.2 provides the load baseline for all users. User No. 4 and user No. 11, respectively, have two and three load sectors, as shown in the table. The total load of users before LM is 760 MW, leading to the load setting of 620 MW

Table 5.2 Data of IEEE 14-bus system.

No.	Neighbor	Baseline	Weights	No.	Neighbor	Baseline	Weights
1	2, 5	0	20	5	1, 2, 4, 6	60	10
2	1, 3, 4, 5	0	20	7	4, 8, 9	70	10
3	2, 4	0	20	12	6, 13	80	10
6	5, 11, 12, 13	0	20	13	6, 12, 14	90	10
8	7	0	20	10	9, 11	100	1
4	2, 3, 5, 7, 9	50(10, 15, 25)	20	11	6, 10	120(40, 80)	1
9	4, 7, 10, 14	150	20	14	9, 13	40	1

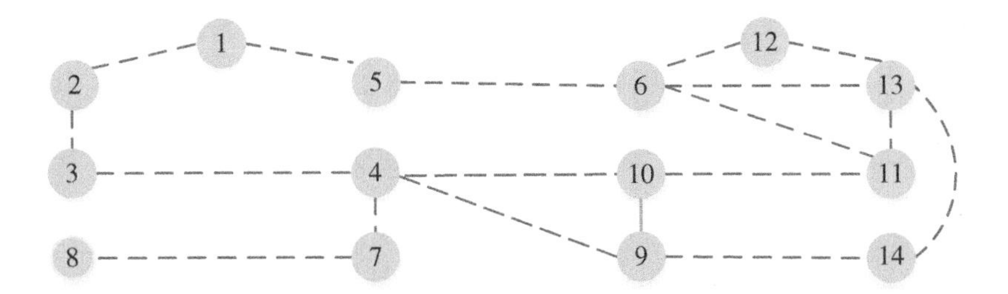

Figure 5.5 Network topology of IEEE 14-bus system.

for LM event. Figure 5.5 shows the graph of the communication network for agent communication, which is the same as that of the topology of connection of the physical power network.

Normal Operating Conditions Figure 5.6 shows the update of utility for agents during optimization. The number of iterations for the algorithm to converge is only 14, with the final obtained utility being 7120.The earned incentive of this coalition for fulfilling the load reduction is $\$0.5 * 140 * 10^3 = \$70\,000$ for an hour. Table 5.3 presents the optimized states of loads (sectors). Figure 5.7a,b presents the updates of load settings of selected loads (loads No. 4, No. 10, No. 11, and No. 14) at two of chosen agents, i.e. agents No. 10 and agent No. 11. The optimized load settings for users No. 4, No. 10, No. 11, and No. 14 are 50 MW, 0 MW, 120 MW, and 0 MW, respectively. Note that the agent of a specific user initializes its own load setting with its load baseline, with other the settings of other agents being zero. Yet, after the algorithm is converged, the optimized states at agents are the same. Hence,

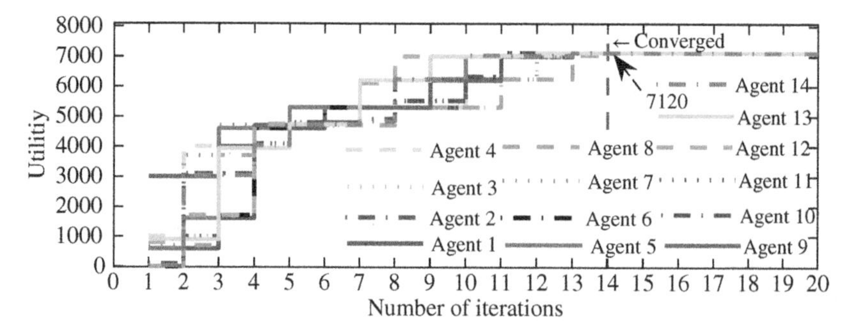

Figure 5.6 Update of utility for agents during optimization.

Table 5.3 Optimized states of loads.

Load	1	2	3	4	5	6	7
Status	ON	ON	ON	(ON,ON,ON)	ON	ON	ON
Load	8	9	10	11	12	13	14
Status	ON	ON	**OFF**	(ON,ON,ON)	ON	ON	**OFF**

the proposed algorithm for DDP guarantees the consistency of optimal solutions obtained by all the distributed agents.

Abnormal Operating Conditions Three abnormal operating conditions during the process of optimization will be tested to evaluate the robustness of the proposed LM solution. These conditions include the loss of communication link, the disconnection of load, and the loss of agent.

Loss of Communication Link For this scenario, we assume that communication links between agents No. 9 and No. 14 and agents No. 12 and 13 stop working after the fifth iteration. However, as shown in Figure 5.9, the graph of a communication network with loss of communication is still a connected graph, and this indicates that condition 1 for convergence introduced previously still holds.

Figure 5.8 shows the update of utility under this scenario. The final converged utility is 7120, which is the same value of that without loss of communication links. For this scenario, the algorithm takes 15 iterations to converge, and the number of iterations only increases by 1. Figure 5.9 shows the update of load settings at agent No. 14. It can be observed that with loss of communication links, the load setting of agent No. 14 changes at the 14th iteration, while this change takes place at 13th

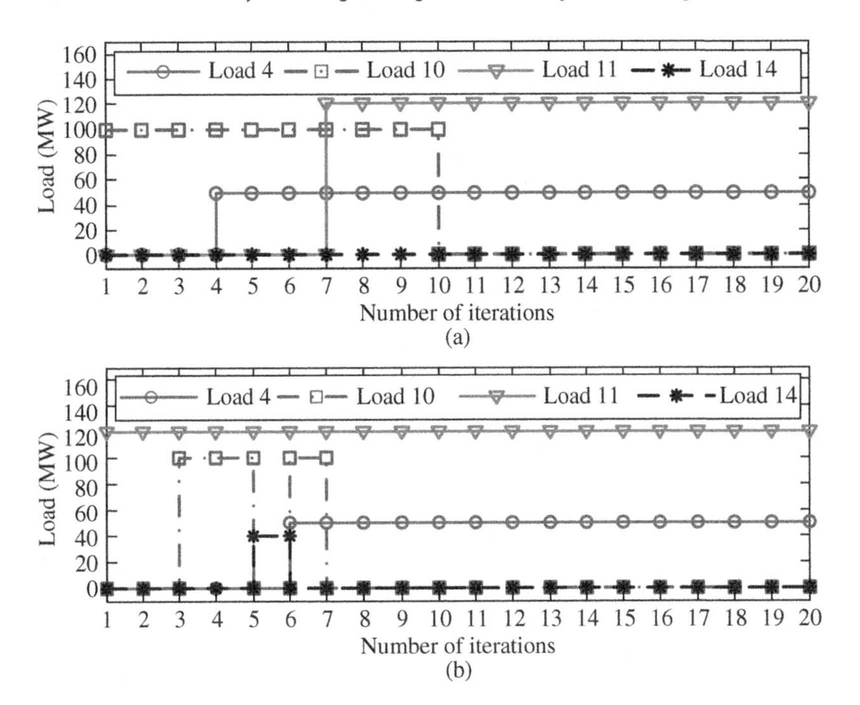

Figure 5.7 Update of load settings during optimization. (a) Update of load settings at agent No. 10. (b) Update of load settings at agent No. 11.

Figure 5.8 Update of utility for agents with loss of communication links.

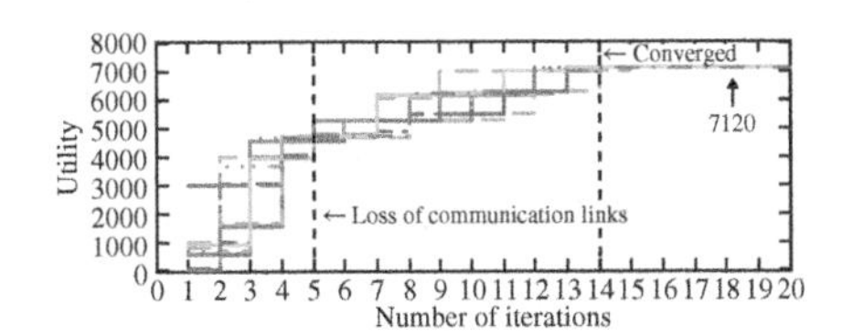

Figure 5.9 Comparison of load settings at bus No. 14.

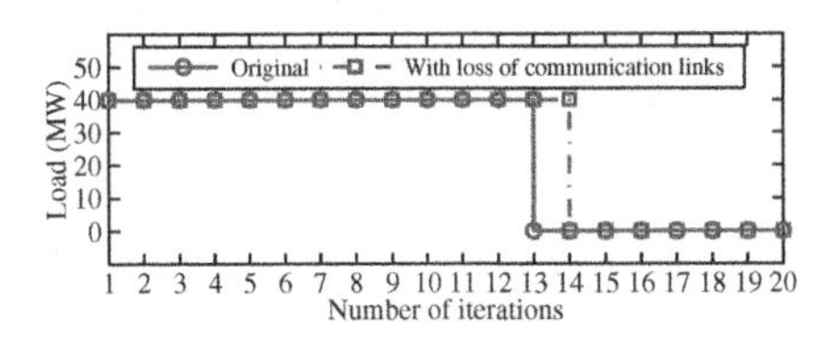

iteration with the original communication settings. For this case, we can see that the loss of the communication links only slows down the overall converging speed slightly. In addition, we are still able to find the feasible and optimal solution as long as the graph corresponding to the communication network is a connected graph.

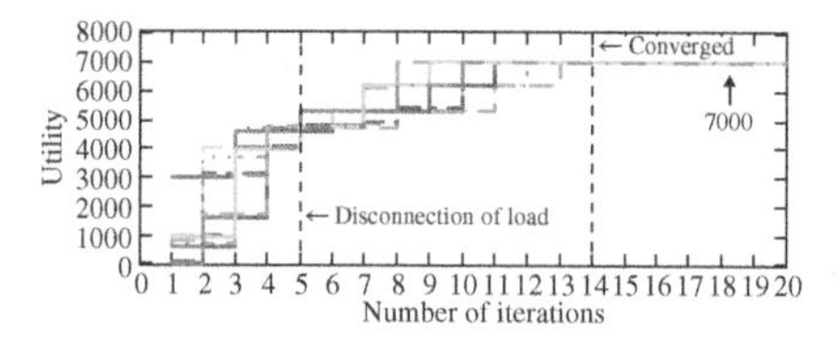

Figure 5.10 Update of utility for agents with disconnection of load.

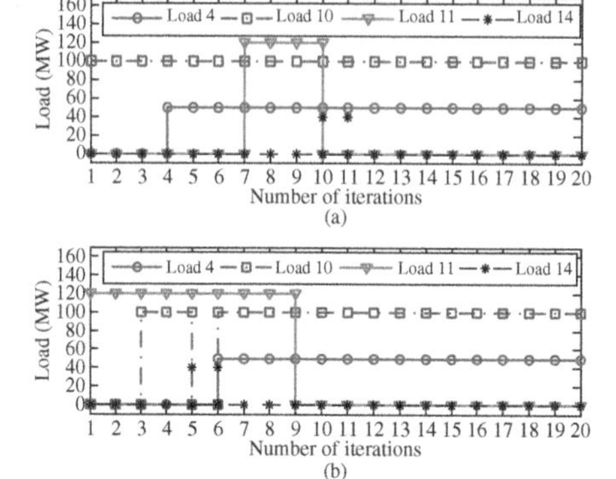

Figure 5.11 Update of load settings for agents with disconnection of load. (a) Load settings at agent No. 10. (b) Load settings at agent No. 11.

Disconnection of Load It is assumed that the disconnection of load occurs at the fifth iteration, with the load at bus No. 10 being disconnected.

Figure 5.10 presents the update of utility during this process. For this scenario, the final converged utility reduces to 7000, compared to the value of 7120 for the case without load disconnection. This is due to two causes. On the one hand, the utility of the load at bus No. 10 is not applicable because it is disconnected. On the other hand, the total load reduction increases to 160 MW (20 MW more than the required), which also results in a decrease in the overall utility.

Figure 5.11a,b presents profiles of load setting of agents No. 10 and No. 11 during optimization. As can be seen in the figure, the load setting for load No. 10 is fixed at a virtual value of 100 MW after its disconnection, which indicates that load No. 10 is excluded from participating in LM response. After the algorithm is converged, the load at No. 11 is optimized to shed 120 MW load to meet the requirement of LM. It is worthy pointing out that the designed LM system can still work properly with the occurrence of the load disconnection.

Loss of Agent Another abnormal scenario is the loss of the agent in the LM system. Here, we assume that the agent No. 10 malfunctions after five times of iteration.

Because of the loss of agent No. 10, the communications between agent No. 10 and its neighboring agents are no longer available. As a consequence, agent No. 10 does not participate in the remaining optimization process with its load setting set to be fixed at 100 MW after fifth iteration. The optimization process carries

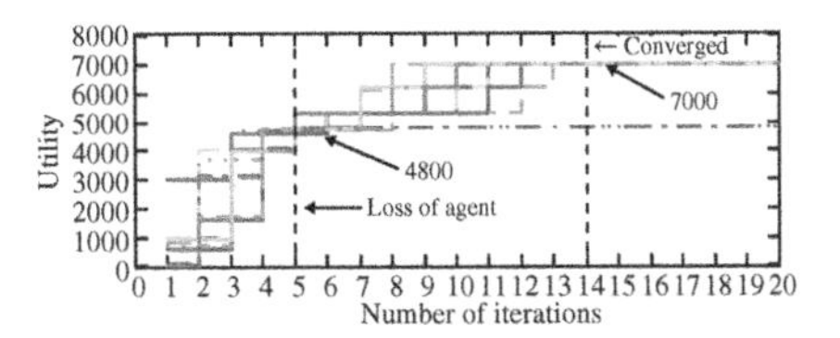

Figure 5.12 Update of utility for agents with loss of agent.

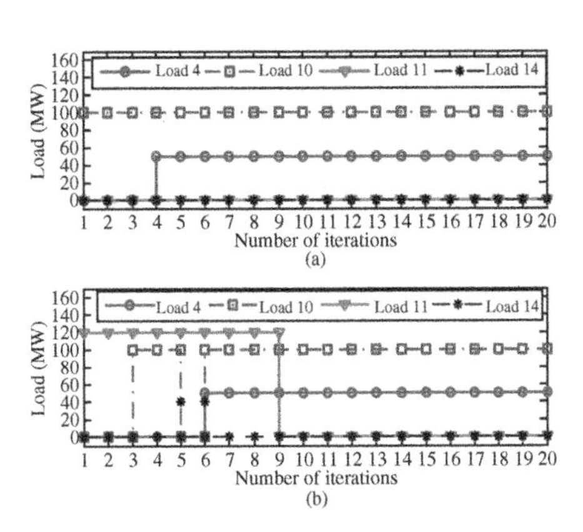

Figure 5.13 Update of load settings for agents with loss of agent. (a) Load settings at agent No. 10. (b) Load settings at agent No. 11.

on because all of the remaining agents still work properly. Note that the obtained optimal solution for this scenario is similar to that with the disconnection of load, as can be observed in Figures 11.12 and 11.13.

As can be observed from results with abnormal operating conditions, the proposed LM system still yields the optimal solution with the loss of the communication network if the graph of the communication network is still a connected graph. With load disconnection or loss of agent, the designed LM system can still obtain the comprised optimal solution because the disconnected load or agent is excluded for further optimization.

With Dynamic Incentives To evaluate the performance of the proposed LM solution under consecutive LM events, a case with a dynamic incentive mechanism is tested here. We assume that the incentive mechanism is given by the system operator as $I_c = I_c^* + 0.15 * \Delta P$, and this setting is derived based on an industrial DR program. Here, I_c^* refers to the incentive trigger point, which is set to \$75/MWh and ΔP denotes the part of shed load that is larger than 75 MW. Figure 5.14a presents the load reduction command and incentives in a similar day within five consecutive hours (10:00 a.m.–3:00 p.m.). The LM event will be broadcasted by the operator to energy user every other hour. Figure 5.14b shows the corresponding utility and earned payment for this LM coalition. As can be seen in the figures, with the increase of required load reduction, the earned payment for this

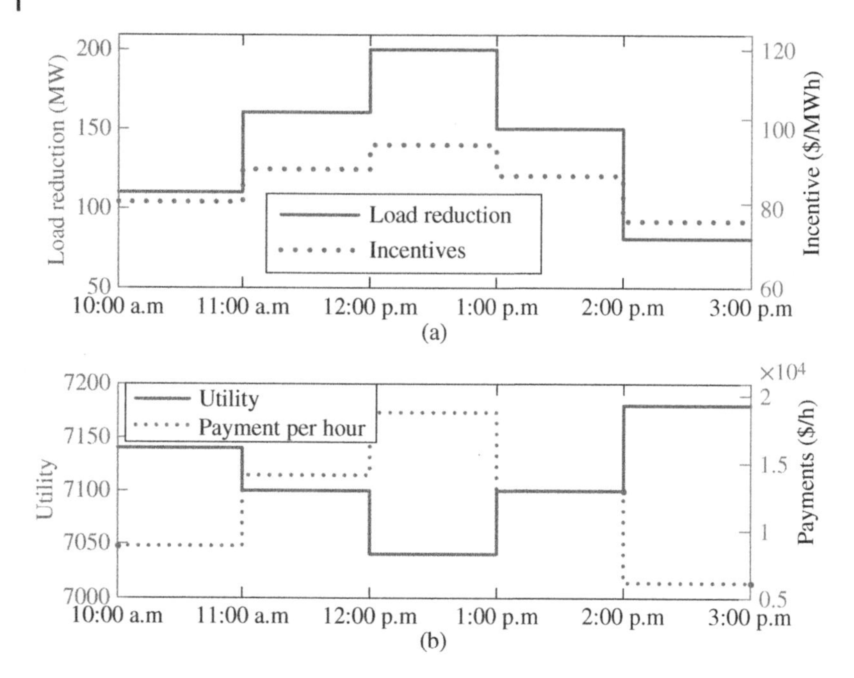

Figure 5.14 LM events in a similar day. (a) Load reduction and incentive settings. (b) Overall utility and earn payment of users.

coalition increases along with the decrease of the total utility. When participating in the LM program, the energy users can always allocate low-preference load or non-vital load with low weights to earn payment while maximizing their total utility. Figure 5.14b shows the earned payment of this coalition during the whole period of LM response (12:00 p.m.–1:00 p.m.), and it can be seen that the payment reaches as high as $18 750 considering that only 200 MW of load is shed for this LM event.

5.1.6.2 Large Test Systems

Here, we test three systems with different sizes to evaluate the performance of the proposed LM solution method. Table 5.4 summarizes the configuration of three tested systems. Here, n_c refers to the total number of communication links and n_{cp} denotes the number of average communication links per agents.

Figure 5.15 shows the converged utility of the tested system, with the corresponding test results of them being summarized in Table 5.5. According to our implementation, the average time for one round of agent communication based on JADE platform is about 3 ms [31]. As for the centralized solution, there is only one agent (centralized controller), the time for agent communication is not applicable here. As can be seen from the test results, the converged utilities of the proposed

Table 5.4 Configuration of test systems.

Test system	n_c	n_{cp}	P_G (MW)	P_R (MW)	I_c (\$\MWh)
14-Agent	20	1.43	760	140	500
162-Agent	284	1.75	15387	1585	750
590-Agent	908	1.54	18707	1169	750
1062-Agent	1635	1.54	34053	1651	750

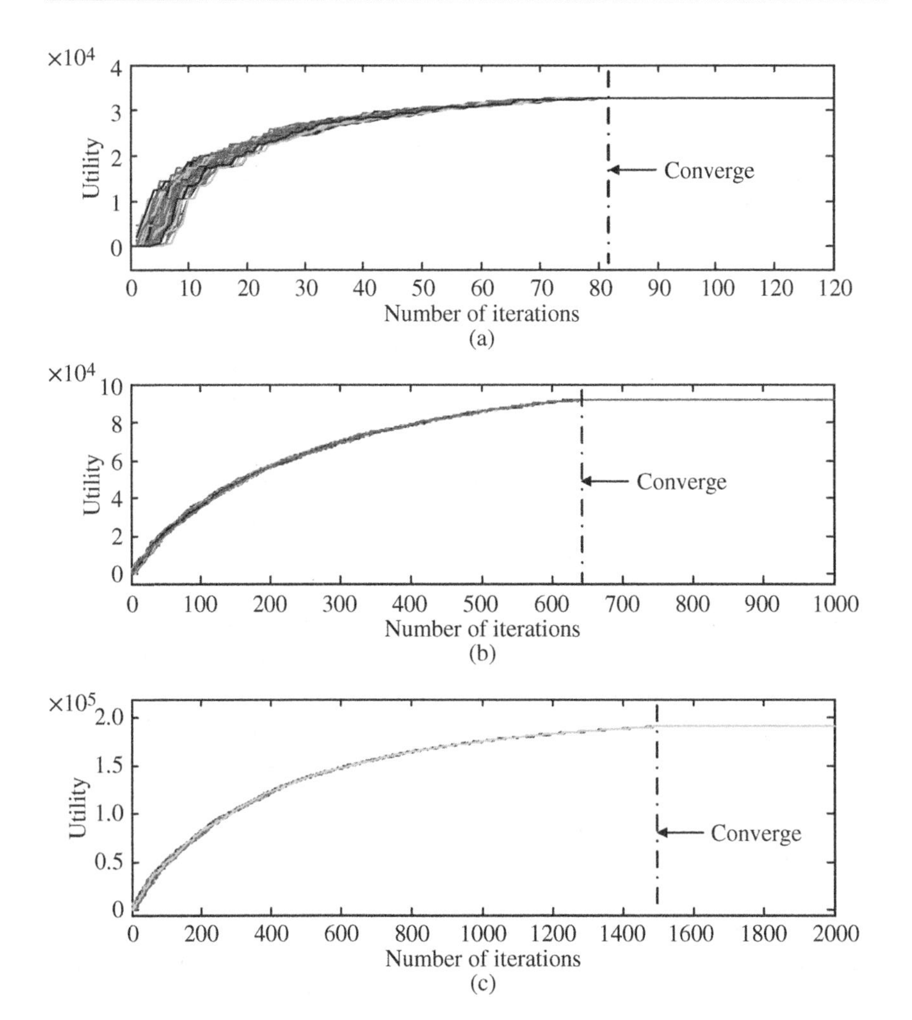

Figure 5.15 Utility update process for agents with large test systems. (a) 162-Agent system. (b) 590-Agent system. (c) 1062-Agent system.

Table 5.5 Comparison between centralized and distributed solutions.

System		Utility	Time/Iter. (ms)		Iter.	Total(ms)	I_p
			ID	SU			
14-Agent	Cen.	7120	—	31	1	31	1
	Dis.	7120	3	<1	14	77	0.49
162-Agent	Cen.	33768	—	7920	1	7920	1
	Dis.	32663	3	<1	82	320	23.94
590-Agent	Cen.	101700	—	34897	1	34897	1
	Dis.	91950	3	2	640	3200	9.86
1062-Agent	Cen.	203386	—	107874	1	107874	1
	Dis.	191174	3	3	1 470	9050	11.23

distributed solution may be less than that of the centralized method. Yet, the maximum deviation is only less than 10%, which is acceptable for industrial practice. It can also be observed that the time consumed by centralized solution increases dramatically with the increase of the scale of the system as can be seen in Table 5.5 for a small-scale system with 14 agents, the distributed solution does not perform better than the centralized solution, with the performance index being only 0.49. However, for a large-scale system with 162 agents or more, the distributed solution results in a high value of performance index (9.86 or higher). With the distributed solution, no control center is needed to acquire the data from all distributed energy users; instead, the agents of energy users only communicate with their neighboring agents to exchange necessary information, via an asynchronous communication protocol. Consequently, the time needed for data acquisition is reduced significantly. In addition, the DDP algorithm we used for implementation distributes the computation efforts among all agents, which dramatically reduced the time for computation. As demonstrated in the simulation, for the large system such as the 1062-bus system, the algorithm can converge within 10 seconds, whereas the centralized algorithm takes as long as 100 seconds even without taking the time used for data acquisition into consideration. Thus, it is safe to say that our proposed LM system can respond in a timely manner.

5.1.6.3 Variable Renewable Generation

In this test case, we assume that the 1062-bus system is under stress condition wherein the spinning reserve of the conventional generators has been depleted. The system operator has to utilize the LM program to support the safe operation of the system within a dispatch interval of 15 minutes. Before the trigger of the

LM event, the power shortage is 15 707 MW. Wind power can compensate for the part of the power shortage; however, it is not reliable because of its intermittency. Figure 5.16a shows the profiles of power shortage as well as the wind power during this period. Figure 5.16b shows the carried out load reduction profiles of both centralized and distributed solution. It can be observed that the centralized method failed to respond in a timely way because it cannot track the power shortage fast enough. As can be seen in Figure 5.16c, with the centralized scheme, the system frequency nadir reaches to a value of 59.79 Hz, which falls in the under

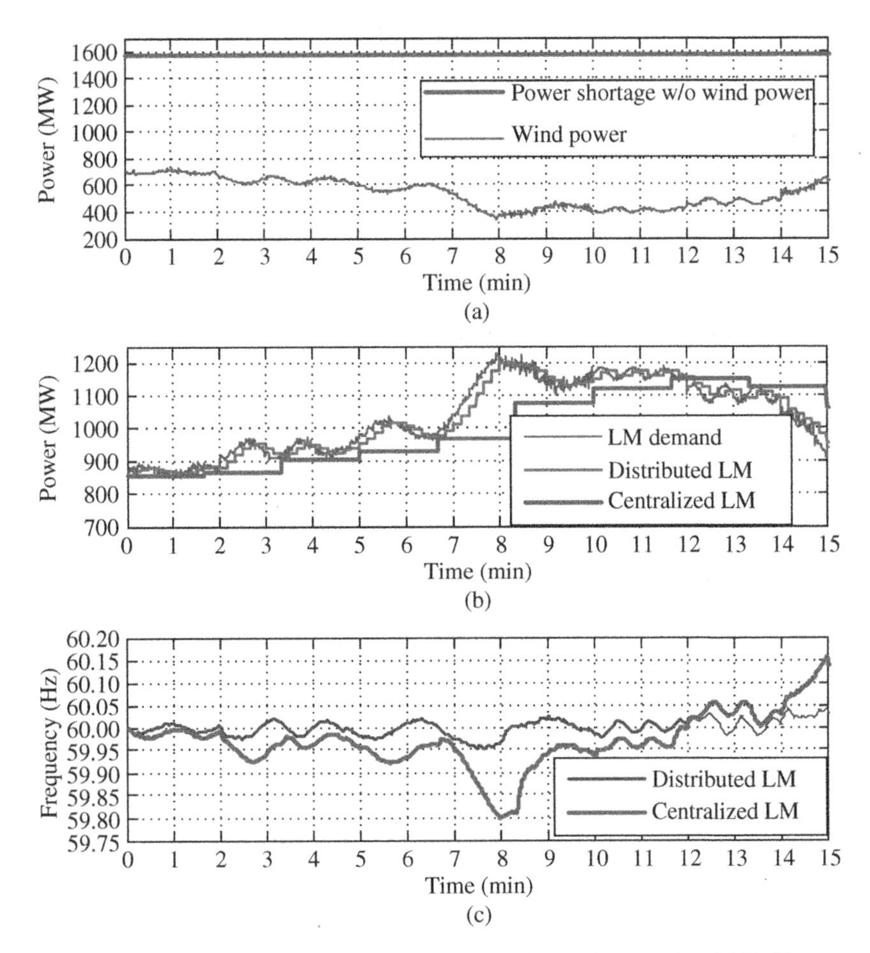

Figure 5.16 Centralized/distributed LM with variable wind generation. (a) Profiles of power shortage and wind power. (b) Load reduction with different LM solutions. (c) Frequency response with different LM solutions.

frequency zone while the frequency peak is reaching 60.16 Hz, being very close to over frequency zone [32].

Compared with the centralized solution, the frequency fluctuation of the distributed solution always falls in the normal operating range (± 0.05 Hz) as the proposed solution can reach the full deployment of the load reduction with 10 seconds. Note that the convergence of the proposed solution will not be influenced by the variation of the renewable generation, such as the wind power in this case. However, faster change of these renewable resources calls for the system operator to trigger the LM event in a more frequent manner. Our proposed LM solution track the LM demand in a very fast way, e.g. less than 10 seconds for the large-scale system with 1062 buses, which can guarantee the decent frequency performance of the overall power grid. Accordingly, the proposed LM solution can be applied to the power grids that are prone to experience fast operating condition changes.

5.1.6.4 With Time Delay/Packet Loss

We continue to investigate the proposed DDP under the circumstance with packet loss. Here, the simulation is done by assuming that the probability of packet loss per iteration for each agent is 0.45. It is operated under the situation with packet loss on all agents in each step of iteration under the probability of 0.45. The simulation results of three test systems introduced previously are shown in Figure 5.17. It can be seen that the algorithm still converges without difficulties. This is because that the condition 1 for the convergence can still hold with the packet loss occurs in a way that follows certain specific probability. However, the occurrence of the packet loss does impact the converging speed.

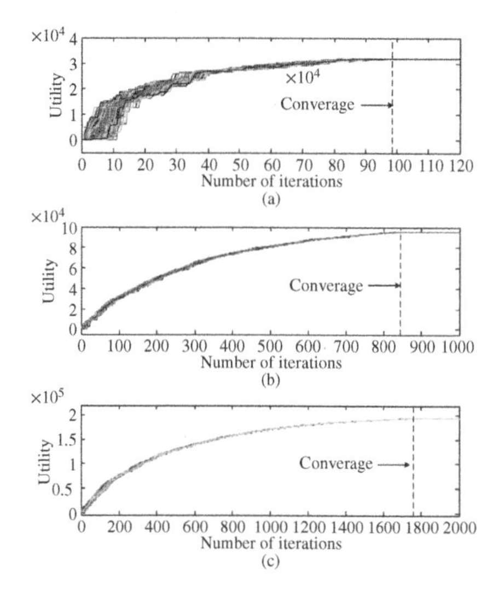

Figure 5.17 The process of utility update for agents with packet losses. (a) 162-Agent system. (b) 590-Agent system. (c) 1062-Agent system.

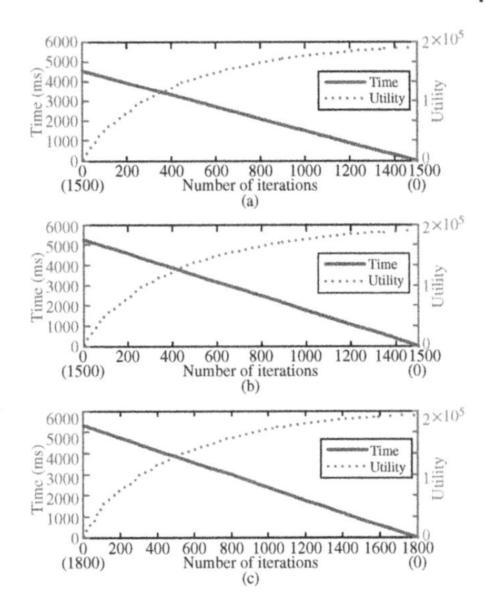

Figure 5.18 Converging time of different scenarios with 1062-bus system. (a) Original. (b) With time-delay. (c) With packet loss.

As the probability of packet loss increases, the number of iterations required for convergence also increases. Another merit of the proposed algorithm is that it does not require the graph corresponding to the communication network to be always connected, which is helpful when the system undergoes abnormalities regarding the communications, thus enables us to adopt the asynchronous communication protocols during implementation.

Time delay of the communication also impacts the converging speed of the algorithm in a similar way. Here, we provide the converging speed of the implemented DDP algorithm with a 1062-bus system under different scenarios. As shown in Figure 5.18, when there is no communication delay or packet loss, the algorithm converges within 4500 ms, as shown in (a). With the average time of communication delay of 0.5 ms, the convergence time increases to 5250 ms. Meanwhile, it takes around 5350 ms for the algorithm to converge for the scenario with the probability of packet loss of the agent per iteration being 0.45. Nevertheless, the proposed DPP algorithm is able to reach convergence without difficulties under these scenarios with communication adversities.

5.1.7 Conclusion and Discussion

In this chapter, we discussed a distributed solution for energy users to participate in the LM program of power grids. To better implement the proposed solution, a DPP algorithm-based framework is designed. To the authors' best knowledge, it is the first time that the DPP algorithm is utilized in the LM problem. In our implementation, the energy user is equipped with an intelligent agent, which is responsible for receiving the load reduction information and the LM incentives. Agents of users only need to collect local information and communicate with neighboring agents. There is no need to send information or control signal back to the system operator. Therefore, the burden of building bidirectional communication networks between the energy users and the system operator for information exchange can be relieved. Additionally, the proposed solution enables the distributed computation among the agents; hence, the central controller is not required. The test results of all these cases demonstrate the excellent robustness and decent performance of the proposed solution.

This work here concentrates on developing an LM solution. We will further evaluate the performance of the LM solution with real-time simulation in our future work. Through incorporating with various renewable energies, for example, PVs and storage, LM can strengthen our energy supply in many ways. As a consequence, to develop a distributed control approach to coordinate these renewable resources is of great importance to ensure our sustainable and secure energy supply.

Plug-in electric vehicles are a promising alternative to conventional fuel-based automobiles. However, a large number of PEVs connected to the grid simultaneously with poor charging coordination may impose severe stress on the power system. To allocate the available charging power, this chapter proposes an optimal charging rate control of PEVs based on consensus algorithm, which aligns each PEV's interest with the system's benefit. The proposed strategy is implemented based on a MAS framework, which only requires information exchanges among neighboring agents. The proposed distributed control solution enables the sharing of computational and communication burden among distributed agents. Thus it is robust, scalable, and convenient for a plug-and-play operation, which allows PEVs to join and leave at arbitrary times. The effectiveness of the proposed algorithm is validated through simulations.

5.2 Optimal Distributed Charging Rate Control of Plug-in Electric Vehicles for Demand Management

The PEV technology has been acknowledged as a promising alternative to tackle the problem of fossil fuel shortage and environmental issue [33]. Despite its popularity, the high penetration of PEVs charging [34, 35] load brings significant challenges to the power system including (i) system efficiency degradation, (ii) worsening the power quality, and even (iii) lack of stability [36].

On the other hand, the flexibility of PEV's charging behavior can be beneficial to the power grid by offering various ancillary services, such as load leveling and frequency regulation [37] if well coordinated. Additionally, because the charging strategies influence the process of charging in terms of the financial cost, the time of charging, and the state of charge (SOC) of the battery of the PEV when the PEV is leaving the charging station, it is important to design proper charging strategies so as to satisfy the consumers. One of the major challenges of PEVs is how to manage the electrical power demand. One effective solution is to properly design an effective coordination control strategy, which could offer the optimal solution with convergence guarantee. The goal is to maximize the benefit of a PEV driver without operational constraints of the power system.

The existing literature introducing the control strategies of coordination for the demand management of PEVs can be mainly divided into three categories, which are control strategies in a centralized manner, fully decentralized manner, and distributed manner. The centralized manner control strategies coordinate all the PEVs by a central controller. Su et al. present a review of the centralized control strategies [38], which include linear programming [39], model predictive control [40], dynamic programming [41, 42], particle swarm optimization [43], etc. The common factor is that the bidirectional communication between every PEV and central controller is essential, which may lead to a heavy burden of a large amount of data collocation and computation. Thus, these methods are considerably costly and susceptible to single-point failures. The second type of control strategy, which is based on the fully decentralized manner, can be achieved by using the local information solely, such as the droop control-based strategy proposed in [44, 45], which is implemented on the vehicle-to-grid (V2G) scheme. This method requires no communication and thus is robust and of low cost. However, because of the broader available information insufficiency [46], the optimal power sharing might be inaccurate because droop control parameters, which are dynamically based on instantaneous operating conditions, need to be adjusted. The third strategy is based on the distributed control strategy, which relies on the local communication network for information exchanging among neighboring units. Because of the diversity of PEV fleet in quantity and features, the corresponding control and management strategies are expected to be convenient, efficient, and cost-effective. A well-designed distributed strategy is a promising solution for the demand management of the PEV fleet because of its flexibility, reliability, scalability, and adaptivity.

The implementation of distributed control has captured wide attention. Recent literature demonstrates that its implementation is potentially enabled by the improvements in the distributed algorithm design, the technology advancement of PEV chargers, and the enhancement in a communication network. A method based on the noncooperative game is proposed to analyze the charging strategies of PEVs. Meanwhile, it is observed that there is a conceptual similarity between the charging games for PEVs and the routing games in networks [47]. In this method, the PEVs are considered as rational agents that tend to minimize the cost concerning the energy price by solving the local optimization problems. The distributed control strategy is achieved by every PEV user receiving data of power consumption from all other users at every iteration, which mitigates the computation burden on the central controller at the cost of a more complex communication network. The framework of the MAS has been widely applied in the decentralized charging control strategies [48, 49, 51], most of which are implemented in hierarchical architecture. The local agent performs charging power regulation according to the external signal, and the computation burden of the central controller is relieved. However, the communication network that builds connection between central and regional is also an indispensable factor for implementation, such as aggregation agents in the divided regions [48], an auctioneer agent serving for sending the price signal [49], and the agent act as a coordinator [50] or aggregator [51] for sending the coordination signal or information to the agents scattered at local areas. To tackle this problem, a peer-to-peer communication network-based fully distributed cooperative algorithm is proposed in [52], which aims to achieve the PEVs charging control and then maximize the SOC level of PEVs at the end of the charging cycle. By approximating the formulated objective functions, the proposed method can achieve a suboptimal solution. It is still a great challenge to design a fully distributed approach that can find the global optimal solution, which is comparable to those of the centralized algorithms.

This section proposes a MAS-based optimal distributed charging rate control strategy of PEV fleet. The proposed approach is aimed at mainly two goals. The first is to meet the dynamic available charging power constraints and the other is to minimize the total charging power loss incurred by the battery internal resistance. It is assumed that all PEVs are at the same charging station, while PEVs in a different charging station can be grouped into another fleet, which can then be regulated independently by using the same control approach. Each recharging socket is assigned with an agent, and each agent makes a prediction about the optimal charging rate based on the SOC, the required time of charging, and other relevant battery parameters. By using a consensus algorithm-based method, which only needs local communication, the optimal charging power can be predicted and then the available charging power can be obtained. Finally, by minimizing the aggregated deviations of available charging power, the global optimal solution is obtained. The proposed solution shows the effectiveness of the proposed method in sharing the communication and computation burden among all the agents. The convergence of the proposed approach is guaranteed by rigorous stability analysis.

5.2.1 Background

Consensus-based algorithms have gained popularity in various applications and fields. A recent published research on this topic can be found in [53]. With only the local information required, the consensus-based algorithm can achieve information sharing in a distributed manner. This can be represented as:

$$x_i[t+1] = \sum_{j=1}^{n} d_{ij} x_j[t] \tag{5.19}$$

where n is the total number of agents, $x_j[t]$ is the state value of the agent i's state variable, which represents the local information exchanged with the agent j at the iteration t, and d_{i_j} is the communication coefficient between agent i and agent j. $x_i[t+1]$ is the update of state variable. The matrix form of Eq. (5.19) can be rewritten as follows:

$$X[t+1] = \mathbf{D}\mathbf{X}[t] \tag{5.20}$$

where $X[t]$ and $X[t+1]$ are the column vectors that represent the discovered information at the iteration t and $t+1$, respectively. D is the Laplacian matrix of the network graph [53].

There are various approaches to determine d_{i_j} with different convergence speeds. Mean metropolis algorithm is distributed, adaptive to communication network topology changes, and able to guarantee the convergence with a nearly optimal convergence speed [15]. This algorithm is adopted in this chapter.

$$d_{i_j} = \begin{cases} 2/(n_i + n_j + 1) & j \in N_i \\ 1 - \sum\limits_{j \in N_i} 2/(n_i + n_j + \Delta) & i = j \\ 0 & \text{otherwise} \end{cases} \tag{5.21}$$

where N_i represent the set of agents communicating with the agent i, Δ is a small number, and n_i and n_j are the number of agents that communicate with the agents i and j, respectively. The designed Laplacian matrix $\mathbf{D}$ can be verified to satisfy the following two conditions:

1. The sum of elements in rows and columns are all equal to one.
2. The eigenvalues λs of $\mathbf{D}$ satisfy $|\lambda_i| \leq 1$, for $i = 1, \ldots, n$. Every x_i will converge to a common value as:

$$[x_i[\infty]] = \frac{1}{n} \sum_{i=1}^{n} x_i[0], \text{ for } \quad i = 1, \ldots, n \tag{5.22}$$

5.2.2 Problem Formulation of the Proposed Control Strategy

Suppose there are several PEVs plan to charge in a charging station, which then negotiates with these PEVs to design their charging plans in the K time slots with a total length ΔT. It is assumed that the charging station process the information of the available charging power during every time slot. PEVs are charging with a constant charging rate and expected to reach desired SOCs before the departure time.

If the PEV charging process is fully controlled by the central controller, the individual choice of the user might be overlooked, and it may also potentially pose a hindrance to the adoption of the V2G technology. A properly designed algorithm, which can better suit the users' choice and the benefit of system in the fully distributed manner, satisfies the charging demand while reducing the charging cost and maintaining the power supply–demand balance.

The users' concerns are mainly focused on the charging service quality, i.e. the battery SOC at the end of the charging cycle and the total charging cost. In this chapter, first, the model and the charging process performance of the PEV battery are investigated. The recent literature on the modeling of electric vehicle battery can be categorized into two types:

1. Electrochemical models [54, 55] are mainly used for the battery design optimization, health characterization, and health-conscious control.
2. Equivalent circuit models [41, 56, 57] are primarily applied to the online estimation and power management. In this chapter, a simple equivalent circuit model as shown in Figure 5.19 is adopted [41, 58]. The model consists of two parts, the one is the constant voltage source and the other is a constant resistance. The two parts are in the series. The discrete time form of this model is listed as follows:

$$V_i(k) = V_{\mathrm{oc},i} + R_i I_i(k) \tag{5.23}$$

$$\mathrm{SOC}_i(k+1) = \mathrm{SOC}_i(k) + \frac{\Delta T}{Q_i} I_i(k) \tag{5.24}$$

$$\mathrm{SOC}_i(K_i) = \mathrm{SOC}_i(0) + \sum_{k=1}^{k=K_i} \frac{\Delta T}{Q_i} I_i(k) \tag{5.25}$$

where V_i, I_i, R_i, $V_{\mathrm{oc},i}$, Q_i, and SOC_i are the terminal voltages, the charging current, the battery equivalent internal resistance, the open-circuit voltage, the charging capacity of the battery, and the battery SOC, respectively. ΔT is the charging time with $k_i = T_i/\Delta \mathrm{T}$, where T_i is the total charging time of $P_{EV,i}$.

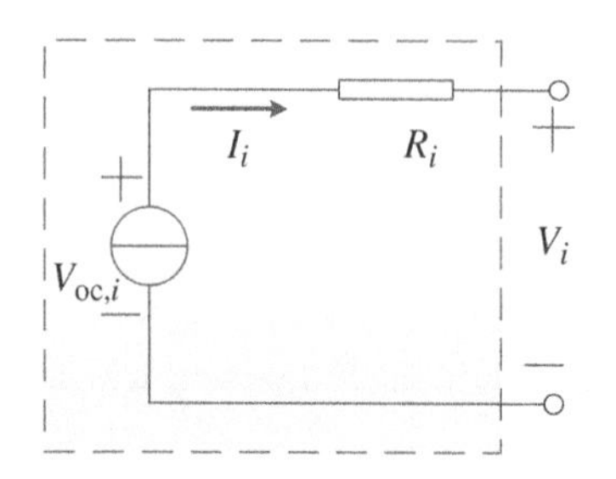

Figure 5.19 PEV battery equivalent circuit model.

The total consumed power $P_{\mathrm{EV},i}$, i.e. the charging power, and the internal power loss of the $P_{EV,i}$ during the charging process are modeled as:

$$P_{\mathrm{EV},i}(k) = V_{oc,i}I_i(k) + R_i I_i^2(k). \tag{5.26}$$

From Eq. (5.26), the charging current can be obtained:

$$I_i(k) = \frac{1}{2R_i}\left(\sqrt{4R_i P_{\mathrm{EV},i}(k) + V_{\mathrm{oc},i}^2} - V_{\mathrm{oc},i}\right) \tag{5.27}$$

The total charging power of all PEVs should be less than the total available power of the charging station:

$$\sum_{i=1}^{n} P_{\mathrm{EV},i}(k) \leq P_C(k) \tag{5.28}$$

where $P_C(k)$ is the total available power.

The charging price p_i of $P_{EV,i}$ is regulated as proportional to the average charging rate to convince the PEV users to make the reasonable charging decisions. The p_i is calculated as follows:

$$p_i = k_p \frac{\mathrm{SOC}_i^* - \mathrm{SOC}_i(0)}{T_i} Q_i \tag{5.29}$$

where SOC_i is the PEV user's desired SOC and k_p is the price. From the PEV user's perspective, assume that the charging price is constant, then the objective is to minimize the charging cost, i.e.

$$\begin{aligned} &\min \quad p_i \cdot \sum_{k=0}^{K_i} P_{\mathrm{EV},i}(k)\Delta T \\ &\mathrm{s.t.} \quad 0 \leq P_{\mathrm{EV},i}(k) \leq P_{\mathrm{EV},i}^{\max} \\ &\sum_{k=1}^{k=K_i} \frac{\Delta T}{Q_i} I_i(k) = \mathrm{SOC}_i^* - \mathrm{SOC}_i(0) \end{aligned} \tag{5.30}$$

From Eq. (5.26), it can be deduced that the total charging power of each PEV can be decomposed into the two parts: the stored energy and the energy loss. The former is stored in the battery and the latter is incurred because of the battery's internal resistance. Therefore, as long as the energy loss is minimized, the total charging cost is also minimized. Then, the objective function can be rewritten as:

$$\min \quad \sum_{k=0}^{K_i} R_i I_i^2(k)\Delta T \tag{5.31}$$

The property of function Eq. (5.31) is listed as follows:

$$\sum_{k=0}^{k=K_i} R_i I_i^2(k)\Delta T \geq \frac{R_i \Delta T}{2}\left[\sum_{k=1}^{k=K_i} I_i(k)\right]^2 = \frac{R_i Q^2}{2\Delta T}[\mathrm{SOC}_i(K_i)\mathrm{SOC}_i(0)]^2 \tag{5.32}$$

During the whole process, *if* the charging current remains constant, the equality holds.

$$I_i^{\text{ref}}(k) = \frac{[\text{SOC}_i(K_i)\text{SOC}_i(0)]Q_i}{K_i \Delta T} \tag{5.33}$$

Therefore, in order to minimize the energy loss, PEVs tend to keep a constant charging rate as shown in Eq. (5.33). Provided the power constraint in Eq. (5.28) varying at different time slots, the constant current charging is left with the little possibility to realize for all PEVs. Therefore, according to the left charging time and the left uncharged capacity, the requested charging current at time slot k is updated as:

$$I_i^{\text{ref}}(k) = \frac{[\text{SOC}_i(K_i)\text{SOC}_i(k)]Q_i}{(K_i - k)\Delta T} \tag{5.34}$$

For the charging rate control of PEVs, the goal is to minimize the deviations between the charging current and the desired charging current while ensuring that the system constraints are not violated

$$\begin{aligned} \min\ L &= \sum_{i=1}^{n} L_i = \sum_{i=1}^{n} w_i(k)[I_i(k) - I_i^{\text{ref}}(k)]^2 \\ &= \sum_{i=1}^{N} \left\{ \frac{P_{\text{EV},i}(k)}{R_i} - \frac{V_{\text{oc},i} + 2R_i I_i^{\text{ref}}(k)}{2R_i^2} \sqrt{4R_i P_{\text{EV},i}(k) + V_{\text{oc},i}^2} \right. \\ &\quad \left. + \frac{V_{\text{oc},i}^2}{2R_i^2} + \frac{2I_i^{\text{ref}}(k)V_{\text{oc},i}}{R_i} + I_i^{\text{ref}^2}(k) \right\} w_i(k) \end{aligned} \tag{5.35}$$

where $w_i(k)$ is a weight coefficient that represents the priority of the $P_{EV,i}$ at time slot k. It is based on the time left to complete the charging process and the remaining charging capacity.

Since the last three terms in Eq. (5.35) are independent of $P_{\text{EV},i}(k)$, thus, the objective function can be simplified as:

$$\sum_{i=1}^{n} w_i(k) \left\{ \frac{P_{\text{EV},i}(k)}{R_i} - \frac{V_{\text{oc},i} + 2R_i I_i^{\text{ref}}(k)}{2R_i^2} \sqrt{4R_i P_{\text{EV},i}(k) + V_{\text{oc},i}^2} \right\} \tag{5.36}$$

$$\text{s.}t.0 \le P_{\text{EV},i}(k) \le P_{\text{EV},i}^{\max}$$

$$\sum_{i=1}^{n} P_{\text{EV},i}(k) \le P_C(k)$$

The $P_{\text{EV},i}(k)$ is the second-order derivative of L_i. It can be verified to be a convex function as $P_{\text{EV},i}(k)$ is positive, i.e.

$$\frac{\partial^2 L_i}{\partial P_{\text{EV},i}(k)^2} = 2w_i(k)(V_{\text{oc},i} + 2R_i I_i^{\text{ref}}(k))(4R_i P_{\text{EV},i}(k) + V_{\text{oc},i}^2)^{-\frac{3}{2}} > 0 \tag{5.37}$$

Function in Eq. (5.37) is also convex because it consists of n convex functions. Then, the constrained convex optimization problem is used to model the coordinated charging rate control problem of PEVs. The incremental cost, which is also the Lagrangian variable of the convex function, is obtained by taking the partial derivative of L_i with respect to the PEV_i

$$r_i = \frac{\partial L_i}{\partial P_{\text{EV},i}} = w_i(k)\left(\frac{1}{R_i} - \frac{(V_{\text{oc},i} + 2R_i I_i^{\text{ref}}(k))}{R_i\sqrt{4R_i P_{\text{EV},i}(k) + V_{\text{oc},i}^2}}\right) \tag{5.38}$$

The global optimal solution to Eq. (5.36), which is the equal incremental cost criterion, is shown in [57]

$$\begin{cases} \left.\dfrac{\partial L_i}{\partial P_{\text{EV},i}}\right|_{P_{\text{EV},i}=P^*_{\text{EV},i}} = r^*, & \text{for } P^{\min}_{\text{EV},i} < P^*_{\text{EV},i} < P^{\max}_{\text{EV},i} \\ \left.\dfrac{\partial L_i}{\partial P_{\text{EV},i}}\right|_{P_{\text{EV},i}=P^*_{\text{EV},i}} < r^*, & P^*_{\text{EV},i} = P^{\max}_{\text{EV},i} \\ \left.\dfrac{\partial L_i}{\partial P_{\text{EV},i}}\right|_{P_{\text{EV},i}=P^*_{\text{EV},i}} > r^*, & P^*_{\text{EV},i} = P^{\min}_{\text{EV},i} \end{cases} \tag{5.39}$$

where r^* is the optimal incremental cost. Then, the optimal charging rate for $P_{EV,i}$ can be obtained:

$$P^*_{\text{EV},i} = \begin{cases} \dfrac{(V_{\text{oc},i}+2R_i I_i^{\text{ref}}(k))^2}{4R_i(1-R_i r^*/w_i(k))^2} - \dfrac{V^2_{\text{oc},i}}{4R_i}, & \text{for } P^{\min}_{\text{EV},i} < P^*_{\text{EV},i} < P^{\max}_{\text{EV},i} \\ P^{\max}_{\text{EV},i}, & \dfrac{(V_{\text{oc},i}+2R_i I_i^{\text{ref}}(k))^2}{4R_i(1-R_i r^*/w_i(k))^2} - \dfrac{V^2_{\text{oc},i}}{4R_i} > P^{\max}_{\text{EV},i} \\ P^{\min}_{\text{EV},i}, & \dfrac{(V_{\text{oc},i}+2R_i I_i^{\text{ref}}(k))^2}{4R_i(1-R_i r^*/w_i(k))^2} - \dfrac{V^2_{\text{oc},i}}{4R_i} < P^{\min}_{\text{EV},i} \end{cases} \tag{5.40}$$

The solution in Eq. (5.40) can also be obtained by a centralized control strategy whose performance highly depends on the capacity of the central controller and the bandwidth of the communication network. It establishes a connection between the central controller and all PEVs. First, the central controller gathers the needed data, such as the users' charging choices and the battery parameters of PEV. Second, the central controller calculates the optimal incremental cost and dispatches to all PEVs. However, the centralized control strategy can coordinate the grid and PEVs. However, the data management issue, and the computation and communication burden, may also be incurred by it. In contrast, a properly designed distributed control strategy shows more flexibility, reliability, and scalability and provides better accommodation to the plug-and-play technology compared to a centralized strategy. The control strategy in the distributed manner has shown a promising future in solving the charging management problem for PEVs.

5.2.3 Proposed Cooperative Control Algorithm

5.2.3.1 MAS Framework

The connections between the PEVs and the charging station are established by the charging sockets, which are assigned a corresponding agent. Agents enable the distributed control and can work round the clock. Only the local information, which is acquired from neighboring agents using the consensus-based algorithm introduced in this section, is required by the agent. As illustrated in Figure 5.20, agent i exchange information with the adjacent agents in an ascending sequence, as $i - n_i/2, \ldots, i-1, i+1, \ldots$ and $i + n_i/2$. The adopted communication topology satisfies the "$N-1$" redundant rule, i.e. the communication link failure, and thus, is robust. However, the independence between the communication system and the physical power network could exist, if carefully designed. Even for a large complex system, the corresponding communication network can be simple and designed based on the cost, the level of convenience, and geographical conditions.

5.2.3.2 Design and Analysis of Distributed Algorithm

Iteratively, the incremental cost of the proposed approach for PEVs can be discovered by the following equations:

$$r_i[t+1] = \sum_{j \in N_i} d_{ij} r_j[t] - \varepsilon \cdot P_D[t] \tag{5.41}$$

$$P_{\mathrm{EV},i}[t+1] = \frac{(V_{\mathrm{oc},i} + 2R_i I_i^{\mathrm{ref}}(k))^2}{4R_i(1 - R_i r_i[t+1]/w_i(k))^2} - \frac{V_{\mathrm{oc},i}^2}{4R_i} \tag{5.42}$$

$$P'_{D,i}[t] = P_{D,i}[t] + (P_{\mathrm{EV},i}[t+1] - P_{\mathrm{EV},i}[t]) \tag{5.43}$$

$$P_{D,i}[t+1] = \sum_{j \in N_i} d_{ij} P'_{D,j}[t] \tag{5.44}$$

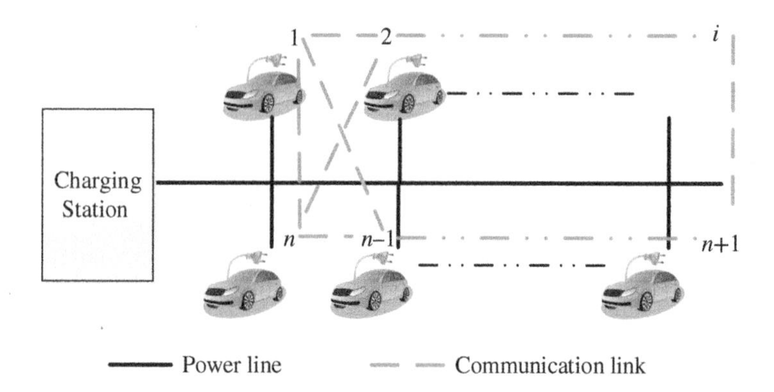

Figure 5.20 The MAS based PEVs charging framework.

where $r_i[t]$ represents the incremental cost during the charging process of $P_{EV,i}$ at iteration t, ϵ represent an adjustable step size that controls the convergence speed, and $P_{D,i}[t]$ represent the local estimation of the global power mismatch. To carry the relative analysis on the properties and convergence of the proposed approach, Eqs. (5.41)–(5.44), which are the updating rules, are rewritten in the matrix form as

$$\mathbf{R}[t+1] = \mathbf{D} \cdot \mathbf{R}[t] - \varepsilon \cdot \mathbf{P}_D[t] \tag{5.45}$$

$$\mathbf{P}_{\mathrm{EV}}[t+1] = \overline{\mathbf{E}} \cdot \mathbf{R}[t+1] \tag{5.46}$$

$$\mathbf{P}_D[t+1] = \mathbf{D} \cdot \mathbf{P}_D[t] + \mathbf{D} \cdot (\mathbf{P}_{\mathrm{EV}}[t+1] - \mathbf{P}_{\mathrm{EV}}[t]) \tag{5.47}$$

where $\mathbf{P}_{\mathrm{EV}}$, $\mathbf{R}$, and $\mathbf{P}_D$ are the column vectors of $P_{\mathrm{EV},i}$, r_i, and $P_{D,i}$, respectively, and the $\overline{E}$ represent the projection operation of R to P_{EV}.

$$\begin{bmatrix} \mathbf{R}[t+1] \\ \mathbf{P}_D[t+1] \end{bmatrix}_{2n\times 1} = \begin{bmatrix} \mathbf{D} & -\varepsilon \mathbf{I}_n \\ \mathbf{D}\bar{\mathbf{E}}(\mathbf{D}-\mathbf{I}_n) & \mathbf{D}+\varepsilon\mathbf{D}\bar{\mathbf{E}} \end{bmatrix}_{2n\times 2n} \cdot \begin{bmatrix} \mathbf{R}[t] \\ \mathbf{P}_D[t] \end{bmatrix}_{2n\times 1} \tag{5.48}$$

Define two matrices as follows: $\mathbf{M} = \begin{bmatrix} \mathbf{D} & 0 \\ \mathbf{D}\bar{\mathbf{E}}(\mathbf{D}-\mathbf{I}_n) & \mathbf{D} \end{bmatrix}$ and $\mathbf{\Delta} = \begin{bmatrix} \mathbf{0} & -\mathbf{I}_n \\ \mathbf{0} & \mathbf{D}\bar{\mathbf{E}} \end{bmatrix}$ where $\mathbf{I}_n$ is an identity matrix of n dimensions. By perturbing matrix $\mathbf{M}$ with $\epsilon\Delta$, the system matrix of Eq. (5.48) can be obtained. The eigenvalues of $\mathbf{D}$ and $\mathbf{M}$ are the same. $|\lambda\mathbf{I}_{2n} - \mathbf{M}| = |\lambda\mathbf{I}_n - \mathbf{D}|^2$. $\mathbf{D}$ is a double stochastic matrix, which means it satisfies $\mathbf{D}\mathbf{1}_n = \mathbf{1}_n$ and $(\mathbf{D}-\mathbf{I}_n)\mathbf{1}_n = \mathbf{0}_n$. The largest eigenvalue of $\mathbf{D}$ is $\lambda_1 = 1$. It can be verified that the corresponding eigenvector to the eigenvalue λ_1 of $(\mathbf{M}+\epsilon\ \Delta)$ is $[\mathbf{1}_n, \mathbf{0}_n]^T$

$$\begin{bmatrix} \mathbf{D} & -\varepsilon \mathbf{I}_n \\ \mathbf{D}\bar{\mathbf{E}}(\mathbf{D}-\mathbf{I}_n) & \mathbf{D}+\varepsilon\mathbf{D}\bar{\mathbf{E}} \end{bmatrix}_{2n\times 2n} \begin{bmatrix} \mathbf{1}_n \\ \mathbf{0}_n \end{bmatrix} = \begin{bmatrix} \mathbf{D}\mathbf{1}_n \\ \mathbf{D}\bar{\mathbf{E}}(\mathbf{D}-\mathbf{I}_n)\mathbf{1}_n \end{bmatrix} = \begin{bmatrix} \mathbf{1}_n \\ \mathbf{0}_n \end{bmatrix} \tag{5.49}$$

The proof in [59] indicates that all the eigenvalues lie inside a unit disk, so the system of equation (5.48) would converge as t approaches infinity to the span $[\mathbf{1}_n, \mathbf{0}_n]^T$. Then, $r_i[t]$ converges to the common values r^* and the optimality is guaranteed.

5.2.3.3 Algorithm Implementation

Assume that at the current time, only one agent, the agent No. 1, knows the real-time available charging power of each charging station. The proposed algorithm can be decomposed, as shown in Table 5.6.

Figure 5.21 demonstrates the operation of agents, whose module of measurement and initialization calculate the desired charging current, as depicted in Eq. (5.33). The estimation of the net power and the initialization of the local

Table 5.6 The Proposed Approach.

I. Initialization (Every ΔT minutes)

$k = 1$;

$I_i^{ref}(k) = \frac{[SOC_i(K_i) - SOC_i(k)]Q_i}{(K_i - k)\Delta T}$, $SOC_i(K_i) = SOC_i(k) + \frac{\Delta T}{Q_i} I_i(k)$

$P_{D,i}[0] = \begin{cases} P_C(k), & i = 1 \\ 0, & \text{otherwise} \end{cases}, r_i[0] = \begin{cases} 0, & k = 1 \\ r_i(k-1), & \text{otherwise} \end{cases}$

II. Consensus coordination

$t = 0$

while

$t \leq N_T$ (N_T is the maximum iteration number)

Each agent communicates with neighboring agents based on the updating rule Eqs. (5.41) and (5.44).

$t = t + 1$

End

$P^*_{EV,i}$ is set according to Eq. (5.40).

$k = k + 1$

III. Go back to I

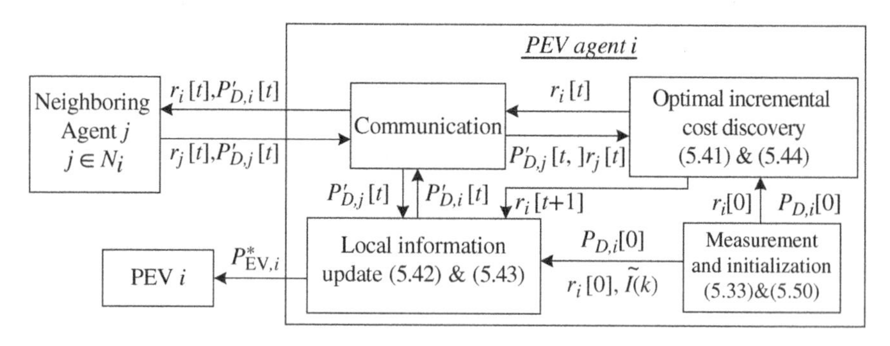

Figure 5.21 Illustration of agent operation.

incremental cost are as follows:

$$\begin{cases} P_{D,i}[0] = \begin{cases} P_C(k), & i = 1 \\ 0, & \text{otherwise} \end{cases} \\ r_i[0] = \begin{cases} 0, & k = 1 \\ r_i(k-1), & \text{otherwise} \end{cases} \end{cases} \tag{5.50}$$

The incremental cost and the estimation of power mismatch are exchanged between an agent and its neighboring agents through the communication module. Using Eqs. (5.41) and (5.44), the relevant information is updated by the module of the optimal cost discovery. According to Eqs. (5.42) and (5.43), the PEV charging power reference and power mismatch estimation are updated by the local information update module, respectively. If $P_{\mathrm{EV},i}$ surpasses the lower/upper limit, it would be set equal to the corresponding limit. This procedure is called projection operation and is aimed to guarantee the convergence of the proposed algorithm [59]. The projection operation can be included in $\bar{E}$. Then, the PEV is charged in accordance with the charging reference.

5.2.3.4 Simulation Studies

Case Study 1

In this section, some case studies on the 14-PEV system, as shown in Figure 5.22, are performed to test the effectiveness of the proposed approach. In case study 1, the number of PEVs and the available power supply for the charging station are all set as constant. In case study 2, the available power supply for the charging station is time-changing. In case study 3, the extendability of the proposed method is validated. In this case study, the charging period is 15 minutes and the available power supply is set to 18 kW. Agents communicate with their neighboring agents and update the data of local incremental cost, the charging power reference, and the allocation of the total power. The update is performed every 0.1 seconds, and the step-size ϵ is set to 0.01 seconds. For $P_{EV,i}$, the priority weight factor can be obtained as follows:

$$w_i(k) = \frac{1}{\mathrm{SOC}_i(k)} \frac{1}{(T_i - \Delta T \cdot k + \varepsilon_t)} \tag{5.51}$$

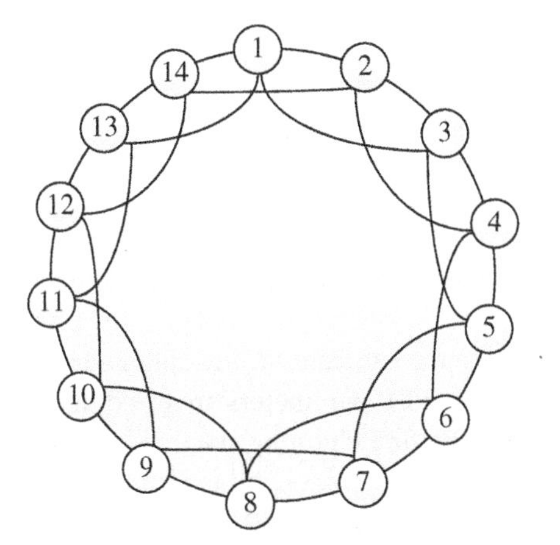

Figure 5.22 Topology of communication network of the 14-PEV scenario.

Table 5.7 Parameters of the battery energy storge systems (BESSs).

Symbol unit	R (Ω)	V_{oc} (V)	Q (A h)	I_{EV}^{max} (A)	SOC (0)	SOC (T)	Time (h)
PEV 1	1.05	280.0	20	10	0.35	0.80	2.5
PEV 2	1.13	313.6	28	10	0.20	0.85	4.0
PEV 3	1.06	308.0	25	10	0.30	0.90	4.0
PEV 4	1.13	319.2	30	10	0.25	0.85	3.5
PEV 5	1.14	294.0	30	10	0.17	0.90	4.0
PEV 6	1.05	302.4	25	10	0.2	0.90	3.5
PEV 7	1.03	305.2	30	10	0.25	0.85	4.0
PEV 8	1.09	274.4	30	10	0.30	0.85	3.0
PEV 9	1.08	277.2	25	10	0.22	0.90	4.0
PEV 10	1.27	280.0	32	10	0.28	0.85	4.0
PEV 11	1.26	285.6	23	10	0.18	0.90	3.0
PEV 12	1.37	291.2	24	10	0.21	0.90	4.0
PEV 13	1.21	285.6	25	10	0.20	0.90	4.0
PEV 14	1.32	288.4	28	10	0.22	0.85	4.0

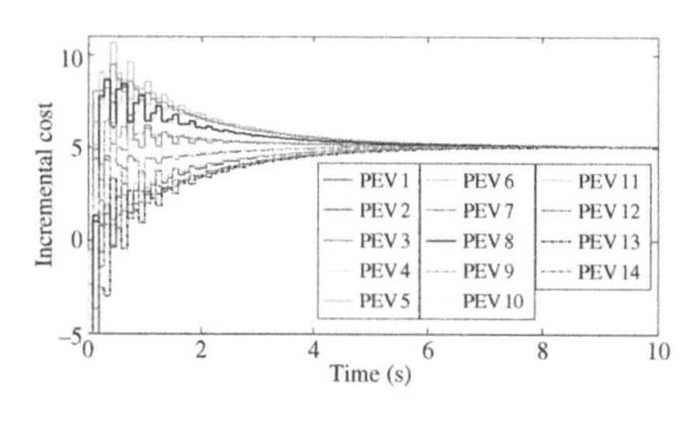

Figure 5.23 The updating process of the incremental cost of PEVs.

where ϵ_t is set to be positive so as to avoid the division by zero. In this case study, $\epsilon_t = 0.05$.

In this case study, the number of adjacent agents of $P_{EV,i}$ are four, whose indices are $i-2$, $i-1$, $i+1$, and $i+2$, respectively. The topology of the communication network consists of 28 edges and the maximum number of edges that the system could possibly have is 91. Then, the graph density is $28/91 = 0.3077$. The communication coefficients d_{ij} are obtained according to Eq. (3.21). The initial conditions and the PEV parameters are listed in Table 5.7.

At the first charging stage, as shown in Figure 5.23, the incremental cost converges to a common value within 10 seconds. Figure 5.24 shows that the power allocation for each PEV converges after seven iterations. Figure 5.25 shows that

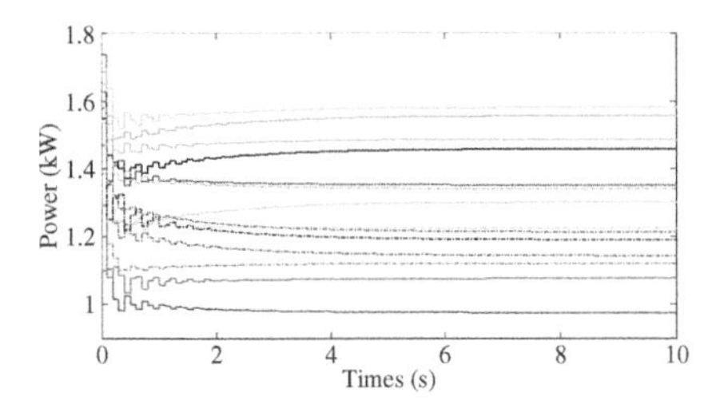

Figure 5.24 The updating process of the allocated power for PEVs.

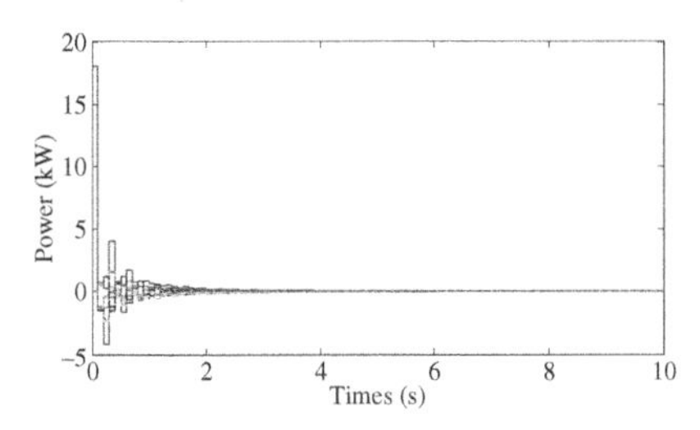

Figure 5.25 The updating process of the supply–demand mismatch estimation for PEVs.

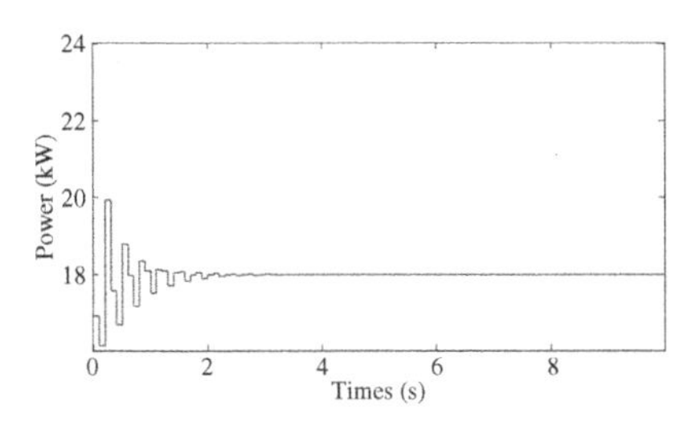

Figure 5.26 The updating process of the total allocated power for PEVs.

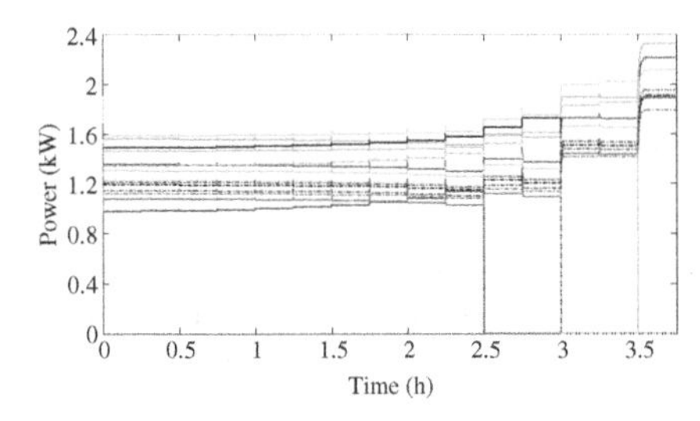

Figure 5.27 The updating process of allocated power profiles for PEVs at the charging process.

the power mismatch estimation converges to zero while the optimal charging rate of every PEV reaches optimality. Figure 5.26 shows that the total charging power converges to 18 kW, which is the assumed total available charging power.

The allocated power of each PEV, the comparison between the desired and actual charging current, the SOC profile of each PEV, and the total allocated power are shown in Figures 5.27–5.30, respectively. As shown in Figure 5.27, the PEV 1 completes its charging process within 2.5 hours, PEV 8 and PEV 11 reach the desired SOC levels within three hours, and that of PEV 4 and PEV 6 are

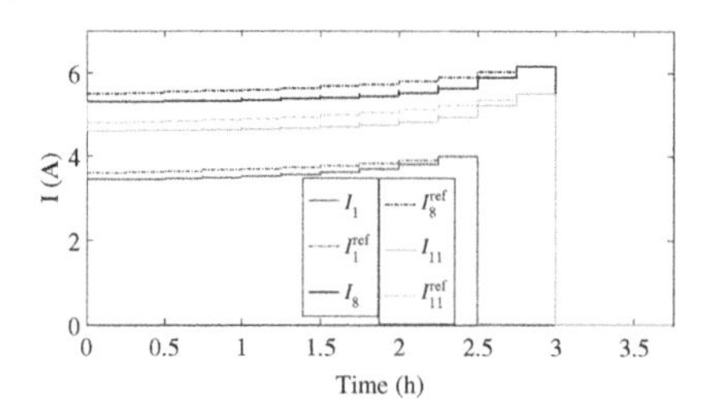

Figure 5.28 The updating process of the desired charging current and actual charging current of PEVs.

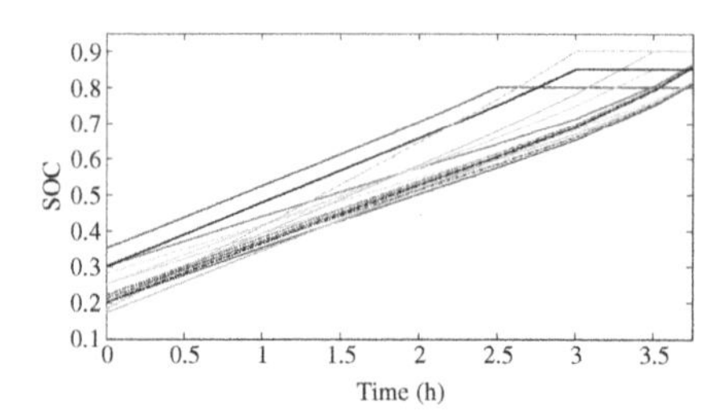

Figure 5.29 The updating process of SOC of PEVs at charging process.

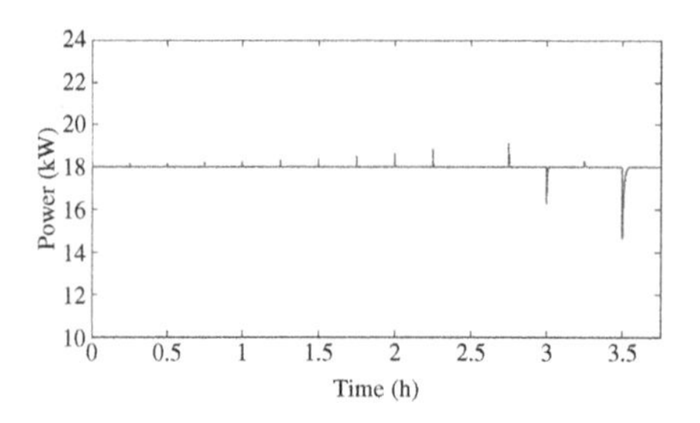

Figure 5.30 The updating process of total allocated power at charging process.

within 3.5 hours as expected. After the departure of PEV 1, the allocated power of the rest of PEVs is increased. One of the major advantages of the proposed approach is that it enables the plug-and-play function. The charging power will drop to zero if a PEV is leaving the charging station. Figure 5.27 shows that with the adaptability of plug-and-play when there are the departures of PEVs from the charging station, the rest of the PEVs can still reach the consensus by exchanging the information of the available charging power. Figure 5.28 illustrates that the charging current that has been allocated and that is referred is very close to each other when it is near the end of the charging cycle. This is because, as shown in Figure 5.29, the priority factor increased when the PEV is near to finish the charging process so that the PEV can be charged to the desired SOC before the departure. Figure 5.30 shows that during the charging, there is a small deviation of the total allocated power. The proposed approach has little impact on the system in terms of frequency disturbances, indicating its promising applications in the isolation system, namely, the autonomous microgrid.

Case Study 2

In order to prove the effectiveness of the proposed approach, this case study uses an isolation system under the condition of the time-varying supply and demand to study its performance. Figure 5.31 shows that the system consists of a synchronous generator, a load that is not a PEV type, and a PEV load. The system frequency change rate can be approximated as [58]:

$$\frac{df}{dt} = \frac{f_0}{2HS_b}\left(P_G - P_{\text{NEV}} - \sum_{i=1}^{n} P_{\text{EV},i}\right) \tag{5.52}$$

where H, S_b, f_0, P_G, and P_{NEV} are the synchronous generator's inertia constant r, the MVA rating used in the system, the nominal value of the system frequency, the generation output, and the non-PEV load, respectively. The key to minimizing the influence on the stability of the system frequency is to keep the net power of the system close to the zero. Therefore, a crucial goal is to use the total PEV charging power to compensate the power mismatch between the generation and the non-PEV loads. The sampling rate (SR) for the measurement of the available charging power, which is set to be five minutes, is aimed to tackle the available power that changes along with the time and to achieve more accurate control. The generation output is set to be 40 kW. Figure 5.32 shows the demand profile of the non-PEV load.

As shown in Figure 5.33, there is a decline in the allocated charging power along with an increment in the non-PEV load 2.5 hours before the PEV leaves the charging station and vice versa. As shown in Figure 5.34, when it is at the end of its charge cycle, the charging currents of PEVs 1, 8, and 11 are very close to their charging current references. PEV 1 finishes the charging process in 2.5 hours, and then PEVs 8 and 11 reach their desired SOC in three hours, as shown in Figure 5.35.

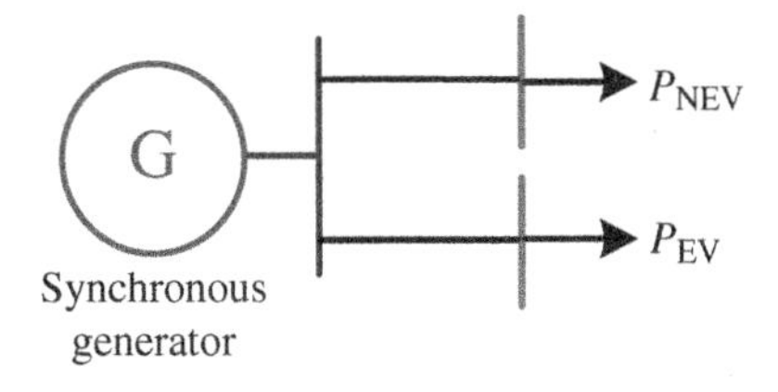

Figure 5.31 The utilization of a single isolated synchronous generator.

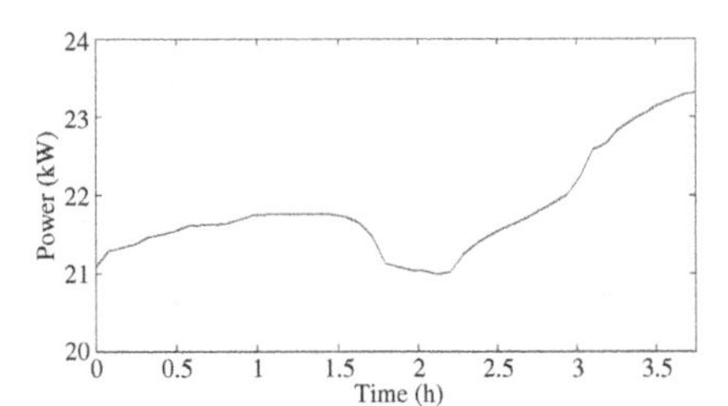

Figure 5.32 The demand profile of non-PEV load in the case 2 scenario.

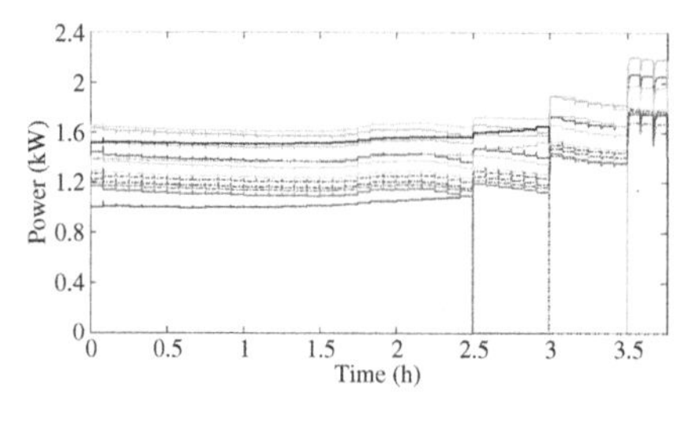

Figure 5.33 The updating profile of allocated power with SR = 5 minutes.

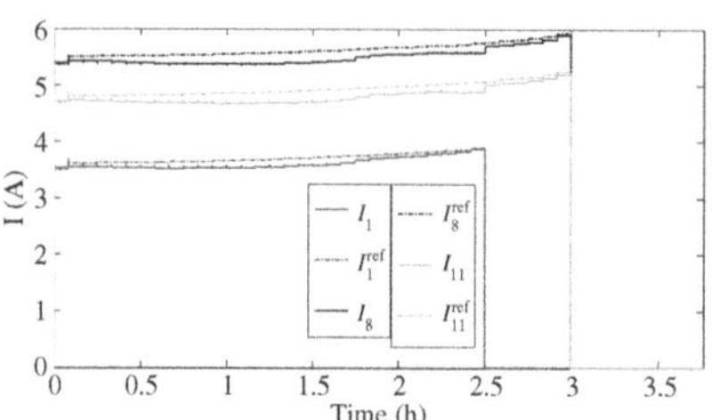

Figure 5.34 The updating profiles of desired and actual charging current with SR = 5 minutes.

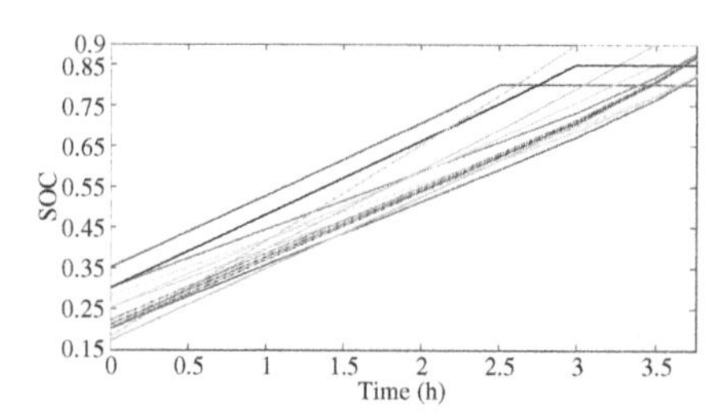

Figure 5.35 The updating profiles of battery SOC with SR = 5 minutes.

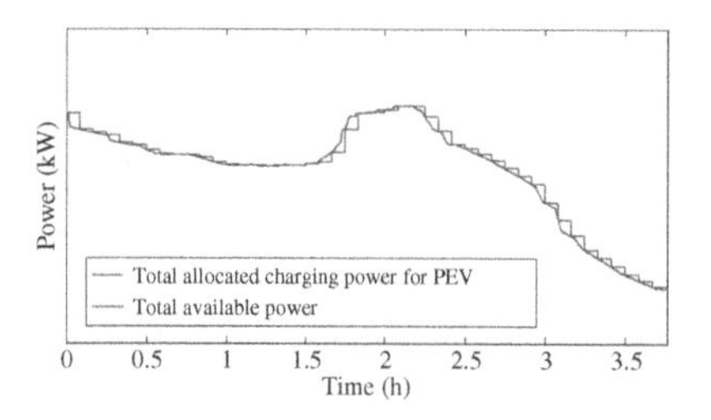

Figure 5.36 The updating profile of available and total allocated charging power with SR = 5 minutes.

Figure 5.36 shows the PEV's profile of the total allocated charging power and the available charging power. The mismatch ratio between the two terms is shown in Figure 5.37. The mismatch ratio is within 2%. The mismatched power is calculated under the assumption that the available charging power of each sample period remains constant. Thus, it can be reduced by using a sampling cycle with a shorter time.

Then, the setting of sample rate is changed from five to two minutes, and the mismatch ratio is reduced to under 1% as shown in Figures 5.38 and 5.39. Therefore, by

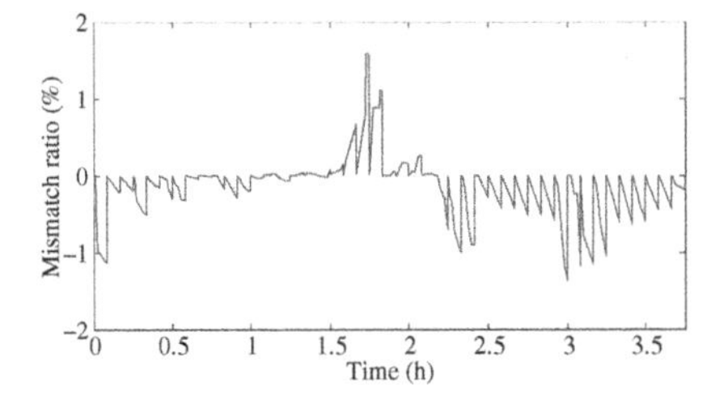

Figure 5.37 The updating profile of mismatch ratio of the available and the total allocated power with SR = 5 minutes.

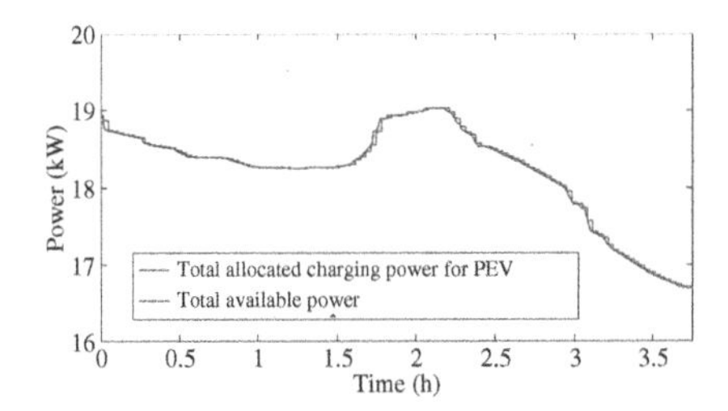

Figure 5.38 The updating profile of available and total allocated charging power with SR = 2 minutes.

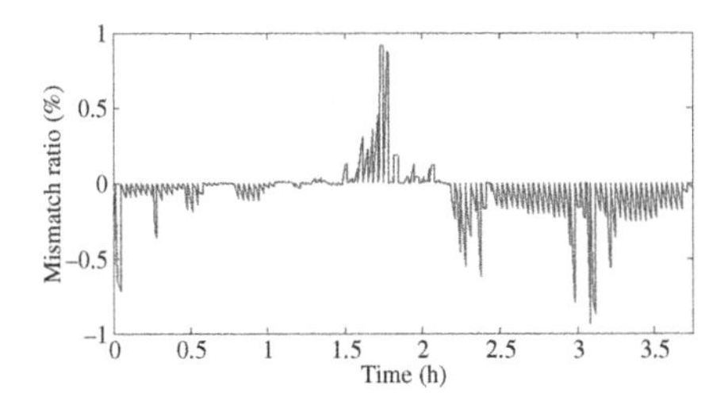

Figure 5.39 The updating profile of mismatch ratio between the available and total allocated power with SR = 2 minutes.

setting a proper sampling rate, the proposed approach can successfully reduce the power mismatch of an isolated system under the condition of fast-changing supply and demand. The simulation results show that the proposed approach has broad application prospects in the autonomous microgrid with intermittent renewable generations.

Case Study 3

In this case, by implementing the proposed approach on two different systems, the 50-PEV and 100-PEV systems, the scalability of the proposed approach is validated. The supporting communication system of the 50-PEV system is similar to that of case study 1. It is assumed that each agent exchange information with 10 neighboring agents. Therefore, the topology of the communication network is formed by the strong connection of 250 edges. Because the maximum number of edges the system could possibly have is 4950, the graph density is 250/1225 = 0.2041. In the 100-PEV system, the agents are assumed to have 20 neighboring agents, and

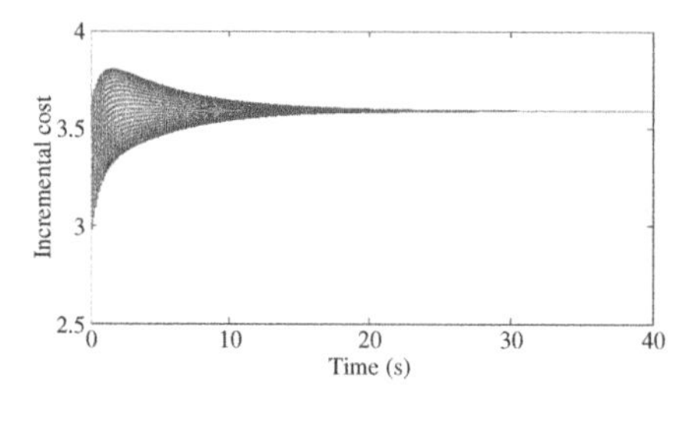

Figure 5.40 The updating profile of incremental cost for 50-PEV systems.

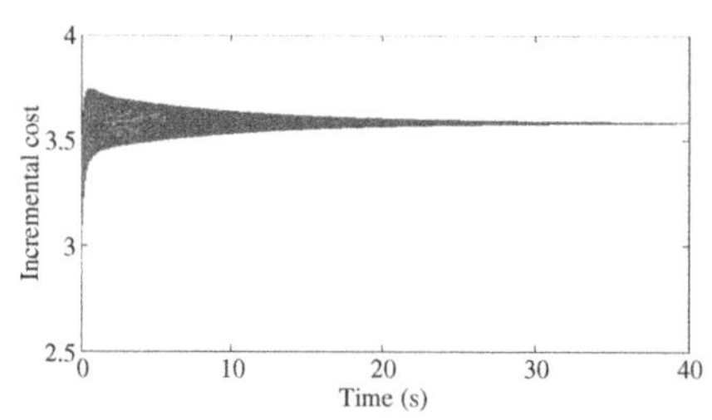

Figure 5.41 The updating profile of incremental cost for 100-PEV systems.

thus, the topology of the communication system is formed by the strong connection of 1000 edges. The graph density is 1000/4950 = 0.2020. The information is updated every 0.1 seconds, and the PEV parameters are from the modification of that in Table 5.7. The available charging power ratings of the 50-PEV and 100-PEV system are set to 60 and 120 kW, respectively.

The proposed approach has been implemented on the 50-PEV and 100-PEV systems. The results show that the convergence is reached within 30 seconds, as shown in Figures 5.40 and 5.41. The simulation results indicated that the density of communication connectivity is the main factor that determines the convergence speed. Thus, despite the large scale of the system, the convergence speed can be maintained or improved by building denser communication connectivity. Hence, the trade-off between the cost of the communication network and the convergence speed should be considered.

5.3 Conclusion

A distributed optimal charging rate control algorithm for PEVs is proposed in this chapter. The proposed approach combines the goal of reducing energy loss and satisfying system constraints. The proposed approach achieves the following four main advantages. The first is to introduce a MAS-based framework in a fully distributed manner and a consensus algorithm that reduces the cost concerning the communication network compared to a centralized algorithm. The second is to consider the charging power loss incurred by the internal resistance of the battery.

This indicates the significant potential to encourage user participation. The third is global optimality. It is obtained without any approximation. The fourth is to verify the feasibility of the proposed approach being applied to different scales of systems, i.e. small systems and large systems. The effectiveness of the proposed approach is validated by the simulation study. It demonstrates the promising applications in the autonomous microgrids with the intermittent renewable generations.

In order to utilize the proposed approach in the field applications, a number of practical issues need to be further elaborated:

1. Consider the influence of the charging terminal on the system frequency and the voltage deviation as well as the power loss incurred by the power line.
2. A PEV battery model with more accuracy is needed.
3. Integrate the predictions and the forecasting technology into the MAS framework so that the local agents could achieve better charging preference settings.
4. Considering the system frequency and the voltage changes, the proposed algorithm is applied to the islanded microgrid system with the intermittent renewable power generation.

References

1 Chauhan, R.K., Rajpurohit, B.S., Hebner, R.E. et al. (2015). Design and analysis of PID and fuzzy-PID controller for voltage control of DC microgrid. *IEEE PES Innovative Smart Grid Technologies (ISGT) Asian Conference.*

2 Kang, Q., Zhou, M., An J., and Wu, Q. (2013). Swarm intelligence approaches to optimal power flow problem with distributed generator failures in power networks. *IEEE Transactions on Automation Science and Engineering* 10 (2): 343–353. https://doi.org/10.1109/TASE.2012.2204980.

3 Abdollahi, A., Moghaddam, M.P., Rashidinejad, M. (2012). Investigation of economic and environmental-driven demand response measures incorporating UC. *IEEE Transactions on Smart Grid* 3 (1): 12–25. https://doi.org/10.1109/TSG.2011.2172996.

4 US Department of Energy (2006). Benefits of demand response in electricity market and recommendation for achieving them. http://www.ee.washington.edu/research/pstca (visited on 30 September 2019).

5 Electric Power Research Institute (2002). *The Western States Power Crisis: Imperatives and Opportunities.* working paper [online]. http://www.epri.com/westernstatespowercrisissynthesis.pdf (accessed 22 July 2020).

6 The Battle Group (2007). *Quantifying demand response benefits in PJM.* Report for PJM interconnection and Mid Atlantic Distributed Resource Initiative(MADRI).

7 David Kathan, D., Daly, C., Gadani, J., et al. (2008). Assessment of demand response and advanced metering. Report for Federal Energy Regulatory Commission.

8 Paterakis, N.G., Erdinç, O., and Catalão, J.P.S. (2017). An overview of demand response: key-elements and international experience. *Renewable and Sustainable Energy Reviews* 69: 871–891.

9 Pacific Gas and Electric Company (2014). Energy management programs: base interruptible programe, demand bidding program. http://www.pge.com (accessed 22 July 2020).

10 Walawalkar, R., Blumsack, S., Apt, J., Fernands, S. (2008). Analyzing PJM' economic demand response program. *Power and Energy Society General Meeting-Conversion and Delivery of Electrical Energy in the 21st Century, 2008 IEEE*. IEEE, pp. 1–9.

11 Albadi, M.H. and El-Saadany, E.F. (2007). Demand response in electricity markets: an overview. *IEEE Power Engineering Society General Meeting* 2007: 1–5.

12 Rahimi, F. (2009). Overview of demand response programs at different ISOs/RTOs. *Power Systems Conference and Exposition, 2009. PSCE'09. IEEE/PES. IEEE*, pp. 1–2.

13 Crane, M. (2012). PJM's revised demand response compensation plan get mixed reviews. *SNL Energy Finance Daily*.

14 Madaeni, S.H. and Sioshansi, R. (2013). Using demand response to improve the emission benefits of wind. *IEEE Transactions on Power Systems* 28 (2): 1385–1394. https://doi.org/10.1109/TPWRS.2012.2214066.

15 Xu, Y. and Liu, W. (2011). Novel multiagent based load restoration algorithm for microgrids. *IEEE Transactions on Smart Grid* 2 (1): 152–161.

16 Zhang, W., Xu, Y., Liu, W. et al. (2013). Fully distributed coordination of multiple DFIGs in a microgrid for load sharing. *IEEE Transactions on Smart Grid* 4 (2): 805–815.

17 Ashok, S. and Banerjee, R. (2001). An optimization mode for industrial load management. *IEEE Transactions on Power Systems* 16 (4): 879–884. https://doi.org/10.1109/59.962440.

18 Fan, Z. (2012). A distributed demand response algorithm and its application to PHEV charging in smart grids. *IEEE Transactions on Smart Grid* 3 (3): 1280–1290. https://doi.org/10.1109/TSG.2012.2185075.

19 Tan, Z., Yang, P., and Nehorai, A. (2014). An optimal and distributed demand response strategy with electric vehicles in the smart grid. *IEEE Transactions on Smart Grid* 5 (2): 861–869. https://doi.org/10.1109/TSG.2013.2291330.

20 Zhang, W., Zhou, S., and Lu, Y. (2012). Distributed intelligent load management and control system. *Power and Energy Society General Meeting, 2012 IEEE*. IEEE, pp. 1–8.

21 Liu, J., Zhao, B., Wang, J. et al. (2010). Application of power line communication in smart power consumption. *2010 IEEE International Symposium on Power Line Communications and Its Applications (ISPLC)*, pp. 303–307. https://doi.org/10.1109/ISPLC.2010.5479945.

22 Bian, D., Pipattanasomporn, M., and Rahman, S. (2014). A human expert-based approach to electrical peak demand management. *IEEE Transactions on Power Delivery* 30 (3): 1119–1127. https://doi.org/10.1109/TPWRD.2014.2348495.

23 Ying-chen, L. and Lu, C. (2010). The distribution electric price with interruptible load and demand side bidding. *2010 China International Conference on Electricity Distribution (CICED)*, pp. 1–6.

24 Qi, W., Liu, J., and Christofides, P.D. (2011). A distributed control framework for smart grid development: energy/water system optimal operation and electric grid integration. *Journal of Process Control* 21 (10): 1504–1516.

25 Bertsekas, D.P. (1975). Monotone mappings in dynamic programming. *1975 IEEE Conference on Decision and Control Including the 14th Symposium on Adaptive Processes*, pp. 20–25. https://doi.org/10.1109/CDC.1975.270641.

26 Bertsekas, D.P. (1982). Distributed dynamic programming. *IEEE Transactions on Automation and Control* 27 (3): 610–616.

27 Bertsekas, D.P. and Yu, H. (2010). Distributed asynchronous policy iteration in dynamic programming. *2010 48th Annual Allerton Conference on Communication, Control, and Computing (Allerton)*. IEEE, pp. 1368–1375.

28 Bellifemine, F.L., Caire, G., and Greenwood, D. (2007). *Developing Multi-Agent Systems with JADE*. Wiley.

29 Zhang, W., Liu, W., Wang, X. et al. (2015). Online optimal generation control based on constrained distributed gradient algorithm. *IEEE Transactions on Power Systems* 30 (1): 35–45. https://doi.org/10.1109/TPWRS.2014.2319315.

30 Xu, Y., Liu, W., and Gong, J. (2011). Stable multi-agent-based load shedding algorithm for power systems. *IEEE Transactions on Power Systems* 26 (4): 2006–2014.

31 Zhang, W. (2013). Mutiagent system based algorithm and their applicaitons in power systems. PhD dissertation. Las Cruces, NM: New Mexico Sate University.

32 Rebours, Y.G., Kirschen, D.S., Trotignon, M., and Rossignol, S. (2007). A survey of frequency and voltage control ancillary services–Part I: Technical features. *IEEE Transactions on Power Systems* 22 (1): 350–357.

33 Voelcker, J. (2011). One million plug-in cars by 2015? [Update]. *IEEE Spectrum* 48 (4): 11–13.

34 Green, R.C. II,, Wang, L., and Alam, M. (2011). The impact of plug-in hybrid electric vehicles on distribution networks: a review and outlook. *Renewable and Sustainable Energy Reviews* 15 (1): 544–553.

35 Fernandez, L.P., Roman, T., Cossent, R. et al. (2011). Assessment of the impact of plug-in electric vehicles on distribution networks. *IEEE Transactions on Power Systems* 26 (1): 206–213.

36 Lopes, J.A.P., Soares, F.J., and Almeida, P.M.R. (2009). Identifying management procedures to deal with connection of Electric Vehicles in the grid. *Powertech,* IEEE Bucharest.

37 Li, C.T., Ahn, C., Peng, H., and Sun, J. (2013). Synergistic control of plug-in vehicle charging and wind power scheduling. *IEEE Transactions on Power Systems* 28 (2): 1113–1121.

38 Su, W., Eichi, H., Zeng, W., and Chow, M.Y. (2012). A survey on the electrification of transportation in a smart grid environment. *IEEE Transactions on Industrial Informatics* 8 (1): 1–10.

39 Jin, C., Jian, T., and Ghosh, P. (2013). Optimizing electric vehicle charging: a customer's perspective. *IEEE Transactions on Vehicular Technology* 62 (7): 2919–2927.

40 Kennel, F., Gorges, D., and Liu, S. (2013). Energy management for smart grids with electric vehicles based on hierarchical MPC. *IEEE Transactions on Industrial Informatics* 9 (3): 1528–1537.

41 Rotering, N. and Ilic, M. (2011). Optimal charge control of plug-in hybrid electric vehicles in deregulated electricity markets. *IEEE Transactions on Power Systems* 26 (3): 1021–1029.

42 Patil, R.M., Kelly, J.C., Filipi, Z., Fathy, H.K. (2013). A framework for the integrated optimization of charging and power management in plug-in hybrid electric vehicles. *IEEE Transactions on Vehicular Technology* 62 (6): 2402–2412.

43 Zhao, J.H., Wen, F., Dong, Z. et al. (2012). Optimal dispatch of electric vehicles and wind power using enhanced particle swarm optimization. *IEEE Transactions on Industrial Informatics* 8 (4): 889–899.

44 Ota, Y. (2012). Autonomous distributed V2G (vehicle-to-grid) satisfying scheduled charging. *IEEE Transactions on Smart Grid* 3 (1): 559–564.

45 Liu, H., Hu, Z., Song, Y., Lin, J. (2013). Decentralized vehicle-to-grid control for primary frequency regulation considering charging demands. *IEEE Transactions on Power Systems* 28 (3): 3480–3489.

46 Mokhtari, G., Nourbakhsh, G., and Ghosh, A. (2013). Smart coordination of energy storage units (ESUs) for voltage and loading management in distribution networks. *IEEE Transactions on Power Systems* 28 (4): 4812–4820.

47 Ma, Z., Callaway, D.S., and Hiskens, I.A. (2012). Decentralized charging control of large populations of plug-in electric vehicles. *IEEE Transactions on Control Systems Technology* 21 (1): 67–78.

48 Karfopoulos, E.L. and Hatziargyriou, N.D. (2013). A multi-agent system for controlled charging of a large population of electric vehicles. *IEEE Transactions on Power Systems* 28 (2): 1196–1204.

49 Kok, K., Roossien, B., MacDougall, P. et al. (2012). Dynamic pricing by scalable energy management systems – field experiences and simulation results using PowerMatcher. *Power & Energy Society General Meeting.*

50 Unda, I., Papadopoulos, P., Skarvelis-Kazakos, S. et al. (2014). Management of electric vehicle battery charging in distribution networks with multi-agent systems. *Electric Power Systems Research* 110: 172–179.

51 Papadopoulos, P., Jenkins, N., Cipcigan, L.M. et al. (2013). Coordination of the charging of electric vehicles using a multi-agent system. *IEEE Transactions on Smart Grid* 4 (4): 1802–1809.

52 Rahbari-Asr, N. and Chow, M.Y. (2014). Cooperative distributed demand management for community charging of PHEV/PEVs based on KKT conditions and consensus networks. *IEEE Transactions on Industrial Informatics* 10 (3): 1907–1916.

53 Olfati-Saber, R., Fax, J.A., and Murray, R.M. (2007). Consensus and cooperation in networked multi-agent systems. *Proceedings of the IEEE* 95 (1): 215–233.

54 Smith, K.A., Rahn, C.D., and Wang, C.Y. (2010). Model-based electrochemical estimation and constraint management for pulse operation of lithium ion batteries. IEEE Transactions on Control Systems Technology 18 (3): 654–663.

55 Santhanagopalan, S., Guo, Q., Ramadass, P., White, R.E. (2006). Review of models for predicting the cycling performance of lithium ion batteries. *Journal of Power Sources* 156 (2): 620–628.

56 Ceraolo, M. (2000). New dynamical models of lead-acid batteries. *IEEE Transactions on Power Systems* 15 (4): 1184–1190.

57 Bashash, S. and Fathy, H.K. (2014). Cost-optimal charging of plug-in hybrid electric vehicles under time-varying electricity price signals. *IEEE Transactions on Intelligent Transportation Systems* 15 (5): 1958–1968.

58 >Doherty, R., Lalor, G., and O'Malley, M. (2005). Frequency control in competitive electricity market dispatch. *IEEE Transactions on Power Systems* 20 (3): 1588–1596.

59 Yang, S., Tan, S., and Xu, J.X. (2013). Consensus based approach for economic dispatch problem in a smart grid. *IEEE Transactions on Power Systems* 28 (4): 4416–4426.

6
Distributed Social Welfare Optimization

Traditionally, economic dispatch and demand response (DR) are considered separately or implemented sequentially, which may degrade the energy efficiency of the power grid. One important goal of optimal energy management is to maximize social welfare through the coordination of the suppliers' generations and customers' demands. Thus, it is desirable to consider the interactive operation of the economic dispatch and DR and solve them in an integrated way. In this chapter, a fully distributed online optimal energy management solution is proposed for the smart grid. The proposed solution considers the economic dispatch of the conventional generators, DR of users, and operating conditions of the renewable generators altogether. The proposed distributed solution is developed based on the market-based self-interest motivation model. This model can realize the global social welfare maximization among system participants. The proposed solution can be implemented with the multi-agent system (MAS), with each system participant assigned an energy management agent (EMA). Based on the designed distributed algorithms for price update and supply–demand mismatch discovery, the optimal energy management among agents can be achieved in a distributed way. Simulation results demonstrate the effectiveness of the proposed solution.

6.1 Introduction

One of the key challenges proposed by the increasing number of renewable generators in the smart grid is related to the economic operation [1]. Because of the existence of multiple traditional and renewable generators in power grids, the smart grids' economic operation requires the optimal management on both demand and supply sides. However, conventional economic dispatch and DR are operated, respectively.

The economic dispatch is implemented to distribute the power generation among generators economically with consideration of the physical restrictions

Distributed Energy Management of Electrical Power Systems, First Edition.
Yinliang Xu, Wei Zhang, Wenxin Liu, and Wen Yu.

Published 2021 by John Wiley & Sons, Inc.

of the power system on the generation side [2]. Conventionally, the load requirements are considered as constant. DR is now becoming an integral part of both the power system and market operations on the demand side [3]. Through the operation of DR, customers have the ability to make an acquainted decision with a consideration of their energy consumption, which promotes the reduction of the entire peak load, reconstruction of demand profiles, and the improvement of grid sustainability. DR applications are currently implemented based on presupposed load curves, schedules, or price patterns [4, 5].

On the one hand, the outputs of the traditional generators require to be adjusted based on operating conditions, for instance, diversification of the renewable generation. The specific approach of adjustment among them requires to be re-dispatched in an economical manner. On the other hand, customers will adjust their demands, responding to the variations in the market price for the purpose of maximizing the profits [6]. The execution of economic dispatch or DR will impact the other users because both of them participate in the energy market. For example, the implementation of DR will reconstruct the load profiles. Thus, the original optimal point acquired via economic dispatch will become less useful because the generations will stray from it. Thus, the economic dispatch requires to be operated again. The implementation of economic dispatch will influence the market price and stimulate DR in turn. By and large, the implementation of economic dispatch and DR should be coupled and designed simultaneously while operating online energy management (OEM). Yet, the interactive process will consume a fairly long time to converge on condition that the economic dispatch and DR be operated independently and sequentially. As a consequence, considering the interactive implementations of economic dispatch and DR, and settle both of them integrally, is of great significance.

In general, the centralized solutions are stubborn, and the single-point failures could influence the solutions easily [7–13]. To operate such methods under a complex communication network, an effective central controller needs to collect global information and to handle a large amount of data. Consequently, when encountering a rigorous operating situation, such methods cannot satisfy the response requirements in a timely manner, especially during the changes caused by the intermittency and uncertainty of renewable energy resources. Therefore, such methods require to be improved in terms of response speed.

For the purpose of addressing the above mentioned disadvantages of centralized solutions, a variety of solutions have been proposed. Some scholars proposed DR approaches on the basis of utility maximization [14]. These scholars designed the bidding process in a distributed way, but these approaches need an alpha controller to assemble the required information from all consumers on-demand side in order to obtain the market price. To solve the economic dispatch in a distributed manner, an incremental cost consensus algorithm-based method is suggested by some scholars. However, the DR was not considered by them [15]. For the dispatch of the distributed generators in a smart grid, some scholars also proposed a population dynamics approach, which also needs an auctioneer agent acting as a centralized coordinator [16]. All of these proposed methods have not considered the impact of intermittent renewable generation. As the power system needs the integration of more renewable generators and more user participation into the energy market, an advanced distributed algorithm with considerations of both generators and energy users for the integrated optimal energy market is required.

In this chapter, a distributed solution for OEM is proposed to maximize the total social welfare of the whole power system. Such a solution is applied with a MAS, designing on the ground of the self-interest motivation market model. MAS has been implemented to diversified problems, such as the active power and the reactive power control [17–19]. Solution based on a well-designed MAS performs more flexibly, reliably, and can be easily implemented. Moreover, such a solution has advantages of robustness against single-point failures.

In terms of the proposed solution for OEM, we use one EMA to denote a single unit in the power system. In order to achieve the market equilibrium, an EMA is designed to communicate with its surrounding EMAs within its communication range and participates in the price negotiation. Accordingly, for the OEM problem, its optimal solution is the acquired market equilibrium. The main highlights of this chapter are illustrated as follows:

1. An OEM problem is formulated by considering the interactive operation of economic dispatch and DR. Both of the social welfare maximization models and market-based self-interest motivation models are introduced in formulating the problem. This chapter also discusses the relationship between these two models.
2. The OEM problem is solved by utilizing a distributed approach, and this approach is designed on the basis of distributed price updating and supply–demand mismatch discovery algorithm, with the market-based self-interest motivation model being applied.
3. The designed OEM solution is implemented by using MAS architecture, which simply requires the local information exchange among neighboring agents. Therefore, such a solution overcomes the disadvantages of centralized solutions.

The rest of the chapter is structured as follows. Section 6.2 proposes two OEM optimization models and introduces relationships between these two models. Section 6.3 introduces the specifics of the establishment and application of the proposed distributed algorithms. Section 6.4 presents two cases with the 6-bus power system and the IEEE 30-bus system.

6.2 Formulation of OEM Problem

The renewable generation could not be assumed as dispatchable because of its intermittent nature. Hypothetically, in this chapter, renewable generation is unable to satisfy all loads. Thus, it is always consumed whenever available. The conventional generation will accommodate the unserved load. The following equation presents the active power balance constraints that should be satisfied in a power system with n_G conventional generators, n_R renewable generators, and n_L loads:

$$\sum_{i=1}^{n_G} P_{G,i} - \sum_{j=1}^{n_L} P_{D,j} + \sum_{k=1}^{n_R} P_{R,k} - P_{\text{loss}} = 0 \tag{6.1}$$

where $P_{G,i}$ is the generation supply of the ith unit, while $P_{D,i}$ is the load demand of the jth units. $P_{R,k}$ is the renewable generation of kth unit. The active power loss of the network is P_{loss}. Typically, social welfare maximization cannot be considered as self-interest maximization. However, in the situation established in this chapter, social welfare maximization equals to the self-interest maximization, which will be later discussed in detail.

6.2.1 Social Welfare Maximization Model

The OEM aims to minimize all generators' production cost and maximize all users' utility for the purpose of maximizing the social welfare [20]:

$$\max\left(-\sum_{i=1}^{n_G} C_{s,i}(P_{G,i}) + \sum_{j=1}^{n_L} C_{u,j}(P_{D,j})\right) \tag{6.2}$$

The production cost of the supplier (generator) i is denoted with $C_{s,i}$, and the utility of the user j is denoted by using $C_{u,j}$. In most countries, the renewable generation is currently consumed whenever available as per the policy and environmental regulations. Therefore, here, the cost for the renewable generation could be considered as zero. The following equations present the production cost for generator i:

$$C_{s,i}(P_{G,i}) = \frac{1}{2}a_i P_{G.i}^2 + b_i P_{G.i} + c_i \tag{6.3}$$

$$P_{G,i}^{\min} \leq P_{G,i} \leq P_{G,i}^{\max} \tag{6.4}$$

where a_i, b_i, and c_i represent the generation cost coefficients of generator i. $P_{G.i}^{\min}$ and $P_{G.i}^{\max}$ are the minimum and maximum output of the generation, respectively.

In this chapter, the utility of a user is defined as converged utilities for loads with varying tasks instead of the utility of a personal device. It is rational to consider that the utility function is nondecreasing as more consumed power enables us to accomplish more tasks. The form of the user's utility cost is typically a quadratic function. User js utility function is denoted as [21, 22]:

$$C_{u,j}(P_{D,j}) = \frac{1}{2}\alpha_i P_{D,j}^2 + \omega_j P_{D,j} \tag{6.5}$$

The coefficients of the utility function are denoted as α_i and ω_i. Typically, α_i is negative, while ω_i is positive. The loads can be divided based on the two classification criteria, i.e. the loads that are controllable and the must-run loads [22]. The loads that could be paused, regulated, or shifted are considered as controllable loads, e.g. airconditioners and PHEVs. In contrast, must-run loads could not be adjusted; for instance, refrigerators are unable to react to price variations and thus are must-run loads. The minimum power demand of a user is decided by its must-run loads. Obviously, users' demand is limited by its rating capacity.

$$P_{D,j}^{\min} \le P_{D,j} \le P_{D,j}^{\max} \tag{6.6}$$

Therefore, the entire OEM problem for social welfare maximization is established as:

$$\begin{cases} \max\left(-\sum_{i=1}^{n_G} C_{s,i}(P_{G,i}) + \sum_{j=1}^{n_L} C_{u,i}(P_{D,j})\right) \\ \text{subject to } (6.1), (6.4) \text{ and } (6.6) \end{cases} \tag{6.7}$$

The Lagrangian function to solve (6.7) is denoted as:

$$\begin{aligned} L = &-\sum_{i=1}^{n_G} C_{s,i}(P_{G,i}) + \sum_{j=1}^{n_L} C_{u,j}(P_{D,j}) \\ &+ \lambda\left(\sum_{i=1}^{n_G} P_{G,i} - \sum_{j=1}^{n_L} P_{D,i} + \sum_{k=1}^{n_R} P_{R,k} - P_{\text{loss}}\right) \\ &+ \sum_{i=1}^{n_G} \overline{\mu}_{G,i}(P_{G,i}^{\max} + P_{G,i}) + \sum_{i=1}^{n_G} \underline{\mu}_{G,i}(P_{G,i} - P_{G,i}^{\max}) \\ &+ \sum_{j=1}^{n_L} \overline{\mu}_{D,j}(P_{D,j}^{\max} - P_{D,j}) + \sum_{j=1}^{n_L} \underline{\mu}_{D,j}(P_{D,j} - P_{D,j}^{\min}) \end{aligned} \tag{6.8}$$

where λ is the Lagrangian multiplier for constraint (6.1), $\overline{\mu}_{G,i}$, $\underline{\mu}_{G,i}$, and $\overline{\mu}_{D,j}$, $\underline{\mu}_{D,j}$ are the non-negative Lagrangian multipliers for constraints (6.4) and (6.6), respectively. The following conditions determine the optimal solution of (6.7):

$$\begin{cases} Ms_i(P_{G,i}) = \lambda(1-\gamma_i) - \overline{\mu}_{G,i} + \underline{\mu}_{G,i} \\ Mu_j(P_{D,j}) = \lambda(1+\gamma_j) + \overline{\mu}_{D,j} - \underline{\mu}_{D,j} \\ \sum_{i=1}^{n_G} P_{G,i} - \sum_{j=1}^{n_L} P_{D,j} + \sum_{k=1}^{n_R} P_{R,k} - P_{\text{loss}} = 0 \\ \overline{\mu}_{G,i}(P_{G,i}^{\max} - P_{G,i}) = 0, \underline{\mu}_{G,i}(P_{G,i} - P_{G,i}^{\min}) = 0 \\ \overline{\mu}_{D,j}(P_{D,j}^{\max} - P_{D,j}) = 0, \underline{\mu}_{D,j}(P_{D,j} - P_{D,j}^{\min}) = 0 \end{cases} \tag{6.9}$$

The following equation defined γ_i and γ_j in (6.9), where γ_i is the coefficient of generator i's power loss, while γ_j is the coefficient of generator j's power loss [23]:

$$\gamma_i = \frac{\partial P_{\text{loss}}}{\partial P_{G,i}}, \gamma_j = \frac{\partial P_{\text{loss}}}{\partial P_{D,j}} \tag{6.10}$$

The following equation defines $Ms_i(P_{G,i})$ and $Mu_j(P_{D,j})$, which represents supplier i and user j's marginal cost and utility:

$$Ms_i(P_{G,i}) = \frac{d(C_{s,i}(P_{G,i}))}{dP_{G,i}} = a_i P_{G,i} + b_i \tag{6.11}$$

$$Mu_i(P_{D,j}) = \frac{d(C_{u,j}(P_{D,j}))}{dP_{D,j}} = \alpha_j P_{D,j} + \omega_j \tag{6.12}$$

Equation (6.9) could be solved by many centralized approaches [24, 25]. The iterative way to acquire a solution of (6.9) by applying a gradient method could be given as follows:

$$(Ms_i(P_{G,i}) + \overline{\mu}_{G,i} - \underline{\mu}_{G,i})/(1-\gamma_i) = \lambda[t] \tag{6.13}$$

$$(Mu_j(P_{D,j}) - \overline{\mu}_{D,j} + \underline{\mu}_{D,j})/(1+\gamma_j) = \lambda[t] \tag{6.14}$$

$$\lambda[t+1] = \lambda[t] - \Delta P[t]\varepsilon \tag{6.15}$$

In (6.15), $\Delta P[t]$ is the supply–demand mismatch at the tth iteration. In fact, the Lagrangian function's gradient with respect to λ is $\Delta P[t]$, and ε is the stepsize. $\Delta P[t]$ could be calculated according to (6.8):

$$\Delta P[t] = \sum_{i=1}^{n_G} P_{G,i}[t] - \sum_{j=1}^{n_L} P_{D,j}[t] + \sum_{k=1}^{n_R} P_{R,k}[t] - P_{\text{loss}}[t] \tag{6.16}$$

The power loss could be calculated as follows, considering the active power loss resulted by the transmission of suppliers' generation and users' consummation:

$$P_{\text{loss}}[t] = \sum_{i=1}^{n_G} \gamma_i P_{G,i}[t] + \sum_{j=1}^{n_L} \gamma_j P_{D,j}[t] + \sum_{k=1}^{n_R} \gamma_k P_{R,k}[t] \tag{6.17}$$

Here, γ_k denotes the power loss coefficient of the renewable generator. Therefore, the supply–demand mismatch could be expressed as follows:

$$\Delta P[t] = \sum_{i=1}^{n_G}(1-\gamma_i)P_{G,i}[t] - \sum_{j=1}^{n_L}(1+\gamma_j)P_{D,j}[t] + \sum_{k=1}^{n_R}(1-\gamma_k)P_{R,k}[t] \tag{6.18}$$

The unique optimal solution of (6.7) is bound to a particular unique Lagrange multiplier λ^*, which is the so-called system λ [26]. This particular λ^* is be obtained when $\Delta P[t]$ approaches zero, namely, the supply and demand are balanced out.

6.2.2 Market-Based Self-interest Motivation Model

The self-interest motivated OEM problem is designed in a deregulated market. In this chapter, the suppliers refer to the conventional generators and the users refer to loads. The market equilibrium could be achieved through negotiation on prices between suppliers and users. Because the cost of the renewable generator is zero, it could be considered as selfless. However, as discussed above, renewable generators could still influence the final market-clearing price. The supplier i's predicted profit function could be represented as follows based on a given selling price r_i:

$$U_{s,i} = r_i P_{G,i}(1-\gamma_i) - C_{s,i}(P_{G,i}) \tag{6.19}$$

As provided in (6.3), $C_{s,i}(P_{G,i})$ is the generation cost function, and the power loss parameter γ_i is given in (6.10). Likewise, user j's anticipated utility function can be defined as:

$$U_{u,j} = C_{u,j}(P_{D,j}) - r_j P_{D,j}(1+\gamma_j) \tag{6.20}$$

Here, user j's bidding price is denoted as r_j. As given in (6.5), user j's utility function is $C_{u,j}(P_{D,j})$ and user j's power loss coefficient γ_j is defined in (6.10). Here, we assume that all suppliers are selfish. Thus, when marginal cost equals the selling price, supplier i will accomplish the adjustment of its generation, achieving the optimal strategy, on condition that the selling price r_i is given. The following equation presents the solution of maximizing the profit function:

$$\frac{dU_{s,i}}{dP_{G,i}} = r_i(1-\gamma_i) - Ms_i(P_{G,i}) = 0 \tag{6.21}$$

To maximize the profit, a specific user tends to purchase until the marginal utility reaches the bidding price:

$$\frac{dU_{u,j}}{dP_{D,j}} = Mu_j(P_{D,j}) - r_j(1+\gamma_j) = 0 \tag{6.22}$$

Here, $Ms_i(P_{G,i})$ represents the marginal utility of user i, which is defined in (6.12). By taking the constraints given in (6.4) and (6.6) into consideration, the suppliers and users will determine their generation and load according to (6.23) and (6.24), respectively:

$$P_{G,i} = \begin{cases} P_{G,i}^{\min} & \text{if } r_i(1-\gamma_i) < Ms_i^{\min} \\ P_{G,i}^{\max} & \text{if } r_i(1-\gamma_i) > Ms_i^{\max} \\ \frac{r_i(1-\gamma_i)-b_i}{a_i} & \text{otherwise} \end{cases} \tag{6.23}$$

$$P_{D,j} = \begin{cases} P_{D,j}^{\min} & \text{if } r_j(1+\gamma_j) < Mu_j^{\min} \\ P_{D,j}^{\max} & \text{if } r_j(1+\gamma_j) > Mu_j^{\max} \\ \frac{r_j(1+\gamma_j)-\omega_j}{\alpha_i} & \text{otherwise} \end{cases} \tag{6.24}$$

where $Ms_i^{\min} = Ms_i(P_{G,i}^{\min})$, $Ms_i^{\max} = Ms_i(P_{G,i}^{\max})$, and $Mu_j^{\min} = Mu_j(P_{D,j}^{\min})$, $Mu_j^{\max} = Mu_j(P_{D,j}^{\max})$.

The following equations show how the suppliers and users tend to change the selling prices or the bidding prices based on the supply–demand mismatch in the deregulated market.

$$\begin{cases} r_i[t+1] = r_i[t] - \Delta P[t]\varepsilon_i \\ r_j[t+1] = r_j[t] - \Delta P[t]\varepsilon_j \end{cases} \tag{6.25}$$

The supply–demand mismatch $\Delta P[t]$ here is calculated by using (6.16), and ε_i is the incentive factor that inspire producer i to improve its generation to achieve market equilibrium, whereas ε_j signifies user j's willingness in improving its bidding price to meet its demand.

Here, we considered that the market is comparatively stable and competitive, which indicates that selling and bidding price can meet at a point where the market is cleared. Thus, the equilibrium price is represented as follows:

$$r_i = r_j = r^*, \quad \text{for } i = 1, 2, \dots, n_G,\ j = 1, 2, \dots, n_L \tag{6.26}$$

The r^* in (6.26) is the frequently mentioned market-clearing price, which clears the market on condition where the supply–demand balance is achieved, as discussed in (6.1).

6.2.3 Relationship Between Two Models

By dropping the subscripts i, j in (6.25), the price updating formula can be written in a simpler form:

$$r[t+1] = r[t] - \varepsilon\Delta P[t] \tag{6.27}$$

By replacing r with λ, one can easily find that (6.27) is precisely the same as (6.15), which indicates that updating the rule of the market price is the same as that of the system λ. Such a situation demonstrates that both of them converge when supply and demand reach a balance.

For the social welfare maximization model, one of the conditions for convergence is that $\lambda[t+1] = \lambda[t] = \lambda^*$. The supplier's generation and the user's demand under this system λ can be determined by the following equations according to (6.13) and (6.14):

$$Ms_i(P_{G,i}) + \overline{\mu}_{G,i} - \underline{\mu}_{G,i} = \lambda^*(1-\gamma_i) \tag{6.28}$$

$$Mu_j(P_{D,j}) - \overline{\mu}_{D,i} + \underline{\mu}_{D,j} = \lambda^*(1+\gamma_j) \tag{6.29}$$

The $\overline{\mu}_{G,i}$, $\underline{\mu}_{G,i}$ and $\overline{\mu}_{D,j}$, $\underline{\mu}_{D,j}$ are Lagrangian multipliers for constraints (6.4) and (6.6), respectively. These multipliers are set to zero when the generator or load is operating within its bounds. For instance, $P_{G,i}$ can be calculated as $P_{G,i} = \frac{\lambda^*(1-\gamma_i)-b_i}{a_i}$ using (6.11) on condition that the generator is operated within its bounds. If the generator reaches its maximum limit, then $P_{G,i} = P_{G,i}^{\max}$. Moreover, $\overline{\mu}_{G,i} > 0$, $\underline{\mu}_{G,i} = 0$, and $Ms_i^{\max} + \overline{\mu}_{G,i} = \lambda^*(1-\gamma_i)$ in this scenario, which yields $\lambda^*(1-\gamma_i) > Ms_i^{\max}$. When a generator hits its lower bound, then $P_{G,i} = P_{G,i}^{\min}$. Under this circumstance, $\overline{\mu}_{G,i} = 0$, $\underline{\mu}_{G,i}$ and $Ms_i^{\min} - \underline{\mu}_{G,i} = \lambda^*(1-\gamma_i)$, accordingly $\lambda^*(1-\gamma_i) < Ms_i^{\min}$. Therefore, the generation of the generator could be obtained by using the following equations:

$$P_{G,i} = \begin{cases} P_{G,i}^{\min} & \text{if } \lambda^*(1-\gamma_i) < Ms_i^{\min} \\ P_{G,i}^{\max} & \text{if } \lambda^*(1-\gamma_i) > Ms_i^{\max} \\ \frac{\lambda^*(1-\gamma_i)-b_i}{a_i} & \text{otherwise} \end{cases} \tag{6.30}$$

The demand of the user could be calculated according to the similar derivation:

$$P_{D,j} = \begin{cases} P_{D,j}^{\min} & \text{if } \lambda^*(1+\gamma_j) < Mu_j^{\min} \\ P_{D,j}^{\max} & \text{if } \lambda^*(1+\gamma_j) > Mu_j^{\max} \\ \frac{\lambda^*(1+\gamma_j)-\omega_j}{\alpha_i}, & \text{otherwise} \end{cases} \tag{6.31}$$

When the algorithm converges, $\Delta P = 0$, that leads to:

$$\begin{aligned} \Delta P(\lambda_*) = & \sum_{i=1}^{n_G}(1-\gamma_i)P_{G,i}(\lambda^*) - \sum_{j=1}^{n_L}(1+\gamma_j)P_{D,j}(\lambda^*) \\ & + \sum_{k=1}^{n_R}(1-\gamma_k)P_{R,k} \end{aligned} \tag{6.32}$$

Substituting (6.30) and (6.31) into (6.32), we can have:

$$\Delta P(\lambda_*) = \sum_{i\in B_G}(1-\gamma_i)P^B_{G,i} + a_s\lambda^* - b_s - \left(\sum_{j\in B_D}(1+\gamma_j)P^B_{D,j} + \alpha_D\lambda^* - \omega_D\right) + \sum_{k=1}^{n_R}(1-\gamma_k)P_{R,k} = 0 \tag{6.33}$$

where B_G and B_D represent generators that reach their limitations and users that touch their demand bounds, respectively. $a_s = \sum_{i\notin B_G}\frac{(1-\gamma_i)^2}{a_i}$, $b_s = \sum_{i\notin B_G}\frac{b_i(1-\gamma_i)}{a_i}$, $\alpha_D = \sum_{i\notin B_D}\frac{(1+\gamma_j)^2}{\alpha_j}$ and $\omega_D = \sum_{i\notin B_D}\frac{\omega_j(1+\gamma_j)^2}{\alpha_j}$. Therefore, the converged system λ^* can be calculated as follows:

$$\lambda^* = \frac{1}{a_s - \alpha_D}\left(b_s - \omega_D - \sum_{i\in B_G}(1-\gamma_i)P^B_{G,i} + \sum_{j\in B_D}(1+\gamma_j)P^B_{D,j} - \sum_{k=1}^{n_R}(1-\gamma_k)P_{R,k}\right) \tag{6.34}$$

As for the self-interest motivation model, the equilibrium price satisfies $r_i = r_j = r^*$ when the market is cleared. Consequently, based on the equilibrium price, suppliers' generation and users' demand can be calculated based on (6.23) and (6.24), respectively. The equations given in (6.23) and (6.24) are the same as (6.30) and (6.31) provided that r in (6.23) and (6.24) is replaced with λ^*. The price that balances supply and demand is the market-clearing price. Consequently, (6.33) still holds if one replaces λ^* with r^*. Then:

$$\lambda^* = r^* = \frac{1}{a_s - \alpha_D}\left(b_s - \omega_D - \sum_{i\in B_G}(1-\gamma_i)P^B_{G,i} + \sum_{j\in B_D}(1+\gamma_j)P^B_{D,j} - \sum_{k=1}^{n_R}(1-\gamma_k)P_{R,k}\right) \tag{6.35}$$

Here, we demonstrate that the converged Lagrangian multiplier for the social welfare maximization model λ^* is, in fact, the market-clearing price for self-interest model r^*. It proves that these two models yield the same solutions. At the equilibrium of the market, one can acquire the maximum of the social welfare, and this phenomenon has been discussed in the existing literature [20].

Solving the self-interest motivation model actually decomposes the original social welfare maximization problem into multiple local optimization problems. Thereby, it can accomplish the task of solving the OEM problem in a distributed manner. The following section will introduce the details of distributively solving the OEM for the optimal solution.

6.3 Fully Distributed MAS-Based OEM Solution

Several problems need to be addressed for the purpose of solving the self-interest motivation model-based OEM problem for a distributed solution. First, suppliers and users require to figure out the supply–demand mismatch (ΔP) in order to update the prices (selling or bidding). In the centralized market environment, the system operators are in charge of providing the information regarding the supply–demand mismatch ΔP, with an incentive factor ε being also provided by them. However, for a distributed scheme, this information is expected to be acquired in the distributed approach. In addition, for all the system participants, the selling and bidding prices defined in (6.25) are expected to be adjusted in a distributed manner for the purpose of achieving a balance of the market. Accordingly, distributed algorithms are proposed in this section for solving the OEM problem to yield a fully distributed solution.

6.3.1 Distributed Price Updating Algorithm

As discussed above, the process of obtaining the optimal generation/demand sets can be viewed as a process of discovering the equilibrium for market clearing. As the designed distributed OEM method, each unit, for instance, a supplier or a user, has been allocated an EMA that takes charge of the price negotiation. Then, each participant communicates with its surrounding EMAs within its communication range to modify its settings regarding the price and generation/load during the process of negotiation. The ith EMA resets its selling or bidding price according to the following formula:

$$r_i[t+1] = \sum_{j=1}^{n} d_{ij} r_j[t] - \varepsilon_i \Delta P[t] \tag{6.36}$$

Here, $r_i[t]$ denotes the price of EMA i at step t, while $r_i[t+1]$ denotes the price for coming step. d_{ij} represents the weights related to agents i and j. n refers to the amounts of EMAs. Similar to (6.25), ε_i denotes incentive factor. $\Delta P[t]$ is the supply–demand mismatch with prices sta quo, whereas the $\varepsilon_i \Delta P[t]$ is defined as the power mismatch incentive. Here, we use i or j to denote the index of an EMA; it can represent either a supplier or a user. As shown in (6.36), the formula for price update includes two terms. The first term is used to estimate the market-clearing price. In general, for a market with multiple sellers and users, the market-clearing price always moves toward the mean price of all sellers and buyers. The second term is used to drive the market toward the supply–demand balance.

In order to guarantee the information exchange among EMAs, the weights d_{ij}s need to be set appropriately. In this chapter, we calculate d_{ij}s by applying the mean metropolis algorithm in [27] as it is thoroughly distributed and adaptive to alterations in the communication network topology, and the weights are determined as follows:

$$d_{ij} = \begin{cases} \frac{2}{n_i+n_j}, & j \in N_i \\ 1 - \sum_{j\in N_i} \frac{2}{n_i+n_j}, & i = j \\ 0, & \text{otherwise} \end{cases} \tag{6.37}$$

Here, n_i refers to the amounts of EMAs that communicate with EMA i, and n_j is defined similarly. N_i denotes the indices of EMAs that can communicate with EMA i for information exchange. According to the proposed method, for each agent, it can only exchange information with the agents, which are permitted to exchange information with, and these agents are also referred to as the neighboring agents of that agent. If two agents are not neighbored, they will not exchange information. For example, the weight d_{ij} that denotes the weight between agent i and j equals to zero implies that there is no communication between agents i and j.

For convenience, we use the matrix form to represent the entire price updating process in (6.36).

$$\mathbf{R}[t+1] = \mathbf{D} \cdot \mathbf{R}[t] - \Delta P[t] \cdot \mathbf{E} \tag{6.38}$$

The weight matrix with the elements being d_{ij}s is denoted by $\mathbf{D}$, while $\mathbf{R}[t] = [r_1[t], r_2[t], \dots, r_n[t]]^T$ and $\mathbf{E} = [\varepsilon_1, \varepsilon_2, \dots, \varepsilon_n]^T$. Letting t approach infinity will result in $r_i[t+1] = r_i[t] = r^*$, which is the equilibrium of the system represented by (6.38). Moreover, adding both sides of (6.38) together yields:

$$\sum_{i=1}^{n} r_i^* = \sum_{i=1}^{n}\sum_{j=1}^{n} d_{ij} r_j^* - \Delta P[\infty] \sum_{i=1}^{n} \varepsilon_i \tag{6.39}$$

where $r_i^* = r_j^* = r^*$ and $\Delta P[t] = 0$ when the market is cleared. Moreover, from (6.37) we can have:

$$\sum_{j=1}^{n} d_{ij} = 1 \tag{6.40}$$

Additionally, from (6.39), we can also have $\sum_{i=1}^{n} r_i^* = \sum_{j=1}^{n} r_j^* = \sum_{j=1}^{n}(\sum_{i=1}^{n} d_{ij}) r_j^*$, which yields:

$$\sum_{i=1}^{n} d_{ij} = 1 \tag{6.41}$$

From (6.40) and (6.41), we can conclude that 1 is both the left and right eigenvalue of the weight matrix $\mathbf{D}$. Note that both (6.40) and (6.41) are satisfied when the matrix **D** is set in the way presented by (6.35). Moreover, for the purpose of simplifying the design of the matrix **D**, we also set it in a symmetric way such that:

$$\begin{cases} d_{ij} = d_{ji}, \quad \text{for all } i, j \\ \sum_{j=1}^{n} d_{ij} = \sum_{i=1}^{n} d_{ji} = 1 \end{cases} \tag{6.42}$$

Because **D** is symmetric, the summation of all rows or columns equals to 1, hence, $\sum_{j=1}^{n} d_{ji} = 1$. Accordingly, we can have

$$\sum_{i=1}^{n}\sum_{j=1}^{n} d_{ij} r_j^* = \sum_{i=1}^{n}\sum_{j=1}^{n} d_{ji} r_i^* = \sum_{i=1}^{n}\left(r_i^* \sum_{j=1}^{n} d_{ij}\right) = \sum_{i=1}^{n} r_i^* \tag{6.43}$$

Since $\varepsilon_i \neq 0$, therefore

$$\Delta P[\infty] = 0 \tag{6.44}$$

Here, we show that the converged market price can ensure the balance between supply and demand after the algorithm converges, as shown in (6.44).

With $r_i[t+1] = r_i[t]$, substituting (6.44) into (6.38) yields:

$$\mathbf{R}^* = \mathbf{D}\mathbf{R}^* \tag{6.45}$$

where $\mathbf{R}^* = [r_1^*, r_2^*, \dots, r_n^*]^T$. According to [27], the solution to (6.45) has the following form:

$$\mathbf{R}^* = r^* \cdot \mathbf{1}, \text{with } \mathbf{1} = [1, 1, \dots, 1]^T \tag{6.46}$$

Thus, the common market-clearing price could be acquired because all providers and users will reach an agreement on this particular price, i.e.

$$r_i[\infty] = r_j[\infty] = r^* \tag{6.47}$$

As a consequence, when both (6.44) and (6.47) hold, the price negotiation process also reaches the overall market equilibrium. Apparently, the designed market price update method presented in (6.36) is not similar to the conventional scheme given in (6.25). Nevertheless, both these two mechanisms can reach the unique market equilibrium by driving the suppliers and users via price negotiation.

6.3.2 Distributed Supply–Demand Mismatch Discovery Algorithm

As shown in (6.36), in order to yield a thoroughly distributed price update algorithm, the global supply–demand mismatch (ΔP) requires to be gained via a distributed manner. The calculation of ($\Delta P[t]$) requires the information of both suppliers and users, as shown in (6.16). We propose to discover $\Delta P[t]$ in a distributed manner by applying the following iterative formula:

$$P_i^{k+1}[t] = \sum_{j=1}^{n} d_{ij} P_j^k[t] \tag{6.48}$$

The formula presented in (6.48) is the update rule of average-consensus algorithm, and one can refer to [28] for more details. According to this algorithm, all P_i^ks in Eq. (6.48) will converge to the same value as follows:

$$P_i^{\infty}[t] = \frac{1}{n}\sum_{i=1}^{n} P_i^0[t] \text{ for } i = 1, 2, \dots, n \tag{6.49}$$

For EMA i, $P_i^0[t]$ is initialized as:

$$P_i^0[t] = n \times [(1-\gamma_i)(P_{G,i}[t] + P_{R,i}[t]) - (1+\gamma_i)P_{D,i}[t]] \tag{6.50}$$

$P_{R,i}[t]$ and $P_{D,i}[t]$ in (6.50) are set to zero if an EMA corresponds to a traditional generator. Similarly, if an EMA relates to a load, $P_{G,i}[t]$ and $P_{R,i}[t]$ are set to zero and the RG EMAs can set $P_{G,i}$ and $P_{D,i}$ in a similar manner.

Such rules are also adoptive to the renewable generator EMAs and loads. In terms of Eqs. (6.48) and (6.49), we can obtain the following equations after the algorithm converges:

$$\begin{aligned} P_i^{\infty}[t] &= \frac{1}{n}\sum_{i=1}^{n} n((1-\gamma_i)(P_{G,i}[t] + P_{R,i}[t]) - (1+\gamma_i)P_{D,i}[t]) \\ &= \Delta P[t] \end{aligned} \tag{6.51}$$

It can be shown that the algorithm used here can obtain information regarding the global supply–demand mismatch in a distributed manner and thus overcomes the difficulties of developing a distributed solution method for OEM. It should be pointed out that the convergence speed of the average-consensus algorithm can be investigated by checking the characteristics of the matrix **D**. As discussed in [29] and [27], the maximum number of iterations required for convergence K, can be determined as:

$$K = \frac{-1}{\log_E\left(\frac{1}{|\lambda_2|}\right)} \tag{6.52}$$

Here, λ_2 is the second largest eigenvalue of **D** with the error tolerance being denoted with E.

6.3.3 Implementation of MAS-Based OEM Solution

The implementation of the proposed OEM solution is shown in Figure 6.1. We assign an EMA to each bus for OEM. Meanwhile, we design the communication

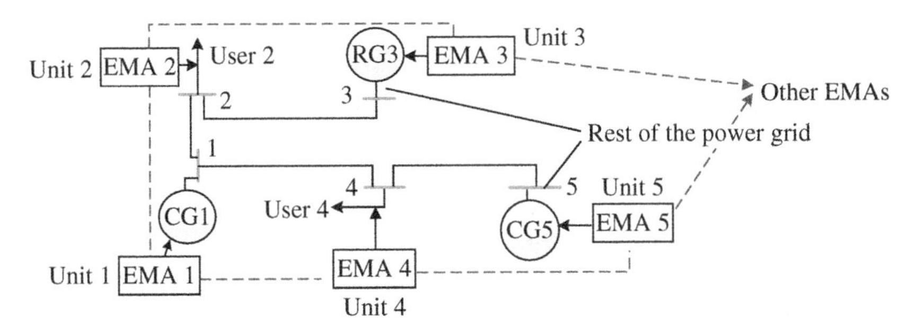

Figure 6.1 MAS-based implementation of the distributed OEM solution.

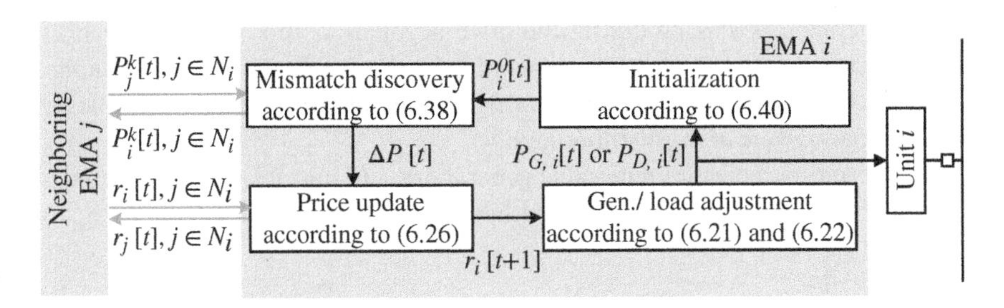

Figure 6.2 Block diagram of the energy management agent.

network topology of EMAs in accordance with the power system's topology such that each agent can communicate with each other if they have an electrical connection. Such a design enables us to use power line communication techniques [30]. However, the communication network can be designed differently with different factors such as cost, convenience, etc., being taken into consideration.

In order to accomplish the task of OEM, each EMA is designed to have two functions, i.e. price update and supply–demand mismatch discovery. Figure 6.2 shows the implementation of an EMA during the price negotiation.

The following three steps show the overall process for distributed OEM.

1. Initialize $P_i^0[t]$ and calculate $(\Delta P[t])$ in terms of (6.50) and (6.48);
2. Based on discovered $\Delta P[t]$, modify selling or bidding price r_i in terms of (6.36);
3. Regulate generation or load in terms of (6.23) and (6.24), separately.

These three steps are operated consecutively until the market equilibrium is obtained. Because the renewable generator keeps outputting its maximum available power, it is unnecessary for the EMA of a renewable generator to adjust the generation. However, the variation in the renewable generation leads to the change in the traditional generation. Thus, it results in a variation in the final market price.

6.4 Simulation Studies

This section will first test the proposed solution with a 6-bus system, and then, the IEEE 30-bus system will also be tested to verify the solution's performance.

6.4.1 Tests with a 6-bus System

As shown in Figure 6.3, the 6-bus system include three loads (L1–L3), two traditional generators (CG4 and CG5), and one renewable generator (RG). CG4 and CG5 could offer essential voltage/reactive power support for the system because both of them are configured with a self-driven voltage regulation system. The supply–demand mismatch during the price negotiation and other disturbances can be compensated since CG4 is operating at the frequency regulation mode. Table 6.1 lists the parameters of the traditional generators and loads, and Table 6.2 lists the parameters of the distribution lines.

Two scenarios, constant renewable generations and unstable renewable generation, are tested during simulations. Although the first scenario is not realistic, its

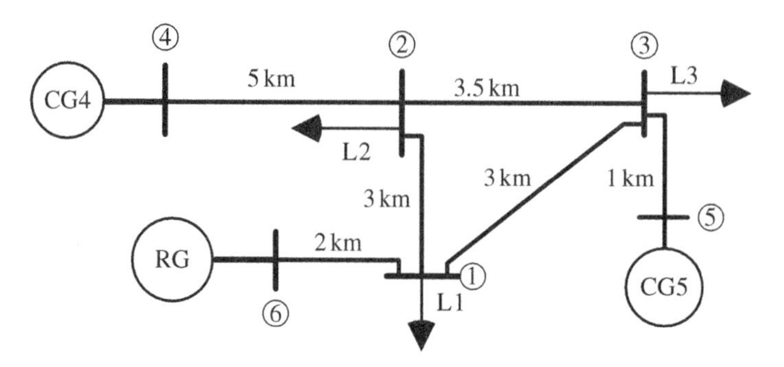

Figure 6.3 The 6-bus test system.

Table 6.1 Parameters of the five units.

Supplier	a_i	b_i	c_i	$P_{G,i}^{min}$ (MW)	$P_{G,i}^{max}$ (MW)
G4	0.0016	4.26	40	180	250
G5	0.0017	4.54	60	150	350
User	α_j	ω_j		$P_{D,j}^{min}$ (MW)	$P_{D,j}^{max}$ (MW)
L1	−0.065	15.86		100	160
L2	−0.061	17.45		150	200
L3	−0.056	19.35		180	250

Table 6.2 The line parameters of the 6-bus system.

From	To	Resistance (Ω)	Reactance (Ω)	Line length (km)
2	4	0.0636	1.9320	5.0000
2	3	0.0446	1.3524	3.5000
1	6	0.0255	0.7728	2.0000
1	2	0.0382	1.1592	3.0000
1	3	0.0382	1.1592	3.0000
3	5	0.0127	0.3864	1.0000

Table 6.3 The weights settings for agent communication.

Agent	Neighboring agents	n_i	n_j	$d_{ij}s$
1	2,3,6	3	3,2,1	$d_{12} = \frac{2}{6}, d_{13} = \frac{2}{5}, d_{16} = \frac{2}{4}, d_{11} = -\frac{7}{30}$
2	1,3,4	3	3,2,1	$d_{21} = \frac{2}{6}, d_{23} = \frac{2}{5}, d_{24} = \frac{2}{4}, d_{22} = -\frac{7}{30}$
3	2,5	2	3,1	$d_{32} = \frac{2}{5}, d_{35} = \frac{2}{3}, d_{33} = -\frac{1}{15}$
4	2	1	3	$d_{42} = \frac{2}{4}, d_{44} = \frac{1}{2}$
5	3	1	2	$d_{53} = \frac{2}{3}, d_{55} = \frac{1}{3}$
6	1	1	3	$d_{61} = \frac{2}{4}, d_{66} = \frac{1}{2}$

simulation results are useful for illustrating the proposed solution because of its simplicity. The simulation of the latter scenario enables one to thoroughly comprehend how the designed method reacts under the change in operating condition.

The average-consensus algorithm is applied for discovering the supply–demand mismatch $\Delta P[t]$. In addition, here, we design the topology of the communication network the same as the topology of the electrical network of the 6-bus system, as shown in Figure 6.3. According to (6.37), the weights $d_{ij}s$ can be computed and the results are shown in Table 6.3. Figure 6.4 demonstrates an example of a supply–demand mismatch discovery process for this 6-bus system.

The generations of CG4, CG5, and RG are set to be 200, 220, and 150 MW, separately. The values of load are initialized with 160, 180, and 230 MW, separately. The initial values for CG4, CG5, RG, L1, L2, and L3 are 1200, 1320, 900, −960, 1080, and −1380 for $\Delta P[t]$ discovery on condition that all $\gamma_i s$ in (6.50) are set to zero. The obtained supply–demand mismatch is $(1200 + 1320 + 900 - 960 - 1080 - 1380)/6 = 0$ after the algorithm is converged. As can be seen from Figure 6.4, the algorithm can converge within 40

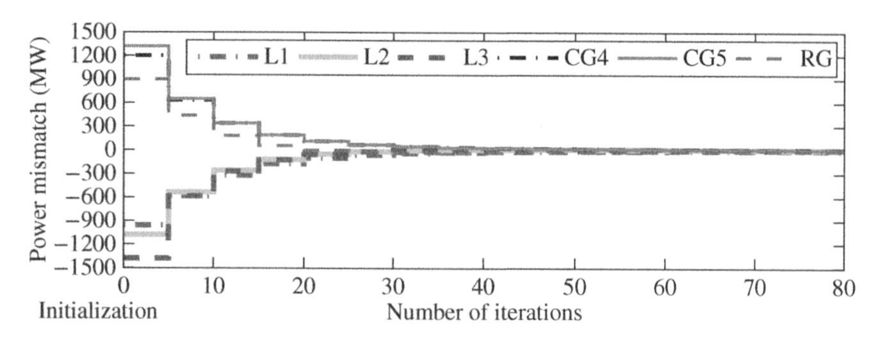

Figure 6.4 Supply–demand mismatch discovery for 6-bus system.

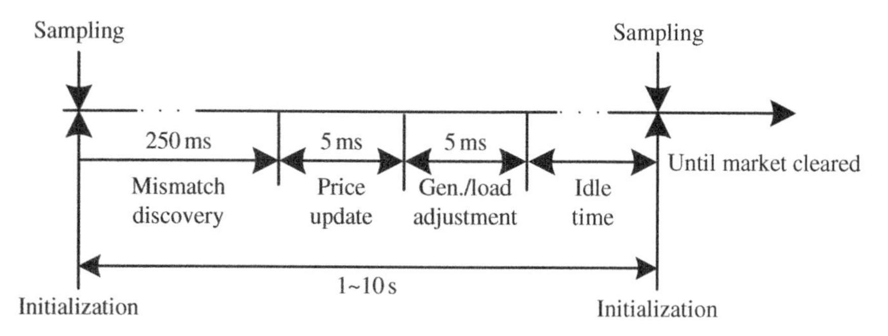

Figure 6.5 Time frame of implementation for the proposed OEM solution.

iterations. With our designed MAS based on Java Agent Development (JADE) framework, each iteration only requires around 6 ms [19]. Therefore, the total time cost is about 250 ms for finishing the ΔP discovery process.

Figure 6.5 presents the time frame of the OEM solution's operation. It should be noted that the price update interval should be chosen by considering the parameters of power grids such as inertia of generators, the response time of users, and etc., which is typically set 1–10 seconds.

6.4.1.1 Test Under the Constant Renewable Generation

In this test, the initial load demands of L1, L2, and L3 are set to be 160, 180, and 230 MW, respectively, while the initial generations of CG4 and CG5 are set to be 220 and 206 MW, separately. The output of RG is assumed to be a constant, i.e. 150 MW. The network power loss is 6 MW for this generation/load condition; thus, the initial supply and demand are balanced for this initial condition. The power loss coefficients (defined in (6.10)) for L1, L2, L3, CG4, CG5, and RG are 0.017, 0.022, 0.013, −0.025, 0.016, and 0.010, respectively.

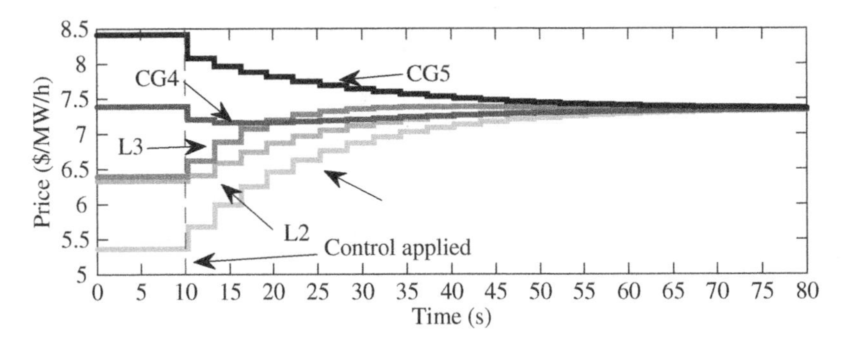

Figure 6.6 Updates of prices during negotiation.

The designed solution is activated at $t = 10$ seconds. The price update interval for our test is set to three seconds. The process of updating prices via distributed negotiation is presented in Figure 6.6. As can be seen from the figure, the generators (CG4 and CG5) hold comparatively higher margin costs before the deployment of the OEM solution. On the other hand, the loads (L1–L3) hold relatively lower margin utilities under such a case. The deployed solution drives suppliers and users to reach a new market equilibrium by dynamically regulating their selling or bidding prices. Figure 6.6 shows that the process merely takes approximately 60 seconds (20 iterations) to achieve the equilibrium. Figure 6.7 presents the generation/demand profiles during this process.

Because RG at bus #6 always outputs its maximum available power (150 MW), it does not react to the market price during the negotiation process. The reference and actual demand of #L1 gauged at bus #1 during dynamic optimization is provided in Figure 6.8a, while the reference and the actual output of CG#5 are shown in Figure 6.8b. Note that within each update interval (three seconds), both generator and load can track their references generated by corresponding EMAs. It is

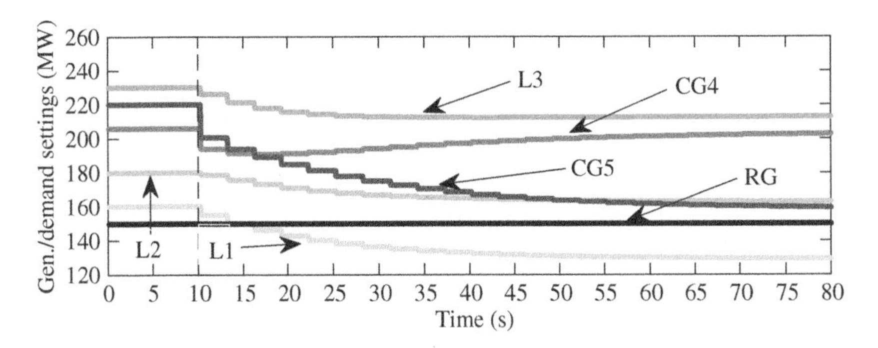

Figure 6.7 Generation/demand adjustment in response to price update.

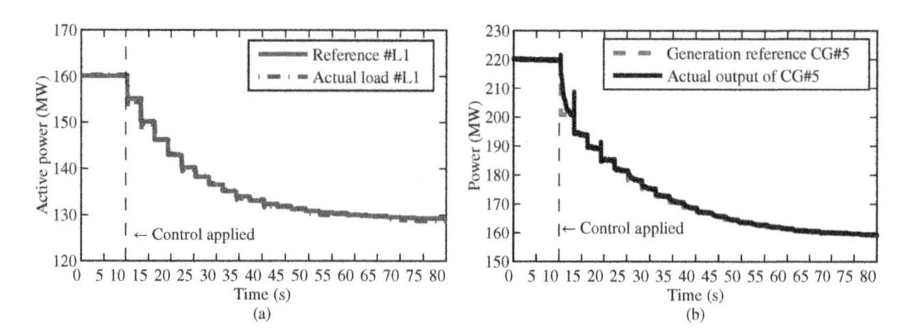

Figure 6.8 Reference tracking for load#1 and generator#5. (a) The profiles of load#1. (b) Power output of CG#5.

worthy to point out that we set the price update interval to a small value of three seconds since the ΔP discovery can be accomplished within 0.25 seconds. However, excessively frequent updates are difficult for load or generator to track and may even result in the instability of the system. Therefore, the price update interval for deploying the proposed solution should be carefully chosen by taking into the response speed of the system.

The change of benefits of a user (L1) and a provider (CG#5) during OEM are shown in Figure 6.9a,b, respectively. As shown in the figures, during OEM, both the supplier's and user's profits are increased. Figure 6.10 demonstrates that global social welfare is maximized when all users and providers achieve their maximum benefits. Such an interesting phenomenon verifies the previous analysis that the maximum of social welfare can be achieved when the equilibrium is reached.

The system's frequency and voltage responses during OEM are presented in Figure 6.11a,b. As a consequence of the generation/demand adjustments, slight frequency and voltage oscillations could be observed. The system frequency will be stabilized at 60 Hz and the supply–demand mismatch ΔP approaches zero after the algorithm converges because the active power redistribution has no significant impact on the voltage. The change in bus voltage is relatively small during

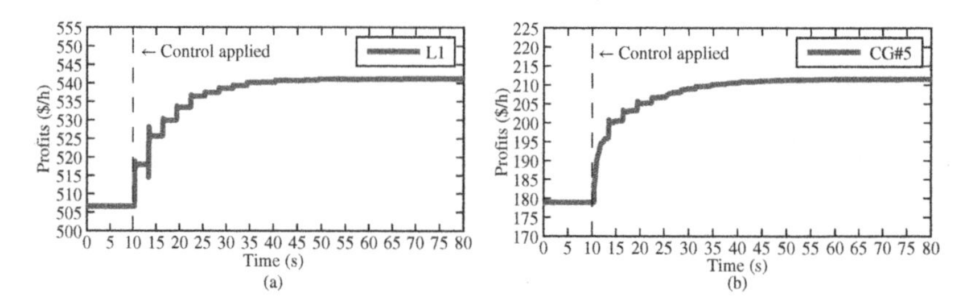

Figure 6.9 Trajectory of profits of load#1 and generator#5. (a) The change of profits of load#1. (b) The change of profits of CG#5.

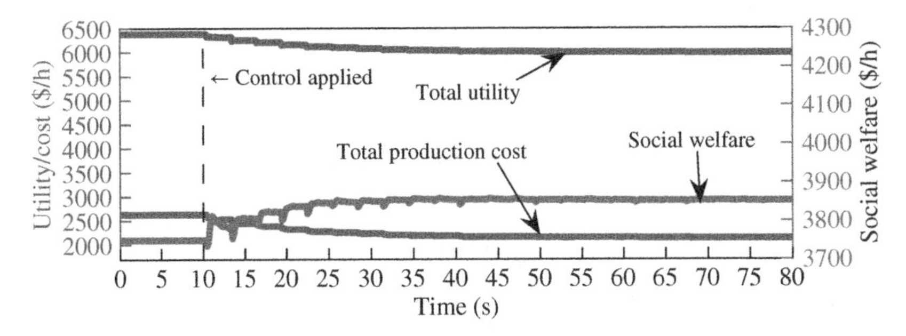

Figure 6.10 Evolution of social welfare during OEM.

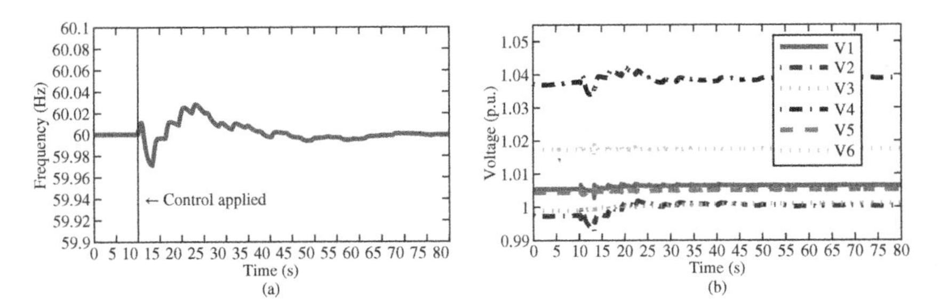

Figure 6.11 Frequency and voltage response during OEM. (a) Frequency response during OEM. (b) Voltage profiles of buses during OEM.

the generation/load regulation, as shown in Figure 6.11b. As can be seen from Figures 6.8–6.11, the stable and optimal operation of the power grid can be secured by deploying the proposed OEM solution.

6.4.1.2 Test Under Variable Renewable Generation

The performance of the proposed OEM solution is further tested via simulating variable renewable generation. Initially, the system is operating at the optimal point acquired after deploying the OEM solution of the previous test, wherein the market-clearing price is 7.3416 $/MW/h. In this test case, the generation of RG is assumed to be increased by 50 MW at t = 10 seconds, reaching 200 MW. The price updates under this circumstance are shown in Figure 6.12, while the corresponding generation and load settings during this process are provided in Figure 6.13. As discussed before, the increase in renewable generation will cause a decrease in the traditional generation. Thus, the overall market price since the cost of renewable generation is zero. In addition, because of the decrease in price, the flexible load demand (L1–L3) will correspondingly increase. Consequently, social welfare will increase in this case because the decrease in average generation cost results in an increase in renewable generation (Figure 6.14).

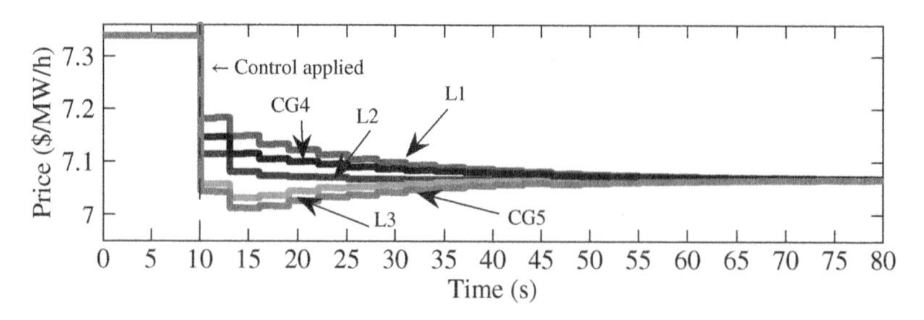

Figure 6.12 Updates of prices under variable renewable generation.

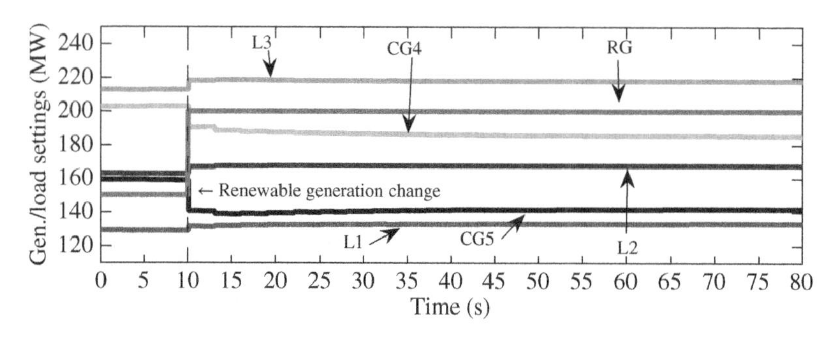

Figure 6.13 Generation/load adjustment under renewable generation change.

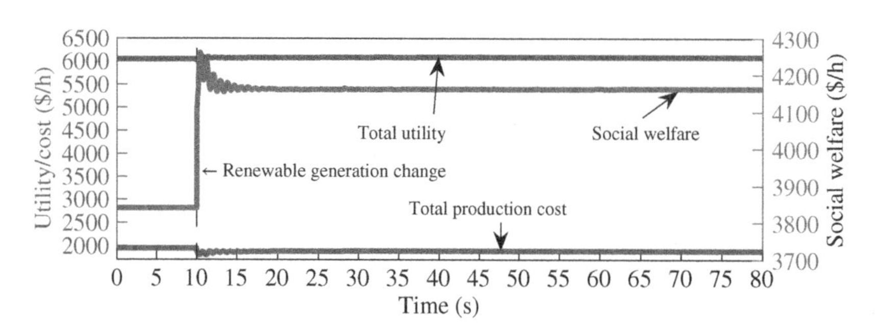

Figure 6.14 Evolution of social welfare with renewable generation change.

6.4.2 Test with IEEE 30-bus System

The designed solution is also implemented on the IEEE 30-bus system to evaluate its performance with larger scale systems. The configuration of the system and network parameters can be found in [31]. There is a total of 30 buses and 6 generators in this IEEE 30-bus system. Here, we select six loads to participate in the DR. Table 6.4 presents the parameters for generators' outcome cost and users' utility functions.

Table 6.4 Parameters for generators and users.

Generator #	Bus #	a_i	b_i	c_i	$P_{G,i}^{min}$	$P_{G,i}^{max}$	P_G^0
1	1	0.0075	2.0	0	50	200	171.27
2	2	0.0350	1.75	0	20	80	53.75
3	5	0.1250	1.00	0	15	50	25.63
4	8	0.0167	3.25	0	10	35	23.87
5	11	0.0500	3.00	0	10	30	19.09
6	13	0.0500	3.00	0	12	40	23.32
User #	Bus #	α_j	ω_j	—	$P_{D,j}^{min}$	$P_{D,j}^{max}$	$P_{D,j}^0$
1	4	−0.0550	5.7	—	22.00	44.00	44.00
2	7	−0.0595	6.20	—	23.51	47.02	47.02
3	10	−0.0487	5.10	—	20.00	40.00	40.00
4	12	−0.0800	6.15	—	20.00	40.00	40.00
5	15	−0.1370	7.25	—	16.50	33.00	33.00
6	30	−0.0912	5.25	—	12.50	25.00	25.00

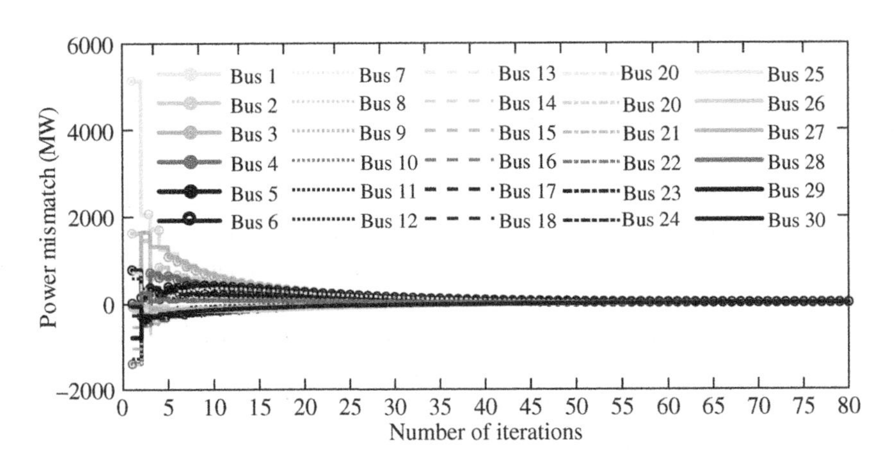

Figure 6.15 Supply–demand mismatch discovery for IEEE 30-bus system.

The designed solution is configured at $t = 10$ seconds. The price update interval is deployed at three seconds. Figure 6.15 shows one round of supply–demand mismatch discovery process with 30 bus (corresponding 30 agents). In this case, it takes only 50 iterations for the algorithm to converge in the process of discovering

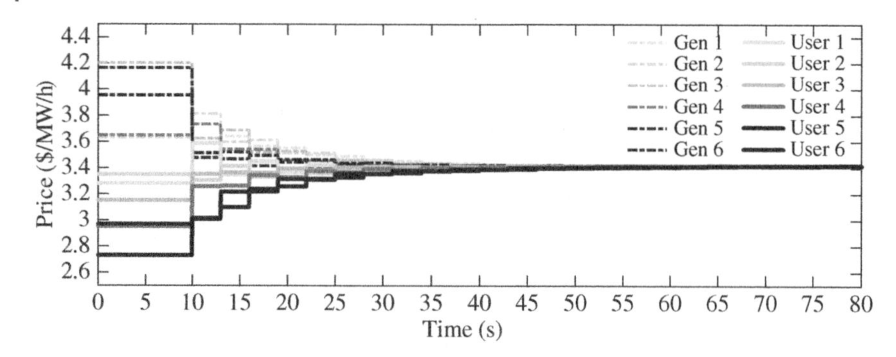

Figure 6.16 Supply-demand mismatch discovery for IEEE 30-bus system.

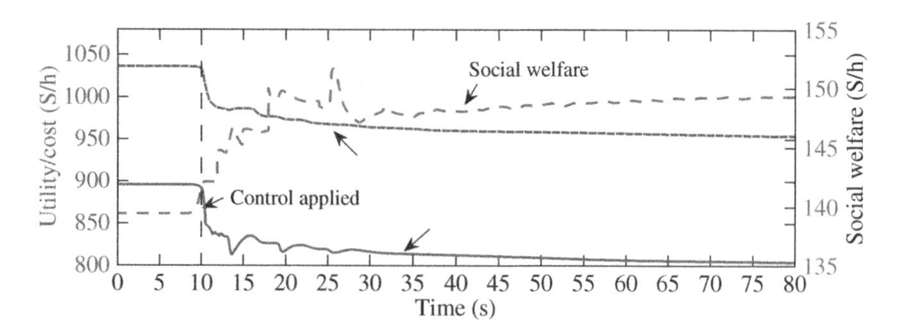

Figure 6.17 Evolution of social welfare during OEM with IEEE 30-bus system.

the supply–demand mismatch, with 30 agents being involved. Apparently, although the number of agents is increased, the algorithm's convergence speed does not increase exponentially.

The price update during the optimization is presented in Figure 6.16. During the process, the profiles of the corresponding social welfare are shown in Figure 6.17. At $t = 80$ seconds (about 23 iterations), the new market equilibrium will be achieved with the entire social welfare being maximized.

The distributed OEM solutions for these two test systems are designed by applying the market-based self-interest motivation model, which can obtain the optimal solution by maximizing the system participants' benefits. Notice that the centralized social welfare maximization model can obtain the same optimal solutions as well. However, it is difficult to implement them in a distributed manner as they require centralized algorithms such as the lambda iteration method introduced in [2].

6.5 Conclusion

A fully distributed OEM solution for smart grids was presented in this chapter. Both the economic dispatch of generators and DR of users can be considered by using the proposed solution, which can maximize social welfare effectively and support the online optimization as well. Deployed using MAS, the proposed solution enables better consumer participation and efficient load curve shaping with fast response. For the proposed solution, the utilization of a formidable central controller handling a massive amount of data or a complex communication network is avoided. Case studies verify the effectiveness of the proposed solution and show that it is promising for the future optimal energy management for smart grids.

References

1 Ipakchi, A. and Albuyeh, F. (2009). Grid of the future. *IEEE Power and Energy Magazine* 7 (2): 52-62. https://doi.org/10.1109/MPE.2008.931384.

2 Chowdhury, B.H. and Rahman, S. (1990). A review of recent advances in economic dispatch. *IEEE Transactions on Power Systems* 5 (4): 1248-1259. https://doi.org/10.1109/59.99376.

3 Medina, J., Muller, N., and Roytelman, I. (2010). Demand response and distribution grid operations: opportunities and challenges. *IEEE Transactions on Smart Grid* 1 (2): 193-198. https://doi.org/10.1109/TSG.2010.2050156.

4 Logenthiran, T., Srinivasan, D., and Shun, T.Z. (2012). Demand side management in smart grid using heuristic optimization. *IEEE Transactions on Smart Grid* 3 (3): 1244-1252. https://doi.org/10.1109/TSG.2012.2195686.

5 Zhang, W., Zhou, S., and Lu, Y. (2012). Distributed intelligent load management and control system. *2012 IEEE Power and Energy Society General Meeting*, pp. 1-8. https://doi.org/10.1109/PESGM.2012.6345457.

6 Chen, L., Li, N., Low, S.H., Doyle, J.C. (2010). Two market models for demand response in power networks. *2010 First IEEE International Conference on Smart Grid Communications*, pp. 397-402. https://doi.org/10.1109/SMARTGRID.2010.5622076.

7 Wood, A.J., Wollenberg, B.F., Sheblé, G.B. (2013). *Power Generation, Operation and Control*. 3rd edition. Wiley-Interscience.

8 Lin, C.E., Chen, S.T., and Huang, C.L. (1992). A direct Newton-Raphson economic dispatch. *IEEE Transactions on Power Systems* 7 (3): 1149-1154.

9 Mohsenian-Rad, A. and Leon-Garcia, A. (2010). Optimal residential load control with price prediction in real-time electricity pricing

environments. *IEEE Transactions on Smart Grid* 1 (2): 120-133. https://doi.org/10.1109/TSG.2010.2055903.

10 Chiang, C.-L. (2007). Genetic-based algorithm for power economic load dispatch. *IET Generation, Transmission and Distribution* 1 (2): 261-269. https://doi.org/10.1049/iet-gtd:20060130.

11 Sinha, N., Chakrabarti, R., and Chattopadhyay, P.K. (2003). Evolutionary programming techniques for economic load dispatch. *International Journal of Emerging Electric Power Systems* 7 (6): 83-94.

12 Kuo, C.C. (2008). A novel coding scheme for practical economic dispatch by modified particle swarm approach. *IEEE Transactions on Power Systems* 23 (4): 1825-1835.

13 Xu, Y., Liu, W., and Gong, J. (2011). Stable multi-agent-based load shedding algorithm for power systems. *IEEE Transactions on Power Systems* 26 (4): 2006-2014.

14 Li, N., Chen, L., and Low, S.H. (2011). Optimal demand response based on utility maximization in power networks. *Power & Energy Society General Meeting*.

15 Zhang, Z. and Chow, M.Y. (2012). Convergence analysis of the incremental cost consensus algorithm under different communication network topologies in a smart grid. *IEEE Transactions on Power Systems* 27 (4): 1761-1768.

16 Pantoja, A. and Quijano, N. (2011). A population dynamics approach for the dispatch of distributed generators. *IEEE Transactions on Industrial Electronics* 58 (10): 4559-4567. https://doi.org/10.1109/TIE.2011.2107714.

17 De Jonghe, C., Hobbs, B.F., and Belmans, R. (2012). Optimal generation mix with short-term demand response and wind penetration. *IEEE Transactions on Power Systems* 27 (2): 830-839. https://doi.org/10.1109/TPWRS.2011.2174257.

18 Xu, Y., Zhang, W., Liu, W., Ferrese, F. (2012). Multiagent-based reinforcement learning for optimal reactive power dispatch. *IEEE Transactions on Systems Man & Cybernetics Part C* 42 (6): 1742-1751.

19 Zhang, W., Liu, W., Wang, X. et al. (2015). Online optimal generation control based on constrained distributed gradient algorithm. *IEEE Transactions on Power Systems* 30 (1): 35-45.

20 Allen, E., Ilic, M., and Younes, Z. (1999). Providing for transmission in times of scarcity: an ISO cannot do it all. *International Journal of Electrical Power & Energy Systems* 21 (2): 147-163.

21 Chen, G., Wang, X., Xu, W., Tajer, A. (2013). Distributed real-time energy scheduling in smart grid: stochastic model and fast optimization. *IEEE Transactions on Smart Grid* 4 (3): 1476-1489.

22 Samadi, P. Mohsenian-Rad, H., Wong, V.W.S., Schober, R. (2013). Tackling the load uncertainty challenges for energy consumption scheduling in smart grid. *IEEE Transactions on Smart Grid* 4 (2): 1007-1016.

23 Mudumbai, R., Dasgupta, S., and Cho, B.B. (2012). Distributed control for optimal economic dispatch of a network of heterogeneous power generators. *IEEE Transactions on Power Systems* 27 (4): 1750-1760.

24 Bertsekas, D.P. (1997). Nonlinear programming. *Journal of the Operational Research Society* 48 (3): 334-334. https://doi.org/10.1057/palgrave.jors.2600425.

25 Boyd, S.P. and Vandenberghe, L. (2004). *Convex Optimization*. Cambridge University Press.

26 Xiao, L. and Boyd, S. (2006). Optimal scaling of a gradient method for distributed resource allocation. *Journal of Optimization Theory and Applications* 129 (3): 469-488. https://doi.org/10.1007/s10957-006-9080-1.

27 Xu, Y. and Liu, W. (2011). Novel multiagent based load restoration algorithm for microgrids. *IEEE Transactions on Smart Grid* 2 (1): 152-161.

28 Olfati-Saber, R., Fax, J.A., and Murray, R.M. (2007). Consensus and cooperation in networked multi-agent systems. *Proceedings of the IEEE* 95 (1): 215-233.

29 Xiao, L. and Boyd, S. (2004). Fast linear iterations for distributed averaging. *IEEE Conference on Decision & Control*.

30 Boyd, J. (2013). An internet-inspired electricity grid. *Spectrum IEEE* 50 (1): 12-14.

31 Lee, K.Y., Park, Y.M., and Ortiz, J.L. (2010). A united approach to optimal real and reactive power dispatch. *IEEE Power Engineering Review* PER-5 (5): 42-43.

7

Distributed State Estimation

As the scale of the power system increases, the state estimation (SE) that covers the entire system consisting of multiple areas becomes increasingly difficult because of the traditional centralized or hierarchical architecture. To address this problem, in this chapter, we will investigate two distributed state estimation approaches that are suitable for multi-area power systems. The first one of distributed estimation approaches is designed based on the consensus algorithm while the second one is developed by using the distributed subgradient algorithm. The consensus algorithm-based approach adopts a two-loop iteration architecture. The inner loop is used to discover the information of gain matrix and gradient vector by applying the consensus confusion technique and the outer loop is used to update the estimation based upon the second-order Newton's method. The distributed algorithm is applied for the second approach to yield a fully distributed integrated solution proposed for multi-area topology identification (TI) and SE problems of power systems. By applying statistical tests, the TI can identify network topology change accurately. The SE estimates the actual states based on the identified network topology. Both approaches investigated in this chapter can be used to implement the distributed state estimation in a fully distributed way. Simulations with variable scales of power systems are also provided for demonstration.

7.1 Distributed Approach for Multi-area State Estimation Based on Consensus Algorithm

The electric power grid is a complex system consisting of multiple regional subsystems, each with its own transmission infrastructure spanning over a huge geographical area [1]. SE is the very basic and most powerful tool for the system operators to use in a monitoring and control system. The goal of state estimation is

Distributed Energy Management of Electrical Power Systems, First Edition.
Yinliang Xu, Wei Zhang, Wenxin Liu, and Wen Yu.

Published 2021 by John Wiley & Sons, Inc.

to obtain a reliable and accurate estimate of the state variables, including bus voltages and phase angles, etc. [2]. State estimation has been traditionally performed at the regional control centers of the corresponding subsystems with limited interaction. However, because of the deregulation of energy markets, large amounts of power are transferred over high-rate, long-distance lines spanning several control regions [3]. As interconnections among regional subsystems become stronger, so the state in a given region needs to be estimated considering the events or decisions in other regions. A system-wide state estimation solution becomes necessary well beyond the extent covered by each control center. Hence, the idea of multi-area state estimation (MASE) methods is gaining renewed interest.

Two computer architectures have been proposed for the MASE problem: the hierarchical and the distributed. In hierarchical MASE, a centralized processor or a coordinator distributes the computation efforts among slave computers performing local area SE and, consequently, coordinates the local estimates. In [4], a distributed state estimation method is proposed for multi-area power systems, wherein each area performs its own state estimation using local measurements and exchanges border information at a coordination state estimator that computes the system-wide state. By incorporating the framework of the convention SE, a two-step hierarchical SE method is proposed in [5], and this method includes two levels: the first local estimator level and the second substation level. The distributed approaches of state estimation are usually based on decomposition techniques or distributed optimization algorithms. Caro et al. proposed a distributed state estimation approach that relies on the decomposition techniques of Lagrangian relaxation, and this method can also be applied to handle bad measurement identification within regions and in border tie-lines [6]. Incorporating the consensus algorithm and innovation approaches, a fully distributed state estimation method is designed for a multi-area power system [7], and Kar et al. also applied a similar design for state estimation as well as energy management for smart grids with distributed generations [8]. The distributed algorithm used in these two papers is actually a kind of variation of distributed subgradient method introduced in [9]. Chavali and Nehorai proposed a distributed algorithm for a power system state estimation with dependencies among the state vectors of neighboring area state vectors at different times being modeled using a factor graph and sum-product message passing algorithm being applying to this graph [10]. The hierarchical architecture for state estimation inevitably needs a centralized controller or coordinator, which may further increase the cost for investment as well as for operation and maintenance. Thus, the distributed architecture is a more preferred solution for MASE. Yet, most of the existing distributed methods are either based on the Lagrangian relaxation-based decomposition method [1, 6] or the distributed subgradient method [7, 8], which on the one hand may face the dilemma of incomplete observability and on the other hand may result in the degraded convergence speed or optimality. In this section, we will investigate a distributed state estimation solution based on the consensus algorithm, which finds a decent trade-off between reducing computational efforts and improving the convergence speed.

7.1.1 Problem Formulation of Multi-area Power System State Estimation

A power system with n buses is partitioned into r nonoverlapping control areas (subsystems). A_i denotes the ith control area with n_i buses and m_i measurements. The areas are connected by tie-lines or transformer branches, as depicted in Figure 7.1 [4]. Each control area is responsible for the measurements in this region and is capable of communicating with its immediate neighboring areas.

The measurement model of the MASE is formulated as

$$\mathbf{z}^i = \mathbf{h}^i(\mathbf{x}) + \mathbf{e}^i i = 1, 2, \ldots, r \tag{7.1}$$

where $(\mathbf{z}^i)^T = [z_1, z_2, \ldots, z_{m_i}]$, is $m_i \times 1$ vector of measurements in area A_i; $\mathbf{x}^T = [x_1, x_2, \ldots, x_n]$, state variable vector, usually including voltage magnitudes and phase angles, n is the number of state variables of the whole system to be estimated; $(\mathbf{h}^i)^T = [h_1(\mathbf{x}), h_2(\mathbf{x}), \ldots, h_{m_i}(\mathbf{x})]^T$, nonlinear vector function with respect to state variables; $(\mathbf{e}^i)^T = [e_1, e_2, \ldots, e_{m_i}]^T$ – error vector corresponding to measurements, which is usually modeled as a Gaussian random vector. The aggregated measurement of all sensors is

$$\mathbf{z} = \mathbf{h}(\mathbf{x})\mathbf{x} + \mathbf{e} \tag{7.2}$$

where $\mathbf{z} = [(\mathbf{z}^1)^T, (\mathbf{z}^2)^T, \ldots, (\mathbf{z}^r)^T]^T$, $\mathbf{h}(\mathbf{x}) = \left[(\mathbf{h}^1(\mathbf{x}))^T \ (\mathbf{h}^2(\mathbf{x}))^T \ \cdots \ (\mathbf{h}^r(\mathbf{x}))^T\right]^T$ and $\mathbf{e} = \left[(\mathbf{e}^1)^T \ (\mathbf{e}^2)^T \ \cdots \ (\mathbf{e}^r)^T\right]^T$. A centralized weighted least square (WLS) estimator

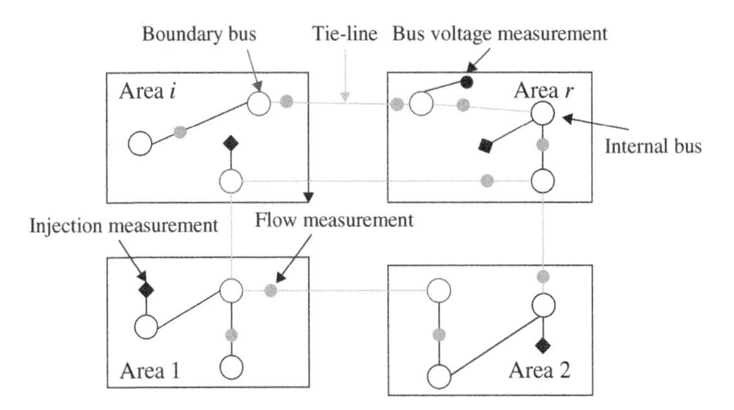

Figure 7.1 Power system with multiple areas. Source: Based on Korres [11].

will minimize the following objective function [12]

$$J(\mathbf{x}) = (\mathbf{z} - \mathbf{h}(\mathbf{x}))^T\mathbf{R}^{-1}(\mathbf{z} - \mathbf{h}(\mathbf{x})) \tag{7.3}$$

where $\mathbf{R} = \mathbf{diag}(\mathbf{R}^1, \mathbf{R}^2, \dots, \mathbf{R}^n)$ is the argument weight matrix. In addition, the following assumptions are customarily made: (i) Error vector has a normal distribution with zero mean, i.e. $E(e_j^i) = 0(i = 1, 2, \dots r; j = 1, 2 \cdots m_i)$; (ii) Measurement errors are independent, i.e. $E(e_j^i e_k^i) = 0(i = 1, 2, \dots, r;\ j, k = 1, 2, \dots, m_i)$. Hence, $\mathrm{cov}(\mathbf{e}^i) = E[\mathbf{e}^i(\mathbf{e}^i)^T] = \mathbf{R}^i = \mathrm{diag}\{\sigma_1^i, \sigma_2^i, \dots, \sigma_{m_i}^i\}$.

In the centralized solution, the control center collecting all the information required for state estimation needs a powerful communication network as well as a processor. In this section, we proposed a distributed state estimation algorithm based on the average consensus algorithm. The algorithm only requires each area to be accessible to local measurements and be able to communicate to its neighboring areas. The algorithm can significantly release the communication burden as well as calculation efforts while still achieves the results comparable to the centralized solution.

7.1.2 Distributed State Estimation Algorithm

To minimize the objective function described in Eq. (7.2), the first-order optimality conditions will have to be satisfied. This can be expressed in the compact form as follows:

$$\mathbf{g}(\mathbf{x}) = \frac{\partial J(\mathbf{x})}{\partial \mathbf{x}} = -\mathbf{H}^T(\mathbf{x})\mathbf{R}^{-1}(\mathbf{z} - \mathbf{h}(\mathbf{x})) = 0 \tag{7.4}$$

where $\mathbf{H}(\mathbf{x}) = \left[\frac{\partial \mathbf{h}(\mathbf{x})}{\partial \mathbf{x}}\right] = [(\mathbf{H}^1(\mathbf{x}))^T, (\mathbf{H}^2(\mathbf{x}))^T, \dots, (\mathbf{H}^r(\mathbf{x}))^T]^T$ and $\mathbf{H}^i(x) = \frac{\partial \mathbf{h}^i(\mathbf{x})}{\partial \mathbf{x}}$. Expanding the nonlinear function into its Taylor series around a state vector $\mathbf{x}^k$ yields

$$\mathbf{g}(\mathbf{x}) = \mathbf{g}(\mathbf{x}^k) + \mathbf{G}(\mathbf{x}^k)(\mathbf{x} - \mathbf{x}^k) + \cdots = 0 \tag{7.5}$$

Neglecting the higher order terms leads to the Gauss–Newton method shown below [1, 4]

$$\mathbf{x}^{k+1} = \mathbf{x}^k - [\mathbf{G}(\mathbf{x}^k)]^{-1} \cdot \mathbf{g}(\mathbf{x}^k) \tag{7.6}$$

where k is the iteration index, $\mathbf{X}^k$ is the solution vector at iteration k, and

$$\mathbf{G}(\mathbf{x}^k) = \frac{\partial \mathbf{g}(\mathbf{x}^k)}{\partial \mathbf{x}} = \mathbf{H}^T(\mathbf{x}^k) \cdot \mathbf{R}^{-1} \cdot \mathbf{H}(\mathbf{x}^k) \tag{7.7}$$

$$\mathbf{g}(\mathbf{x}^k) = -\mathbf{H}^T(\mathbf{x}^k) \cdot \mathbf{R}^{-1} \cdot (\mathbf{z} - \mathbf{h}(\mathbf{x}^k)) \tag{7.8}$$

Assume that $n \leq \sum_{i=1}^{r} m_i$ and matrix $\boldsymbol{H}$ is full rank. Equations (7.7) and (7.8) can be rewritten in the following forms:

$$\mathbf{G}(\mathbf{x}^k) = \sum_{i=1}^{r} (\mathbf{H}^i(\mathbf{x}^k))^T \cdot (\mathbf{R}^i)^{-1} \cdot \mathbf{H}^i(\mathbf{x}^k) \tag{7.9}$$

$$\mathbf{g}(\mathbf{x}^k) = -\sum (\mathbf{H}^i(\mathbf{x}^k))^T \cdot (\mathbf{R}^i)^{-1} \cdot (\mathbf{z}^i - \mathbf{h}^i(\mathbf{x}^k)) \tag{7.10}$$

Notice that computation of the matrix $\mathbf{G}(x^k)$ and vector $\mathbf{g}(x^k)$ can be easily realized in a distributed way based on the average consensus algorithm. Each area has an agent that is responsible for distributed state estimation. For each agent, define local gain matrix $\mathbf{G}^i_{\mathbf{x}^k}(t) \in \mathbf{R}^{n \times n}$ and local gradient vector $\mathbf{g}^i_{\mathbf{x}^k}(t) \in \mathbf{R}^n$ as follows

$$\mathbf{G}^i_{\mathbf{x}^k}(0) = (\mathbf{H}^i(\mathbf{x}^k))^T \cdot (\mathbf{R}^i)^{-1} \cdot \mathbf{H}^i(\mathbf{x}^k) \tag{7.11}$$

$$\mathbf{g}^i_{\mathbf{x}^k}(0) = -(\mathbf{H}^i(\mathbf{x}^k))^T \cdot (\mathbf{R}^i)^{-1} \cdot (\mathbf{z}^i - \mathbf{h}^i(\mathbf{x}^k)) \tag{7.12}$$

and

$$\mathbf{G}^i_{\mathbf{x}^k}(t+1) = a_{ii}(t)\mathbf{G}^i_{\mathbf{x}^k}(t) + \sum_{j \in N_i(t)} a_{ij}\mathbf{G}^j_{\mathbf{x}^k}(t) \tag{7.13}$$

$$\mathbf{g}^i_{\mathbf{x}^k}(t+1) = a_{ii}(t)\mathbf{g}^i_{\mathbf{x}^k}(t) + \sum_{j \in N_i(t)} a_{ij}\mathbf{g}^i_{\mathbf{x}^k}(t) \tag{7.14}$$

According to the consensus algorithm [13], following convergence results can be obtained

$$\lim_{t \to \infty} \mathbf{G}^i_{\mathbf{x}^k}(t) = \frac{1}{r} \sum_{i=1}^{r} (\mathbf{H}^i(\mathbf{x}^k))^T \cdot (\mathbf{R}^i)^{-1} \cdot \mathbf{H}^i(\mathbf{x}^k) \tag{7.15}$$

$$\lim_{t \to \infty} \mathbf{g}^i_{\mathbf{x}^k}(t) = -\frac{1}{r} \sum_{i=1}^{r} (\mathbf{H}^i(\mathbf{x}^k))^T \cdot (\mathbf{R}^i)^{-1} \cdot (\mathbf{z}^i - \mathbf{h}^i(\mathbf{x}^k)) \tag{7.16}$$

Thus, for each agent, the states can be estimated as

$$\mathbf{x}^{k+1} = \mathbf{x}^k - \left[\lim_{t \to \infty} \mathbf{G}^i_{\mathbf{x}^k}(t)\right]^{-1} \cdot \lim_{t \to \infty} \mathbf{g}^i_{\mathbf{x}^k}(t) \tag{7.17}$$

It shows that the overall iterative process is a two-layer nested loop. The inner loop discovers the global information, $\mathbf{H}(x^k)$, based on consensus algorithm while the outer loop updates the estimated state, x^k, through Gauss–Newton method. During the iteration process, the outer loop only updates its state information when the inner loop has converged. The initialization of the gain matrix and gradient vector in Eqs. (7.11) and (7.12) involves the calculation of the Jacobian matrix, which can be developed according to the following development. The relationships between power flow $\mathbf{P}_{ki}$ and $\mathbf{Q}_{ki}$ and the state variables V and δ are given in Eqs. (7.18) and (7.19), respectively.

$$P_{ki} = -V_k^2 G_{ki} + V_k V_i G_{ki} \cos \delta_{ki} + V_k V_i B_{ki} \sin \delta_{ki} \tag{7.18}$$

$$Q_{ki} = V_k^2 B_{ki} + V_k V_i G_{ki} \sin \delta_{ki} - V_k V_i B_{ki} \cos \delta_{ki} + (Y_{ki,s}/2) V_k^2 \tag{7.19}$$

where $G_{ki} + jB_{ki} = [\mathbf{Y}_{\text{bus}}]_{ki}$, $\mathbf{Y}_{\text{bus}}$ is the admittance matrix and $Y_{ki,s}$ is the shunt admittance of the branch. Thus, we have

$$\frac{\partial P_{ki}}{\partial \delta_k} = -V_k V_i G_{ki} \sin \delta_{ki} + V_k V_i B_{ki} \cos \delta_{ki}, \quad \frac{\partial P_{ki}}{\partial \delta_i} = V_k V_i G_{ki} \sin \delta_{ki} - V_k V_i B_{ki} \cos \delta_{ki}$$

$$\frac{\partial P_{ki}}{\partial V_k} = 2V_k G_{ki} + V_i G_{ki} \cos \delta_{ki} + V_i B_{ki} \sin \delta_{ki}, \quad \frac{\partial P_{ki}}{\partial V_i} = V_k G_{ki} \cos \delta_{ki} + V_k B_{ki} \sin \delta_{ki}$$

$$\frac{\partial Q_{ki}}{\partial \delta_k} = V_k V_i G_{ki} \cos \delta_{ki} + V_k V_i B_{ki} \sin \delta_{ki}, \quad \frac{\partial Q_{ki}}{\partial \delta_i} = -V_k V_i G_{ki} \cos \delta_{ki} - V_k V_i B_{ki} \sin \delta_{ki}$$

$$\frac{\partial Q_{ki}}{\partial V_k} = 2V_k B_{ki} + V_i G_{ki} \sin \delta_{ki} - V_i B_{ki} \cos \delta_{ki} + Y_{ki,s} V_k, \quad \frac{\partial Q_{ki}}{\partial V_i} = V_k G_{ki} \sin \delta_{ki} - V_k B_{ki} \cos \delta_{ki}$$

The relationships between bus injection $\mathbf{P}^k$, $\mathbf{Q}^k$, and the state variables are given in Eqs. (7.20) and (7.21), respectively.

$$P_k = V_k \sum_{i=1}^{n} V_i (G_{ki} \cos \delta_{ki} + B_{ki} \sin \delta_{ki}) \tag{7.20}$$

$$Q_k = V_k \sum_{i=1}^{n} V_i (G_{ki} \sin \delta_{ki} - B_{ki} \cos \delta_{ki}) \tag{7.21}$$

Then, we have

$$\frac{\partial P_k}{\partial \delta_i} = \begin{cases} V_k V_i (G_{ki} \sin \delta_{ki} - B_{ki} \cos \delta_{ki}) \text{ if } i \neq k \\ V_k \sum_{j=1, j\neq k}^{n} V_j (-G_{kj} \sin \delta_{kj} + B_{kj} \cos \delta_{kj}) \text{ if } i = k \end{cases}$$

$$\frac{\partial P_k}{\partial V_i} = \begin{cases} V_i (G_{ki} \cos \delta_{ki} + B_{ki} \sin \delta_{ki}) \text{ if } i \neq k \\ \sum_{j=1, j\neq k}^{n} V_j (G_{kj} \cos \delta_{kj} + B_{kj} \sin \delta_{kj}) + 2V_k G_{kk} \text{ if } i = k \end{cases}$$

$$\frac{\partial Q_k}{\partial \delta_i} = \begin{cases} -V_k V_i (G_{ki} \cos \delta_{ki} + B_{ki} \sin \delta_{ki}) \text{ if } i \neq k \\ V_k \sum_{j=1, j\neq k}^{n} V_j (G_{kj} \cos \delta_{kj} + B_{kj} \sin \delta_{kj}) \text{ if } i = k \end{cases}$$

$$\frac{\partial Q_k}{\partial V_i} = \begin{cases} V_i (G_{ki} \sin \delta_{ki} - B_{ki} \cos \delta_{ki}) \text{ if } i \neq k \\ \sum_{j=1, j\neq k}^{n} V_j (G_{kj} \sin \delta_{kj} - B_{kj} \cos \delta_{kj}) - 2V_k B_{kk} \text{ if } i = k \end{cases}$$

For bus voltage magnitude,

$$V_k = V_k \tag{7.22}$$

Obviously, we have $\frac{\partial V_k}{\partial \delta_i} = 0$ and $\frac{\partial V_k}{\partial V_i} = \begin{cases} 1 & \text{if } i = k \\ 0 & \text{if } i \neq k \end{cases}$

If we arrange the measurements vector **Z** in the following way

$$\mathbf{z} = \begin{bmatrix} \mathbf{P}_{\text{line}} \\ \mathbf{P}_{\text{bus}} \\ \mathbf{Q}_{\text{line}} \\ \mathbf{Q}_{\text{bus}} \\ \mathbf{V}_{\text{bus}} \end{bmatrix} \tag{7.23}$$

and state variables are rearranged in Eq. (7.24)

$$\mathbf{x} = \begin{bmatrix} \delta \\ V \end{bmatrix} \tag{7.24}$$

the Jacobian matrix can be expressed in Eq. (7.25)

$$\mathbf{H}^i(\mathbf{x}) = \frac{\partial \mathbf{h}^i(\mathbf{x}^i)}{\partial \mathbf{x}^i} = \begin{bmatrix} \frac{\partial \mathbf{P}^i_{\text{line}}}{\partial \delta} & \frac{\partial \mathbf{P}^i_{\text{line}}}{\partial \mathbf{V}} \\ \frac{\partial \mathbf{P}^i_{\text{bus}}}{\partial \delta} & \frac{\partial \mathbf{P}^i_{\text{bus}}}{\partial \mathbf{V}} \\ \frac{\partial \mathbf{Q}^i_{\text{line}}}{\partial \delta} & \frac{\partial \mathbf{Q}^i_{\text{line}}}{\partial \mathbf{V}} \\ \frac{\partial \mathbf{Q}^i_{\text{bus}}}{\partial \delta} & \frac{\partial \mathbf{Q}^i_{\text{bus}}}{\partial \mathbf{V}} \\ \frac{\partial \mathbf{V}^i_{\text{bus}}}{\partial \delta} & \frac{\partial \mathbf{V}^i_{\text{bus}}}{\partial \mathbf{V}} \end{bmatrix} \tag{7.25}$$

However, for DC flow-based state estimation, usually referred to as the approximate static state estimation model, $\mathbf{h}^i(\mathbf{x})$ can be simplified to a linear function, which can simplify the Jacobian matrix into a constant matrix. In the following section, we will introduce this model and then perform comparison between the results of the approximate and accurate model.

7.1.3 Approximate Static State Estimation Model

The approximate model is based on the following assumptions, which are the same to the DC load flow analysis [14]: (i) For branch i, the reactance X_{ki} is greater than the resistance R_{ki}, i.e. $X_{ki}/R_{ki} \gg 1$, thus $G_{ki} \approx 0$. (ii) Voltage magnitude $V_k \approx V_i \approx 1.0$ and angle $\delta_k \approx \delta_i$. (iii) Active power and reactive power are completely decoupled, i.e. P only relates to voltage angle and Q relates to voltage magnitude. Define $\xi = [V_k, V_i, \delta_{ik}]$. A truncated Taylor series expansion of Eq. (7.18) for a given ξ_0 yields to Eq. (7.26)

$$P_{ki}(\xi) = P_{ki}(\xi)\big|_{\xi=\xi_0} + \left(\frac{\partial P_{ki}(\xi)}{\partial \xi}\bigg|_{\xi=\xi_0} \right)^T (\xi - \xi_0) \tag{7.26}$$

where $\frac{\partial P_{ki}(\xi)}{\partial \xi}$ is a vector of the partial derivative of $P_{ki}(\xi)$ with respect to V_k, V_i, and δ_{ik}, respectively. For the power system, choose $\xi_0 = [110]^T$, then according to the assumptions, Eq. (7.18) becomes (7.27)

$$P_{ki} = B_{ki}(\delta_k - \delta_i) \tag{7.27}$$

A similar expression can also be obtained for Q_{ki}, as shown in Eq. (7.28)

$$Q_{ki} = B_{ki}(V_k - V_i) + Y_{ki,s}V_k - Y_{ki,s}/2 \tag{7.28}$$

For bus injection,

$$P_k = \sum_{i\in k, i\neq k} P_{ki} = \sum_{i\in k, i\neq k} B_{ki}(\delta_k - \delta_i) \tag{7.29}$$

$$Q_k = \sum_{i\in k, i\neq k} Q_{ki} = \sum_{i\in k, i\neq k} B_{ki}(V_k - V_i) + \sum_{i\in k, i\neq k} Y_{ki,s}V_k - \sum_{i\in k, i\neq k} Y_{ki,s}/2 \tag{7.30}$$

Notice that, in Eqs. (7.27) and (7.29), active power only relates to the phase angle; thus, the measurement for active power is utilized to estimate the phase angles, which follows matrix form as (7.31)

$$\mathbf{z}_p = \mathbf{H}_p\mathbf{x}_\delta + \mathbf{e}_p \tag{7.31}$$

where $\mathbf{z}_p$ is the vector for measurement of active power (including line flow and bus injection) and $\mathbf{H}_p$ is the corresponding matrix that relates to network parameters. Here, the function $\mathbf{h}(\mathbf{x}) = \mathbf{H}_p\mathbf{x}_\delta$ is simplified to a linear function; thus, the Jacobian matrix turns out to be $\mathbf{H}_p$, which is a constant. Similarly, for the reactive power, we have

$$\mathbf{z}_Q = \mathbf{H}_Q\mathbf{x}_v + \mathbf{c} + \mathbf{e}_Q \tag{7.32}$$

where $\mathbf{H}_Q$ is also a constant and $\mathbf{c}$ is a vector that relates to shunt admittance $Y_{ki,s}$. For voltage magnitude measurement, we can easily obtain the following form

$$\mathbf{z}_v = \mathbf{H}_v\mathbf{x}_v + \mathbf{e}_v \tag{7.33}$$

Both Eqs. (7.32) and (7.33) are used to estimate the bus voltage; thus, they can be rewritten in a formula given by Eq. (7.34)

$$\mathbf{z}_{Qv} = \mathbf{H}_{Qv}\mathbf{x}_v + \mathbf{e}_v \tag{7.34}$$

Equations (7.31) and (7.34) follow the general form (7.34)

$$\mathbf{z} = \mathbf{Hx} + \mathbf{e} \tag{7.35}$$

Therefore, Eq. (7.3) is simplified as

$$\mathbf{g}(\mathbf{x}) = \frac{\partial J(\mathbf{x})}{\partial \mathbf{x}} = -\mathbf{H}^T\mathbf{R}^{-1}\mathbf{z} + \mathbf{H}^T\mathbf{R}^{-1}\mathbf{Hx} = 0 \tag{7.36}$$

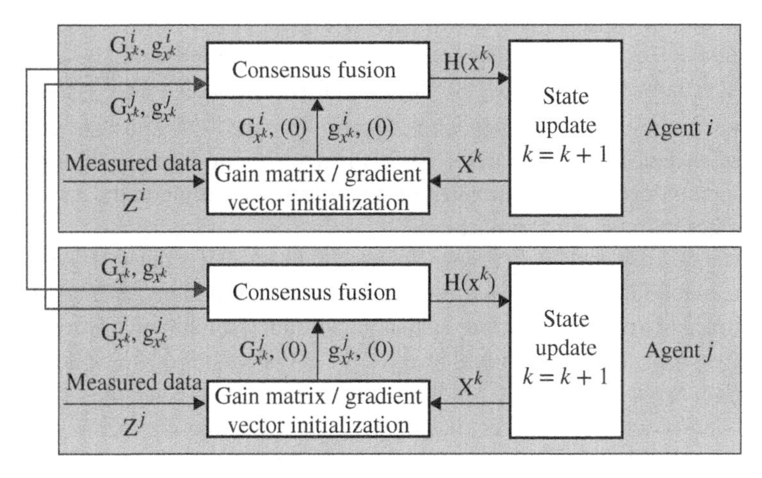

Figure 7.2 Agent Architecture for Distributed State Estimation.

Redefine $\mathbf{G}^i(t)$, $\mathbf{g}^i(t)$ as follows:

$$\mathbf{G}(\mathbf{x}^k) = \sum_{i=1}^{r} (\mathbf{H}^i)^T \cdot (\mathbf{R}^i)^{-1} \cdot \mathbf{H}^i \tag{7.37}$$

$$\mathbf{g}(\mathbf{x}^k) = -\sum_{i=1}^{r} (\mathbf{H}^i)^T \cdot (\mathbf{R}^i)^{-1} \cdot \mathbf{z}^i \tag{7.38}$$

Thus, the estimate stated can be represented as

$$\mathbf{x} = [\lim_{t\to\infty} \mathbf{G}^i(t)]^{-1} \cdot \lim_{t\to\infty} \mathbf{g}^i(t) \tag{7.39}$$

It shows that, for the approximate model, the outer loop iteration described in Figure 7.2 is not necessary. This is because the approximate model is simplified to a linear model. Hence, estimating the state is equivalent to solving the linear equation (7.35); thus, no iteration for an outer loop is needed. Note that the algorithm proposed in [7, 15] is the special case of the algorithm developed in this section. In addition, it also assumes that the function $\mathbf{h}(\mathbf{x})$ is a linear function, which is not applicable when a more accurate analysis is required. In the simulation part, we will provide the test results with the approximate model for comparison.

7.1.4 Regarding Implementation of Distributed State Estimation

Fast decoupled techniques for state estimation: For an accurate model, updating the Jacobian matrix is time-consuming. One way to reduce the computation burden is to maintain the constant Jacobian matrix. Similar to the fast decoupled power flow, we can divide the measurements into two categories; real power measurements and reactive power measurements. Then, we estimate the voltage phase angles and magnitudes alternatively. In addition, for phase angle estimation, ignore $\frac{\partial \mathbf{P}_{\text{line}}}{\partial \mathbf{V}}$ and $\frac{\partial \mathbf{P}_{\text{bus}}}{\partial \mathbf{V}}$, and for voltage magnitude estimation, ignore $\frac{\partial \mathbf{Q}_{\text{line}}}{\partial \delta}$ and $\frac{\partial \mathbf{Q}_{\text{bus}}}{\partial \mathbf{V}}$ and then we would have a constant Jacobian matrix. It is worthy to point out that in each step, we need to update the currently estimated measurements according to the accurate model rather than the approximate model.

Initialization of state variables: Proper setting of the initial values of state variables can reduce the time needed for convergence. Simulations show that for the decoupled state estimation, the voltage magnitude estimation converges faster than the phase angle estimation because the initial values of the voltage magnitudes are closer to the final solutions. Thus, we can predict the state variables first and then use these predicted values as the initial conditions for the current state estimation. This is the same principle as the dynamic state estimation. One of the practical methods to implement a prediction is to use the historical data or the previous state estimation results.

Discriminant handling of internal and boundary variables: Notice that the state vector $\mathbf{x}^i$ either in Eq. (7.7) or 7.28) relates to all buses distributed in all areas because the areas are only connected through the limited number of tie-lines or transmission lines. It is not necessary for a specified area to obtain all the other information except the boundary information. Thus, through reducing the dimension of state vectors for each area, the calculation burden, as well as the communication burden of the whole system, can be significantly released.

7.1.5 Case Studies

In this subsection, two test cases are considered to evaluate the proposed distributed state estimation algorithm. The first test case involves constant operating conditions, while the second considers the variable operating conditions. Both the test cases are conducted on the IEEE 14-bus system [16], and the power system topology and measurement placement configurations are shown in Figure 7.3a. The topology of the communication network is shown in Figure 7.3b. The test techniques adopted in [17] are also utilized. (i) Given the desired network and bus powers, a conventional load flow is calculated to obtain the bus voltages, which are set to be the true values. (ii) According to the measurement placements, types, and accuracy, a set of measurements is simulated by adding errors from a random number generator. (iii) Given the simulated measurement, the state estimations are calculated using the proposed algorithms. (iv) Analyze the obtained results in the sense of residual, errors, etc. The test system is configured to have four control areas, as shown in Figure 7.3a – all the flow and injection measurements meter, both the active and reactive power. Bus voltage measurements are only

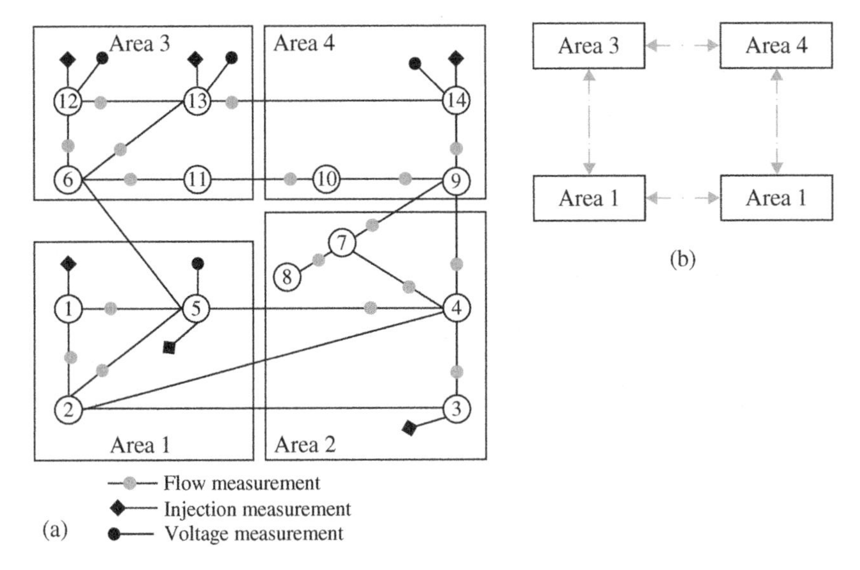

Figure 7.3 Configuration of the IEEE 14-bus system. (a) System topology and measurement placement. (b) Communication topology.

responsible for metering voltage magnitudes. In addition, for PV buses, the bus voltages are assumed to be the pseudo-measurements. All the measurements can be found in Figure 7.3a, and they are assumed to subject to Gaussian distribution with zero mean and equal variance of $\sigma^2 = 0.0001$. $\mathbf{R}$ is set to diagonal, with $\mathbf{R}_{ii} = \sigma^2$. In both test cases, we use the same parameters for the distributed optimization algorithm. The step size is set as a constant with $\alpha = 5 * 10^{-6}$ and weight matrix $\mathbf{A}$ is set to according to the Metropolis [4].

$$\mathbf{A} = \begin{bmatrix} 1/3 & 1/3 & 1/3 & 0 \\ 1/3 & 1/3 & 0 & 1/3 \\ 1/3 & 0 & 1/3 & 1/3 \\ 0 & 1/3 & 1/3 & 1/3 \end{bmatrix} \quad \begin{matrix} \lambda_1 = -0.3333 \\ \lambda_2 = 0.3333 \\ \lambda_3 = 0.3333 \\ \lambda_4 = 1.0000 \end{matrix} \tag{7.40}$$

Instead of setting error for convergence, we use a fixed iteration setting of 1000 for both cases.

7.1.5.1 With the Accurate Model

The voltage magnitudes and angles with the outer iteration loop during the iteration are shown in Figures 7.4 and 7.5, respectively. The corresponding objective function values for each area are shown in Figure 7.6. Note that the algorithm converges quickly, with the number of iterations being less than 10. The outer iteration loop follows the update formula given in (7.6), which is actually the Newton–Raphson method; thus, the order of convergence of the proposed method

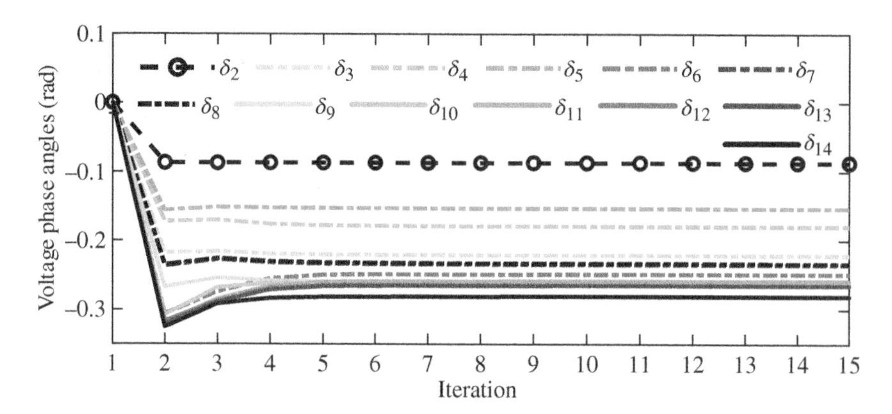

Figure 7.4 The evolution of phase angles.

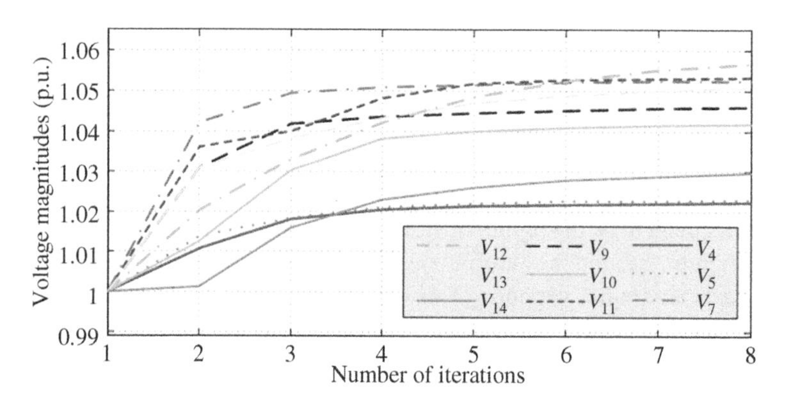

Figure 7.5 The evolution of voltage magnitudes.

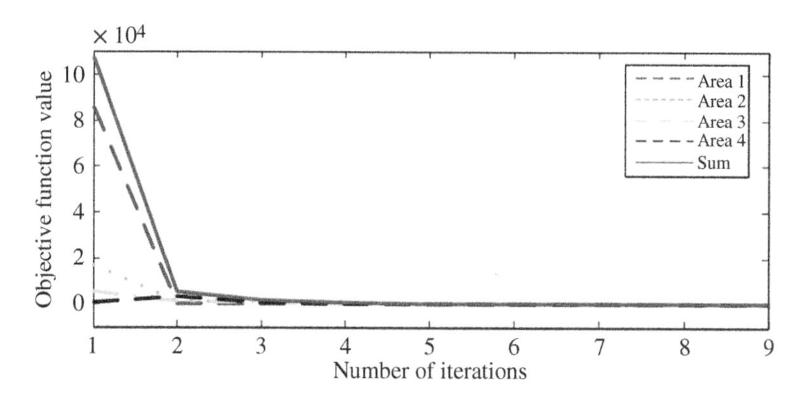

Figure 7.6 The objective function value evolution.

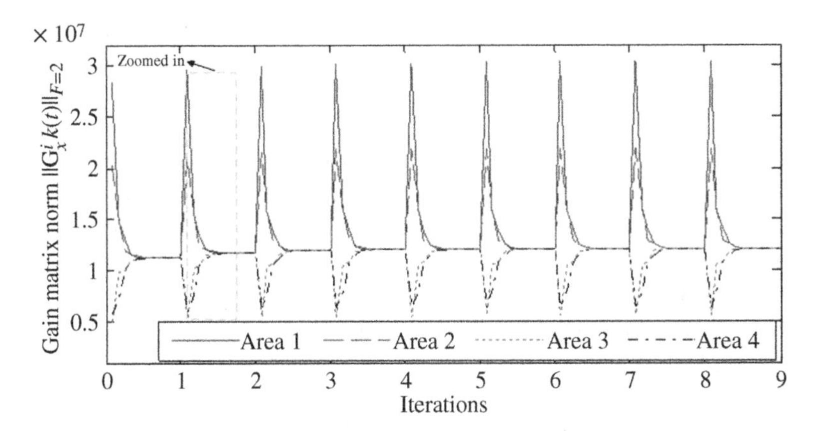

Figure 7.7 Norms of the gain matrix.

is also from the perspective of outer loop iteration. It is worth pointing out here that each outer loop iteration is carried out only when the inner loop consensus algorithm has converged. We use the Frobenius norm [18] of the gain matrix (with $p = 2$) as a tool to show the convergence characteristics of the inner loop iteration. The norm of the gain matrix during the whole solution process is shown in Figure 7.7, with one of the inner loop iteration processes provided in Figure 7.8. It can be shown that in Figure 7.7, the outer loop is indeed initiated when the inner loop iteration process has been completed, i.e. the consensus algorithm has converged. Figure 7.8 shows that the inner loop consensus algorithm takes only 12 iterations to converge, which is also quite faster. Similar observations can also be found in the case of gradient vector. Note that the whole solving process takes

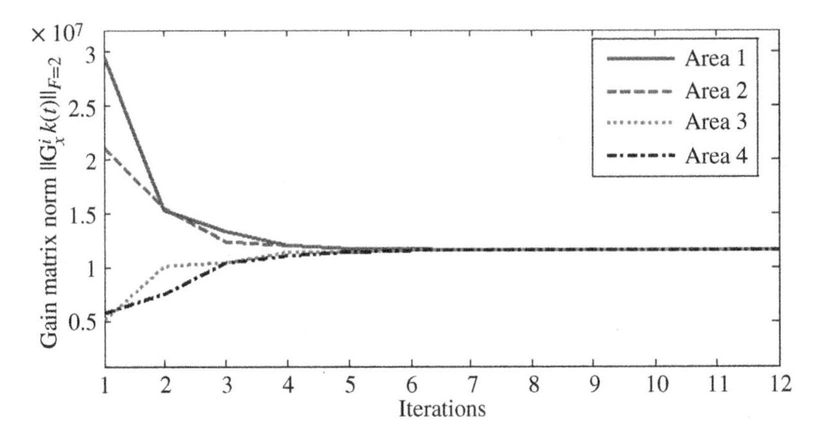

Figure 7.8 Zoomed segment of Figure 7.7.

only 12 * 9 = 108 iterations to converge. Considering that each iteration takes 5 ms, the entire solution process could finish within 0.1, which is sufficient to be used in the MASE applications.

7.1.5.2 Comparisons Between Accurate Model and Approximate Model

With the approximate model, the estimations of voltage phase angles and voltage magnitudes are decoupled and corresponding test results are shown in Figures 7.9 and 7.10, respectively. As can be seen in the figures, the approximate model can obtain the estimations close to the true values. However, the accuracy may not be tolerable for some applications that require accurate estimations, e.g. security-constrained optimal power flow studies, stability analysis, etc. Notice that the proposed approach with an accurate model can obtain the estimations that are almost the same as the true value, especially for the estimation with

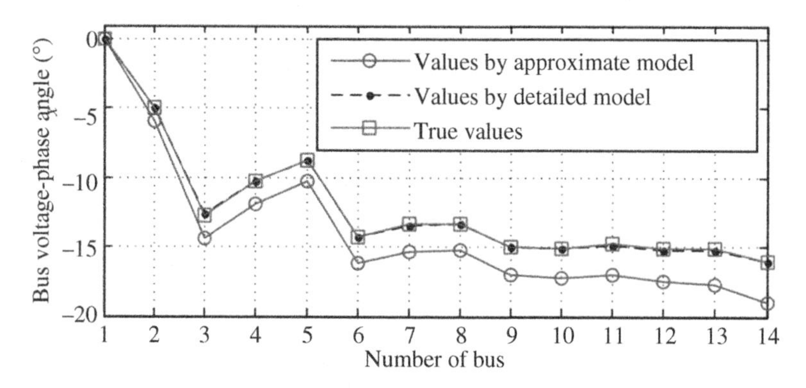

Figure 7.9 Estimated bus voltage phase angles by approximate and accurate model.

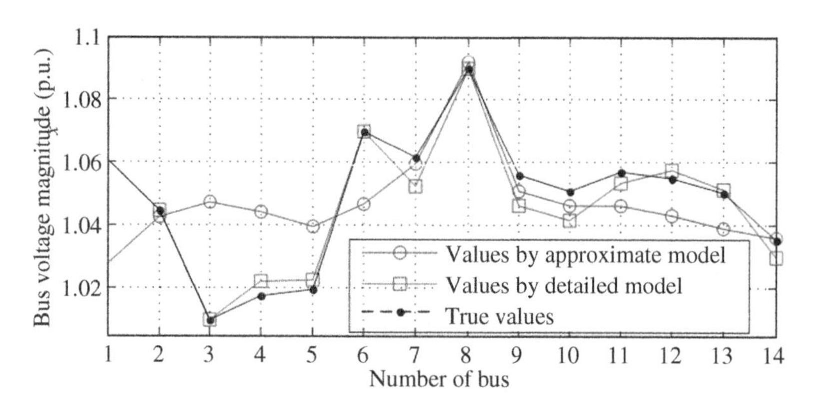

Figure 7.10 Estimated bus voltage phase magnitudes by approximate and accurate model.

voltage phase angles. Because the computation effort with the proposed approach is not large, it can be further applied to any advanced application with accurate estimation results for both voltage phase angles and magnitudes.

7.1.5.3 With Variable Loading Conditions

In order to evaluate the proposed algorithm, we also test the cases with different load patterns. Figure 7.6 shows the load conditions with respect to time, and the load during this period increases from the original level of 1.0 (0–10 minutes) to 1.45 (50–60 minutes). The test results are shown in Figures 7.11– 7.14. The errors used to terminate the iterations are shown in Figure 7.11. Note that in the first pattern, it takes nine iterations to converge, and the patterns follow the convergence in four iterations. Here, for patterns 2–6, we use the pre-obtained solution

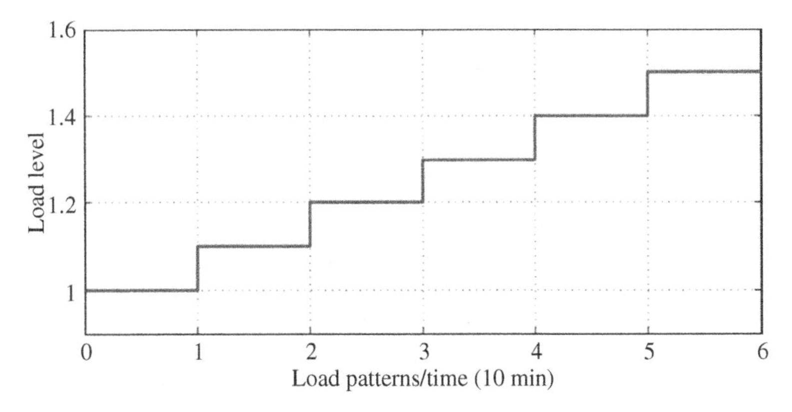

Figure 7.11 Load levels with respect to time.

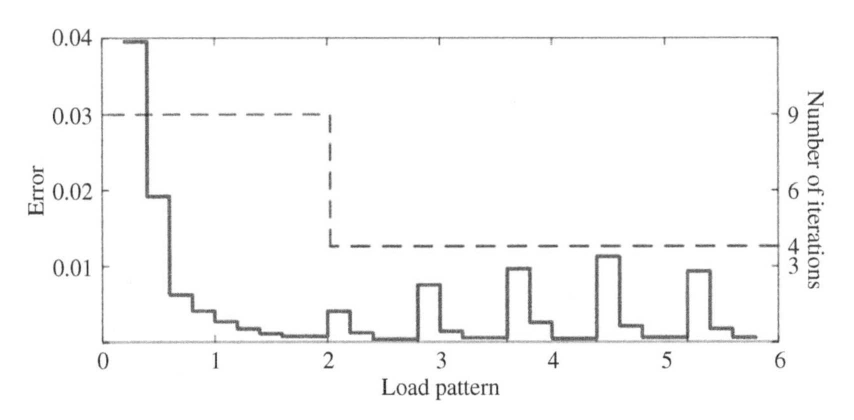

Figure 7.12 Error and number of iterations for state estimation; the solid line denotes the error and dash line denotes the number of iterations.

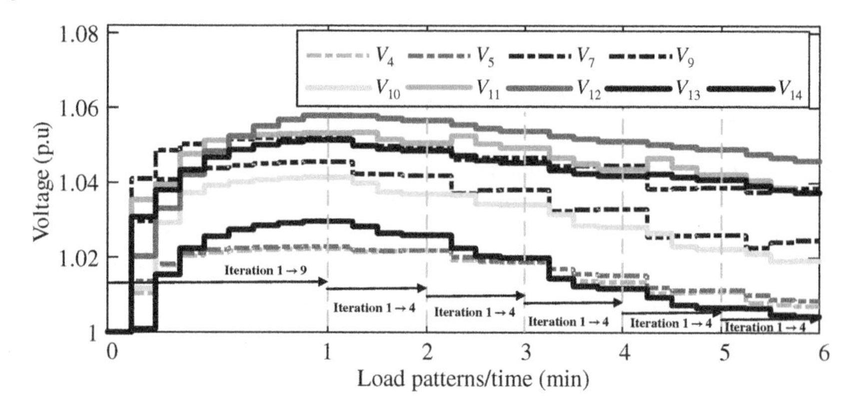

Figure 7.13 The evolution of voltage magnitudes.

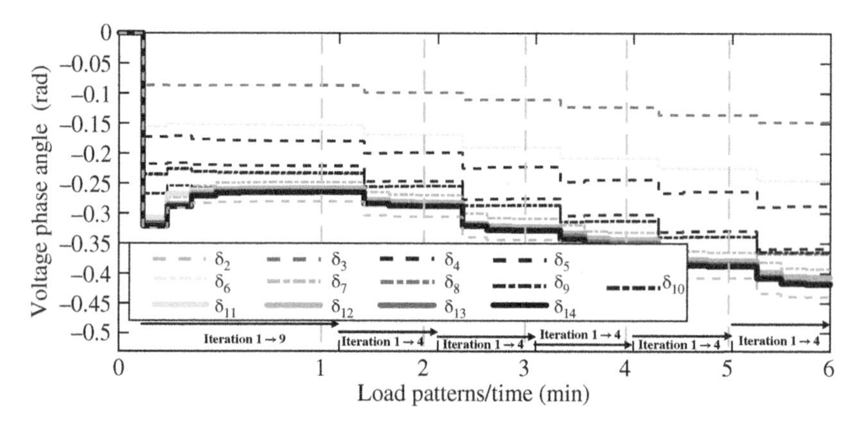

Figure 7.14 The evolution of the voltage phase angles.

of the previous pattern as the initial guess solution, which further increases the convergence speed. The estimated voltage phase angles and magnitudes are shown in Figures 7.15 and 7.16. As can be seen in the figures, the estimated voltage phases for all the load patterns are almost the same as the true values, as discussed previously. We can also observe that the phase angles of nodes decrease as the load increases because the heavier load level requires large power transfer between the generation and the demand node. The estimated bus voltage magnitudes shown in Figure 7.16 are also very close to the true values. However, the error between them is larger than those in the phase angle cases. This is because in this transmission system, the reactive power measurements have a relatively lower value, which results in degraded estimations as the noise may dominate in a few measurements. We can also see that the voltage profile of the system does not change

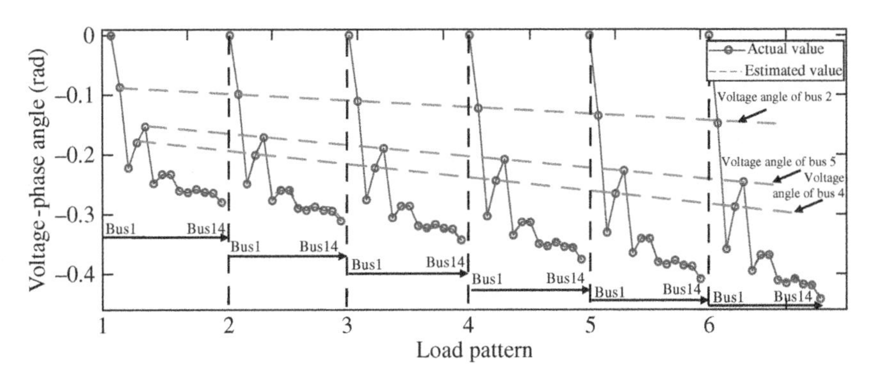

Figure 7.15 Estimated values of bus voltage angles.

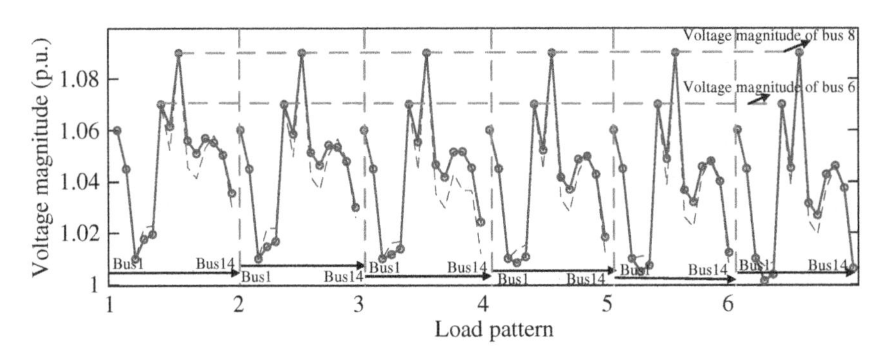

Figure 7.16 Estimated values of bus voltage magnitudes.

a lot. This is because the PV bus node always maintains its bus magnitude as a constant, which helps to keep the voltage profile within the reasonable range.

7.1.6 Conclusion and Discussion

In this section, we presented a distributed state estimation approach for a MASE. By using the consensus algorithm, each area is able to access the information regarding gain matrix and gradient vector of other areas. Through the coordination of the areas in the system, the proposed approach, on the one hand, can converge in a second order; on the other hand, it can obtain the estimations that are decent enough for other advanced power system applications. In addition, the proposed approach does not need a complex communication network to support its implementation. The computational effort and the amount of data that needs to be processed by one area are both significantly reduced. The case studies show that the proposed approach is promising for the MASE.

Appendix

Partial derivative for Taylor series expansion

$$P_{ki}(\xi) = P_{ki}(\xi)\big|_{\xi=\xi_0} + \left(\frac{\partial P_{ki}(\xi)}{\partial \xi}\bigg|_{\xi=\xi_0}\right)^T (\xi - \xi_0) \tag{7.41}$$

$$Q_{ki}(\xi) = Q_{ki}(\xi)\big|_{\xi=\xi_0} + \left(\frac{\partial Q_{ki}(\xi)}{\partial \xi}\bigg|_{\xi=\xi_0}\right)^T (\xi - \xi_0) \tag{7.42}$$

With the assumption that $\xi_0 = \begin{bmatrix}1 & 1 & 0\end{bmatrix}^T$ and $G_{ki} \approx 0$, then we have $P_{ki}(\xi)\big|_{\xi=\xi_0} = 0$, $\frac{\partial P_{ki}(\xi)}{\partial \xi}\bigg|_{\xi=\xi_0} = \begin{bmatrix} \frac{\partial P_{ki}}{\partial V_k}\Big|_{\xi=\xi_0} \\ \frac{\partial P_{ki}}{\partial V_i}\Big|_{\xi=\xi_0} \\ \frac{\partial P_{ki}}{\partial \delta_{ik}}\Big|_{\xi=\xi_0} \end{bmatrix} = \begin{bmatrix} 0 \\ 0 \\ B_{ki} \end{bmatrix}$ and $Q_{ki}(\xi)\big|_{\xi=\xi_0} = Y_{ki,s}/2$, $\frac{\partial Q_{ki}(\xi)}{\partial \xi}\bigg|_{\xi=\xi_0} = \begin{bmatrix} \frac{\partial Q_{ki}}{\partial V_k}\Big|_{\xi=\xi_0} \\ \frac{\partial Q_{ki}}{\partial V_i}\Big|_{\xi=\xi_0} \\ \frac{\partial Q_{ki}}{\partial \delta_{ik}}\Big|_{\xi=\xi_0} \end{bmatrix} = \begin{bmatrix} B_{ki} + Y_{ki,s} \\ -B_{ki} \\ 0 \end{bmatrix}$ $\xi - \xi_0 = \begin{bmatrix} V_k - 1 \\ V_i - 1 \\ \delta_{ki} \end{bmatrix}$ putting these expressions into Eqs.(7.41) and (7.42) yields

$$P_{ki} = B_{ki}(\delta_k - \delta_i) \tag{7.43}$$

$$Q_{ki} = B_{ki}(V_k - V_i) + Y_{ki,s}V_k - Y_{ki,s}/2 \tag{7.44}$$

7.2 Multi-agent System-Based Integrated Solution for Topology Identification and State Estimation

Power systems should be monitored continuously to maintain a normal and safe operation. State estimation (SE) can provide accurate information for power systems, such as power injections, power flows, and voltage magnitudes. The concept of SE was first introduced in the late 1960s and had achieved remarkable development in the past few decades [19].

The implementation of SE is growing into a more challenging problem for the two reasons. First, a large integration of distributed generations greatly increases the complexity of the power system operation [20]. Second, the deregulation of the power systems results in the creation of many local utilities and independent system operators (ISOs), which usually operate independently and thus require the local implementation of SE [11].

Current SE algorithms aim at enhancing the computational speed and accuracy. Several techniques are proposed to increase the computational speed of the state estimators, such as the fast decoupled method and sparse matrix techniques [21]. There have also been extensive efforts on improving the convergence speed of the SE algorithms such as Hachtel's augmented matrix approach and the QR factorization method [22]. Most of these SE algorithms are implemented in a centralized manner, and they are not suitable for multi-area power systems with a large number of local utilities and ISOs because of a large scale of the power grid as well as the difficulty in data acquisition [11].

Distributed SE algorithms are proposed for large-scale power systems. Different decomposition techniques are utilized, such as the Lagrangian relaxation algorithm [6] and the alternating direction method of multipliers [23]. Generally, the performance of the distributed algorithms is dependent on the system structure. For example, the decomposition algorithm proposed in [24] is based on the decomposition of the gain matrix, wherein the off-diagonal elements are neglected. However, this algorithm is not suitable for areas with many interconnection branches. A tie-line constrained distributed SE method is proposed in [25], where the developed solution requires appropriate splitting of the original system and selection of tie-lines. In addition, two-level schemes are proposed [26, 27]. However, this structure normally has extra requirements for the systems. For example, the local states from the lower level are supposed to be fully accessible by the higher level [26], thereby depending on a sophisticated centralized communication network. For the method proposed in [27] for SE, the original system is decomposed into several nonoverlapping subsystems and an interconnection tie-line area, which requires that the entire system and the subsystems are all observable.

The aforementioned distributed SE algorithms require the observability of the subsystems, which normally leads to heavy measurement redundancy. Moreover, a coordination level is required to collect measurement data from the local areas, which increases the difficulty for implementation. In [15, 28], the methods that are used to detect useless data and solve SE problems in a distributed manner are introduced. These methods are developed based on the distributed $\mathcal{LU}$ algorithm. However, these methods are only applicable to linear observation models.

Furthermore, existing TI and SE problems are studied independently. Their implementations are both based on the assumption of a fixed and known power network topology. However, this assumption no longer holds because of the increasing integration of renewable and distributed generations. It may constantly change the topology configuration of the power networks. Consequently, an integrated solution for distributed TI and SE is highly desirable.

In this section, we develop an integrated method for solving the TI and SE problems of multi-area power systems. Both TI and SE are modeled as WLS problems and are solved in a distributed manner using the distributed subgradient algorithm. We utilize the multi-agent system (MAS) framework for the algorithm implementation, which renders a flexible, reliable, and cost-effective

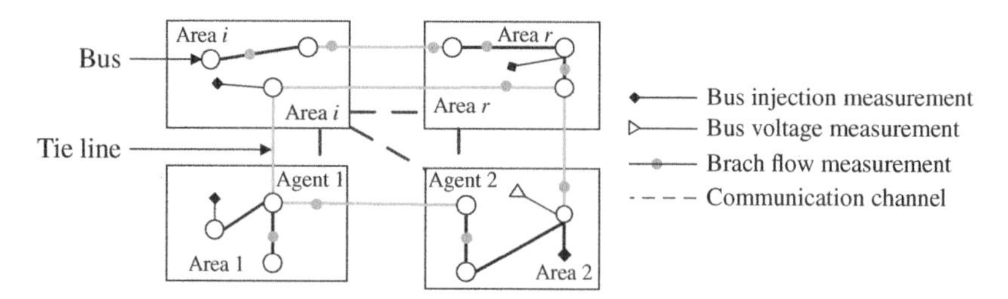

Figure 7.17 Power system with multiple areas.

solution. The proposed solution only requires a simple communication network to realize information exchange. It is also capable of identifying the network topology correctly and obtaining relatively accurate states that are comparable to the centralized solutions.

7.2.1 Measurement Model of the Multi-area Power System

Let us assume that an interconnected power network, with a total of n buses, is divided into r nonoverlapping subsystems or control areas. The ith control area is denoted by A_i with m_i measurements and n_i buses. These subsystems are connected via transformer branches or tie-lines, as portrayed in Figure 7.17 [11].

We use $\mathbf{z}^i = [z_1^i, z_2^i, \dots, z_{m_i}^i]^\top$ to denote the vector of measurements in area A_i, where m_i is the dimension of measurements. Let us denote the global vector of the system states as $\mathbf{x} = [x_1, x_2, \dots, x_{n_s}]^\top$, which includes the magnitudes and phase angles of the voltages. We denote the vector of nonlinear functions of state variables as $\mathbf{h}^i = [h_1^i(\mathbf{x}), h_2^i(\mathbf{x}), \dots, h_{m_i}^i(\mathbf{x})]^\top$. Accordingly, the measurement model for the multi-area TI and SE can be given as follows:

$$\mathbf{z}^i = \mathbf{h}^i(\mathbf{x}) + \mathbf{e}^i, \quad i = 1, 2, \dots, r \tag{7.45}$$

where $\mathbf{e}^i = [e_1^i, e_2^i, \dots, e_{m_i}^i]^\top$ is the vector of measurement errors, which is typically modeled by a Gaussian random vector.

We formulate both the TI and SE of multi-area power systems as an optimization problem. Here, the WLS approach is adopted and the objective of the optimization problem is given as:

$$J(\mathbf{x}) = \sum_{i=1}^{r} (\mathbf{z}^i - \mathbf{h}^i(\mathbf{x}))^\top (\mathbf{R}^i)^{-1} (\mathbf{z}^i - \mathbf{h}^i(\mathbf{x})) \tag{7.46}$$

where $\mathbf{R}^i$ represents the weight matrix of the measurements in area A_i. Unless otherwise specified, the following two assumptions are assumed to hold:

1. The measurement errors conform to a normal distribution with a zero mean, i.e.

$$E(e_j^i) = 0, \quad i = 1, 2, \dots r, \ j = 1, 2, \dots, m_i \tag{7.47}$$

2. The distributions of measurement errors are independent, i.e.

$$E(e_j^i e_k^i) = 0, \quad i = 1, 2, \dots, r; \; j, k = 1, 2, \dots, m_i; \; j \neq k \tag{7.48}$$

thereby

$$\text{cov}(\mathbf{e}^i) = E[\mathbf{e}^i(\mathbf{e}^i)^\top] = \mathbf{R}^i = \text{diag}\{\sigma_1^i, \sigma_2^i, \dots, \sigma_{m_i}^i\} \tag{7.49}$$

In the centralized approach, a central controller collects all the information and relies on a complicated communication network. Our proposed distributed integrated algorithm, on the contrary, only requires an associated agent in each area to collect the local measurements and communicate with the neighboring agents only. The formulated TI and SE problems are solved by a distributed subgradient algorithm, which we will introduce in the following subsection.

7.2.2 Distributed Subgradient Algorithm for MAS-Based Optimization

The distributed subgradient algorithm was first introduced by Tsitsiklis and further developed for distributed multi-agent optimization [29, 30]. Assuming that there are r agents and n_s variables in the system, the goal is to minimize a global function, which is given by

$$f(\mathbf{x}) = \sum_{i=1}^{r} f^i(\mathbf{x}), \quad \text{subject to } \mathbf{x} \in \mathbf{R}^{n_s} \tag{7.50}$$

where $\mathbf{x} = [x_1, x_2, \dots, x_{n_s}]^T$ is the vector of optimization variables, $f^i(x) : \mathbf{R}^{n_s} \rightarrow \mathbf{R}$ is the objective function of agent i, where $f^i(x)$ is only aware by agent i. Agent i aims at minimizing $f^i(x)$ by communicating with other agents.

To solve (7.50), each agent obtains a local initial estimate of the global optimal solution of (7.50), which is denoted by $\mathbf{x}^i(0)$ [30]. At each iteration, each agent obtains an update by exchanging information with its neighbors. The neighbors of agent i are denoted by the set N_i. After acquiring the information $\mathbf{x}(k)$ of agents j, the update of agent i follows the rule:

$$\mathbf{x}^i(k+1) = \sum_{j \in N_i} a_{ij} \mathbf{x}^j(k) + a_{ii} \mathbf{x}^i(k) - \alpha^i \mathbf{d}^i(k) \tag{7.51}$$

where a_{ij} denotes the interinfluence weight of agent j on agent i, a_{ii} is the self-influence weight of agent i and α^i denotes the step size. Interested readers can refer to [29] for methods of deciding the weights and step sizes.

The subgradient of $f^i(\mathbf{x})$ at $\mathbf{x} = \mathbf{x}^i(k)$ is the vector $\mathbf{d}^i(k)$ in (7.51), which is shown as follows:

$$\mathbf{d}^i(k) = \left[\frac{\partial f^i(\mathbf{x})}{\partial x_1}, \frac{\partial f^i(\mathbf{x})}{\partial x_2}, \dots, \frac{\partial f^i(\mathbf{x})}{\partial x_{n_s}} \right]^T$$

Based on (7.51), an agent keeps a local guess for the global optimum and adjusts its estimation based on its local subgradient and the up-to-date estimation of its neighbors. When the algorithm converges, (7.51) satisfies the following two conditions:

1. $\mathbf{d}^i(\infty) = 0, \ \forall i$. It is the "stopped" model discussed in [29].
2. $\mathbf{x}^1(\infty) = \mathbf{x}^2(\infty) = \cdots = \mathbf{x}^r(\infty), \ \forall i$.

The two conditions indicate that the optimal values obtained by the distributed agents can approach the global optimum $\mathbf{x}^*$.

Here, we briefly present the convergence analysis for the subgradient algorithm. Let us define $\mathbf{A}(k)$ as a $r \times r$ matrix with elements a_{ij}, where $a^i(k)$ is the ith column of $\mathbf{A}(k)$. The iterations in (7.51) show that $\forall i, s, k; \ k \geq s$, we have that

$$\begin{aligned}\mathbf{x}^i[k+1] &= \sum_{j=1}^{r} [\mathbf{A}(s)\mathbf{A}(s+1)\cdots\mathbf{A}(k-1)a^i(k)]_j \mathbf{x}^j(s)\\ &\quad - \sum_{j=1}^{r} [\mathbf{A}(s+1)\cdots\mathbf{A}(k-1)a^i(k)]_j \alpha^j(s) d^j(s) - \cdots\\ &\quad - \sum_{j=1}^{r} [\mathbf{A}(k-1)a^i(k)]_j \alpha^j(k-2) d^j(k-2)\\ &\quad - \sum_{j=1}^{r} [a^i(k)]_j \alpha^j(k-1) d^j(k-1) - \alpha^i(k) d^i(k)\end{aligned} \tag{7.52}$$

Define the transition matrix as $\boldsymbol{\Phi}(k,s)$, which is shown as

$$\boldsymbol{\Phi}(k,s) = \mathbf{A}(s)\mathbf{A}(s+1)\cdots\mathbf{A}(k-1)\mathbf{A}(k)$$

The ith column of $\boldsymbol{\Phi}(k,s)$ is

$$[\boldsymbol{\Phi}(k,s)]^i = \mathbf{A}(s)\mathbf{A}(s+1)\cdots\mathbf{A}(k-1)\mathbf{A}(k-1)a^i(k)$$

where the element in the jth row, ith column is given by

$$[\boldsymbol{\Phi}(k,s)]^i_j = [\mathbf{A}(s)\mathbf{A}(s+1)\cdots\mathbf{A}(k-1)\mathbf{A}(k-1)a^i(k)]_j$$

Let us assume that all agents share a common constant step size α. Let $s = 0$, (7.52) can be rewritten as

$$\mathbf{x}^i[k+1] = \sum_{j=1}^{r} [\boldsymbol{\Phi}(k,0)]^i_j \mathbf{x}^j(0) - \alpha \sum_{l=1}^{k} \sum_{j=1}^{r} [\boldsymbol{\Phi}(k,l)]^i_j d^j(l-1) - \alpha d^i(k) \tag{7.53}$$

It is assumed by the stopped" model that agents stop to update $d^j(k)$ after some $\bar{k}$ iterations, such that $d^j(k) = 0$, for all j, k and $k \geq \bar{k}$.

Let $\overline{\mathbf{x}}^i(k),\ i = 1, \ldots, r$ denote the estimations of agent i. Based on (7.53), for all i we have,

$$\overline{\mathbf{x}}^i(k) = \mathbf{x}^i(k), \quad \forall k \le \overline{k}$$

and when $k > \overline{k}$,

$$\overline{\mathbf{x}}^i(k) = \sum_{j=1}^{r} [\boldsymbol{\Phi}(k-1,0)]_j^i \mathbf{x}^j(0) - \alpha \sum_{l=1}^{\overline{k}} \sum_{j=1}^{r} [\boldsymbol{\Phi}(k-1,l)]_j^i d^j(l-1)$$

Let us define $\mathbf{y}(\overline{k})$ as $\lim_{k\to\infty} \overline{\mathbf{x}}^i(k) = \mathbf{y}(\overline{k})$. By virtue of Proposition 2(a) in [29], one can obtain

$$\mathbf{y}(\overline{k}) = \frac{1}{r}\sum_{j=1}^{r} \mathbf{x}^j(0) - \alpha \sum_{l=1}^{\overline{k}} \sum_{j=1}^{r} \frac{1}{r} d^j(l-1) \tag{7.54}$$

Since (7.54) holds for any $\overline{k}$, we re-index $\overline{k}$ with k, which yields:

$$\mathbf{y}(k+1) = \mathbf{y}(k) - \frac{\alpha}{r}\sum_{j=1}^{r} d^j(k) \tag{7.55}$$

When the set of optimal values $\mathbf{x}^*$ is nonempty, by virtue of Lemma 5 in [29], we have

$$\begin{aligned} \mathrm{dist}^2(\mathbf{y}(k+1), \mathbf{x}^*) \le\ & \mathrm{dist}^2(\mathbf{y}(k+1), \mathbf{x}^*) \\ & + \frac{2\alpha}{r}\sum_{j=1}^{r} (\| d^j(k) \| + \| g^j(k) \|) \| \mathbf{y}(k) - \mathbf{x}^j(k) \| \\ & - \frac{2\alpha}{r}[f(\mathbf{y}(k)) - f^*] + \frac{\alpha^2}{r^2}\sum_{j=1}^{r} \| d^j(k) \| \end{aligned} \tag{7.56}$$

where $g^j(k)$ denotes the subgradient of $f^j(\cdot)$ at $\mathbf{y}(k)$.

Suppose that the agent j satisfies the following constraint:

$$\max_{1\le j\le r} \quad \| \mathbf{x}^j(0) \| \le \alpha L$$

Based on Proposition 3(a) in [29], $\forall i \in (1, \ldots, r)$, the upper limit on $\| \mathbf{y}(k) - \mathbf{x}^i(k) \|$ satisfies

$$\| \mathbf{y}(k) - \mathbf{x}^i(k) \| \le 2\alpha L C_1, \quad \forall k \ge 0\ C_1 = 1 + \frac{r}{1-(1-\eta^{B_0})^{1/B_0}} \cdot \frac{1+\eta^{-B_0}}{1-\eta^{B_0}} \tag{7.57}$$

where η satisfies $0 < \eta < 1$ (Assumption 1(a) in [29]) and $B_0 = (r-1)B$, where B is the communication interval limit (Assumption 3, Lemma 4(c) in [29]).

Let us denote the average vectors to estimate the optimal solution as follows [29, 31]:

$$\mathbf{y}(k) = \frac{1}{k}\sum_{h=0}^{k-1}\mathbf{y}(h), \quad \hat{\mathbf{x}}^i(k) = \frac{1}{k}\sum_{h=0}^{k-1}\mathbf{x}^i(h) \tag{7.58}$$

By virtue of Proposition 3 of [29], the upper limits on the cost in the objective $f(\mathbf{y})$ and $f(\mathbf{x})$ are approximated by

$$\begin{aligned} f(\mathbf{y}(k)) &\leq f^* + \frac{r\text{dist}^2(\mathbf{y}(0), \mathbf{x}^*)}{2\alpha k} + \frac{\alpha L^2 C}{2} \\ f(\hat{\mathbf{x}}^i(k)) &\leq f^* + \frac{r\text{dist}^2(\mathbf{y}(0), \mathbf{x}^*)}{2\alpha k} + \alpha L\left(\frac{LC}{2} + 2r\hat{L}_1 C_1\right) \end{aligned} \tag{7.59}$$

where $\hat{L}_1$ is the upper limit of the subgradient of $f^j(\cdot)$ at $\mathbf{x}(k)$ and $C = 1 + 8rC_1$.

It is revealed by (7.57) that the deviation between $\mathbf{y}(k)$ and $\mathbf{x}^i(k)$ is confined by a fixed value that is proportionate to α. Meanwhile, (7.56) suggests that the gap between $\mathbf{y}(k)$ and $\mathbf{x}^*$ is constrained. Consequently, the interval between $\mathbf{x}^i(k)$ generated by (7.51) and $\mathbf{x}^*$ is bounded. The upper limit on $f(\hat{\mathbf{x}}^i(k))$ in 7.59 demonstrates that the inaccuracy from the optimal point f^* includes two parts: one is inversely related to the step size α and reduces to zero at the rate of $1/k$ and the other is correlated to the step size α, L, C, and C_1 [29,30]. As a result, from 7.58 and 7.59, one can obtain that the distributed subgradient algorithm converges with a proper choice of α and $\boldsymbol{\Phi}(k, s)$. Interested readers can refer to [29, 30] for detailed proofs.

In the following section, we introduce the distributed subgradient algorithm to solve the TI and SE problems.

7.2.3 Distributed Topology Identification

TI identifies the changes in the topology by examining the estimated states based on measurements, thereby serving as a preprocessing step for SE. The implementation of TI takes two steps. First, we apply the subgradient algorithm and solve the least square problem to obtain the estimation of the state variables; second, we identify the topology errors via the approach of the statistical test.

7.2.3.1 Measurement Modeling

The general form of the measurement model is given by (7.45). For both the TI and SE problems, we need to derive the models that describe the relationships between measurement and the state variables. In the TI problem, we adopt the decoupled DC power flow model with only power measurements being used. The measurements, as portrayed in Figure 7.18, include

1. the active power flows in lines $k - l$, P_{kl} and P_{lk};

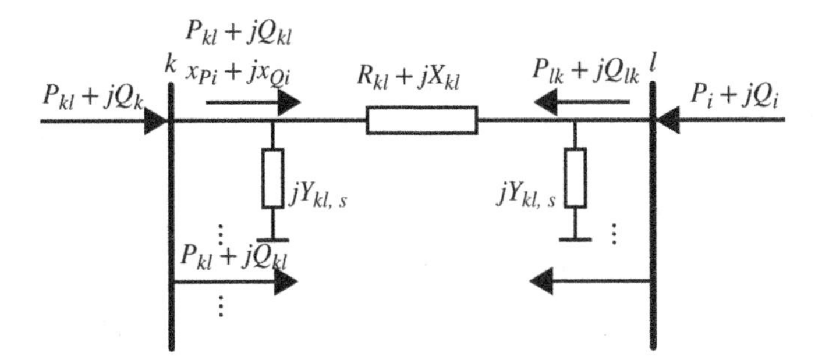

Figure 7.18 π-equivalent circuit of the branch $k - l$.

2. the reactive power flows in lines $k - l$, Q_{kl} and Q_{lk};
3. the active power injections at nodes k and l, P_k and P_l;
4. the reactive power injections at nodes k and l, Q_k and Q_l.

The active and reactive power flows at one ending point of a branch are chosen as the state variables (referred to as the sending end).

Active Power Measurement Modeling

If there is only one active power flow measurement for a branch, this measurement is selected as the state variable corresponding to this branch. If there are more than one measurement, the state variable is chosen from either end of the branch active power flows. Let us denote the active power flow that is chosen as the state variable as the sending end active power flow. The active flow that is not chosen, on the contrary, is defined as the receiving end active power flow.

Assume that node k and node l are connected via a branch. We can represent the sending and receiving ends of the active power flows as follows [32]:

$$P_{kl} = x_{\mathrm{Pi}} + e_{P,kl}, \quad P_{lk} = -x_{\mathrm{Pi}} + P_{\mathrm{L},kl} + e_{P,lk} \tag{7.60}$$

where x_{pi} is the state variable associated with this branch and $e_{P,kl}$ and $e_{P,lk}$ represent the active power measurement errors of both ends.

Let $P_{\mathrm{L},kl}$ in (7.60) denote the active power loss of the line $k - l$, which is shown as

$$P_{\mathrm{L},kl} = G_{kl}(V_k^2 + V_l^2 - 2V_kV_l\cos(\delta_{kl})) \tag{7.61}$$

where V_k and V_L are the voltage magnitudes of buses k and l, respectively, $\delta_{kl} = \delta_k - \delta_l$ represents the difference of the voltage angle on the branch $k - l$, δ_k and δ_l are the phase angles of the voltage at buses k and l, respectively, and G_{kl} is the serial conductance, which is calculated as $G_{kl} = R_{kl}/(R_{kl}^2 + X_{kl}^2)$.

In the DC power flow model, δ_{kl} can be calculated as

$$\delta_{kl} \approx X_{kl}P_{kl} = X_{kl}x_{\mathrm{Pi}} \tag{7.62}$$

Substituting (7.62) into (7.61), setting $V_k = V_l = 1$ yields

$$P_{\mathrm{L},kl} \approx 2G_{kl}(1 - \cos(X_{kl}x_{\mathrm{P}i})) \tag{7.63}$$

Consequently, (7.60) can be rewritten as

$$P_{lk} = -x_{\mathrm{P}i} + 2G_{kl}(1 - \cos(X_{kl}x_{\mathrm{P}i})) + e_{P,lk} \tag{7.64}$$

The injections of active powers at buses k and l are

$$P_k = \sum_{l \in N_k} P_{kl} + e_{P,k}, \quad P_l = \sum_{k \in N_l} P_{lk} + e_{P,l} \tag{7.65}$$

Denote the state variables associated with active power flows with an n_b-dimensional vector $\mathbf{x}_{\mathrm{P}}$, where n_b is the number of branches. Denote the active power measurements with an m_{P}-dimensional vector $\mathbf{z}_{\mathrm{P}}$, where m_P is the number of active power measurements. Therefore, the model for the active power measurement can be written as

$$\mathbf{z}_{\mathrm{P}} = \mathbf{h}_{\mathrm{P}}(\mathbf{x}_{\mathrm{P}}) + \mathbf{e}_{\mathrm{P}} \tag{7.66}$$

where $\mathbf{e}_{\mathrm{P}}$ is an m_P-dimensional vector of random errors that has a zero mean and a known covariance matrix $\mathbf{R}_{\mathrm{P}} = \mathrm{diag}\{\sigma_{\mathrm{P}1}^2, \sigma_{\mathrm{P}2}^2, \dots, \sigma_{\mathrm{P}m_{\mathrm{P}}}^2\}$.

In the TI problem, the simplified DC power flow model performs quite well for the majority of the operating conditions. In some special conditions, e.g. when a branch power flow is extremely small, the DC power flow may imply a wrong TI. Nevertheless, the following state estimation on the basis of the identified topology is still adequate because the power flow is so trivial and can be omitted. In the majority of the state estimation-based applications, this imprecision is acceptable.

Reactive Power Measurement Modeling

The select of the state variables is the same as that of the active power. When the branch has a single reactive flow measurement, the state variable is chosen to associate with this measurement. Otherwise, the state variable is defined to associate with either end of the reactive power flows.

It is shown in Figure 7.18 that the reactive power on both ends of one branch can be defined as [32]

$$Q_{kl} = x_{\mathrm{Q}i} + e_{Q,kl}, \quad Q_{lk} = -x_{\mathrm{Q}i} + Q_{\mathrm{L},kl} + e_{Q,lk} \tag{7.67}$$

where $Q_{\mathrm{L},kl}$ is the loss of the reactive power on line $k - l$, which is shown as follows:

$$Q_{\mathrm{L},kl} = -Y_{kl,s}(V_k^2 + V_l^2) = -B_{kl}(V_k^2 + V_l^2 - 2V_kV_l\cos(\delta_{kl})) \tag{7.68}$$

where B_{kl} is the serial susceptance, i.e. $B_{kl} = -X_{kl}/(R_{kl}^2 + X_{kl}^2)$.

Moreover, for the DC flow model (7.68) can be represented by

$$Q_{\mathrm{L},kl} = -2Y_{kl,s} - 2B_{kl}(1 - \cos(X_{kl}x_{\mathrm{P}i})) \tag{7.69}$$

It should be pointed out that the loss of the reactive power (7.69) relies on $x_{\mathrm{P}i}$, which can be estimated by using $\hat{x}_{\mathrm{P}i}$ from the results of active power estimation. Therefore, we can rewrite (7.69) as

$$\hat{Q}_{\mathrm{L},kl} = -2Y_{kl,s} - 2B_{kl}(1 - \cos(X_{kl}\hat{x}_{\mathrm{P}i})) \tag{7.70}$$

Hence, (7.67) is rewritten as

$$Q_{kl} = x_{\mathrm{Q}i} + e_{\mathrm{Q},kl}, \quad Q_{lk} = -x_{\mathrm{Q}i} + \hat{Q}_{\mathrm{L},kl} + e_{\mathrm{Q},lk} \tag{7.71}$$

Because $\hat{Q}_{\mathrm{L},kl}$ and the state variable $x_{\mathrm{Q}i}$ are independent, it can be obtained when the estimation of the active power $\hat{x}_{\mathrm{P}i}$ is obtained.

The injection of the reactive power at nodes k and l are shown as follows:

$$Q_k = \sum_{l \in N_k} Q_{kl} + e_{\mathrm{Q},k}, \quad Q_l = \sum_{k \in N_l} Q_{lk} + e_{\mathrm{Q},l} \tag{7.72}$$

We present the measurement model for the reactive power in the matrix form as follows:

$$\mathbf{z}_{\mathrm{Q}} = \mathbf{h}_{\mathrm{Q}}(\mathbf{x}_{\mathrm{Q}}) + \mathbf{e}_{\mathrm{Q}} \tag{7.73}$$

where $\mathbf{z}_{\mathrm{Q}}$ is the reactive power measurement vector, which has m_{Q} dimensions and m_{Q} reactive power measurements, $\mathbf{x}_{\mathrm{Q}}$ is the state vector of reactive power flows, which has n_b dimensions, and $\mathbf{e}_{\mathrm{Q}}$ is the vector of random errors, which has m_{Q} dimensions, the mean of $\mathbf{e}_{\mathrm{Q}}$ is zero, and the corresponding covariance is $\mathbf{R}_{\mathrm{Q}} = \mathrm{diag}\{\sigma^2_{\mathrm{Q}1}, \sigma^2_{\mathrm{Q}2}, \ldots, \sigma^2_{\mathrm{Q}m_{\mathrm{Q}}}\}$.

7.2.3.2 Distributed Topology Identification

Because the models of both the active and reactive power given in (7.66) and (7.73) have the same mathematical formulation, we drop the subscripts P and Q and generalize the models as follows:

$$\mathbf{z} = \mathbf{h}(\mathbf{x}) + \mathbf{e} \tag{7.74}$$

For multi-area TI, the measurement model has the same formulation as (7.45). To obtain the estimation of the states (power flows on the lines), the local objective function is given by

$$f^i(\mathbf{x}) = (\mathbf{z}^i - \mathbf{h}^i(\mathbf{x}))^{\mathrm{T}} \mathbf{R}^{\mathbf{i}^{-1}} (\mathbf{z}^i - \mathbf{h}^i(\mathbf{x})) \tag{7.75}$$

For the TI problem that uses the measurements of both the active and reactive flows, the overall objective function can be written as follows:

$$J(\mathbf{x}) = \sum_{i=1}^{r} f^i(\mathbf{x}), \quad \text{subject to } \mathbf{x} \in \mathbf{R}^{n_b} \tag{7.76}$$

Because (7.76) has the same form as (7.50), the distributed subgradient algorithm can be applied to render a distributed solution. The algorithm requires that each agent initializes with an estimate of the global optimal state as follows:

$$x^i(0) = \mathbf{x}_0^i, \quad i = 1, 2, \dots, r \tag{7.77}$$

It should be noticed that $\mathbf{x}_0^i$ can be initialized by the following methods, i.e. by a zero vector or a vector of measurements. The algorithm converges for both methods. However, the second one renders better convergence and is used in our work.

In each iteration, each agent updates its estimate based on (7.51). The subgradient in (7.51) is calculated as

$$\mathbf{d}^i(k) = -2\left(\frac{\partial \mathbf{h}^i(\mathbf{x}^i)}{\partial \mathbf{x}^i}\Big|_{\mathbf{x}^i(k)}\right)^{\mathsf{T}} (\mathbf{R}^i)^{-1}(\mathbf{z}^i - \mathbf{h}^i(\mathbf{x}^i(k))) \tag{7.78}$$

It should be noted that for each agent, the subgradient only requires the local measurements and current estimation of the agent's states. Moreover, the observability of an area is not necessary as long as the entire system is observable. We will illustrate this later.

7.2.3.3 Statistical Test for Topology Error Identification

Upon obtaining the estimates of the state variables $\hat{\mathbf{x}}^i$s, a robust statistical test [32] is adopted for the subsystems to identify the topology configuration.

The basic idea of the robust statistical test is to compare the magnitude of the estimation of the standardized flow in (7.79) to a given cutoff value M_0:

$$M_j^i = \frac{\hat{z}_j^i}{\hat{s}_{z,j}^i} \tag{7.79}$$

Here, $\hat{z}_j^i$ represents the jth estimated power flow of $\hat{\mathbf{z}}^i$ at ith area with $z_j^i = h_j^i(\hat{\mathbf{x}}^i)$. $\hat{s}_{z,j}^i$ is the square root of the jth diagonal entry of the covariance matrix $\text{cov}(\hat{\mathbf{z}}^i)$ of $\hat{\mathbf{z}}^i$, which is given by

$$\text{cov}(\hat{\mathbf{z}}^i) = \mathbf{H}^i(\hat{\mathbf{x}})\text{cov}(\hat{\mathbf{x}})\mathbf{H}^i(\hat{\mathbf{x}})^{\mathsf{T}} \tag{7.80}$$

$\text{cov}(\hat{\mathbf{x}})$ here is calculated as [33]:

$$\text{cov}(\hat{\mathbf{x}}) = \hat{s}_x(\mathbf{H}^i(\hat{\mathbf{x}})^{\mathsf{T}}(\mathbf{R}^i)^{-1}\mathbf{H}^i(\hat{\mathbf{x}}))^{-1} \tag{7.81}$$

where $\hat{s}_x$ is a scale estimate. For Gaussian measurement errors, test results with various scales of systems using Monte Carlo simulations show that $\hat{s}_x = 1.2$ exhibits decent performance [32]; accordingly, this setting is adopted for our method.

If the magnitude of the standardized flow is found to be smaller than M_0, it means that the standardized flow is so small that this line is most likely to be disconnected. Therefore, the associated branch is regarded as disconnected." Otherwise, the corresponding branch is treated as connected." Here, we set M_0 as 2.0. The value of cutoff is designed upon the test experiences via numerous test cases. Our test results show that this setting of 2.0 yields good performance for TI.

It should be pointed out that one area only needs to carry out the statistical test of the related lines in its own area. Therefore, this test can be accomplished in a distributed manner.

7.2.4 Distributed State Estimation

SE focuses on the estimation of the voltage magnitudes and phase angles upon the network configuration and measurements. For simplicity, we assume that the network topology has been identified. The measurements are adopted for SE include active and reactive power flows, active and reactive power injections, and nodal voltage magnitudes.

Notice that the simplified DC-PF model is accurate enough for TI. Though we can use the AC-PF model, it is much more complicated and time-consuming. Therefore, to increase the response speed of TI, we adopt the DC-PF model. Moreover, in order to obtain an accurate SE, the coupled AC-PF model is utilized based on the TI results, which we will introduce, as follows.

Let us denote the vector of active power flow measurements as P_{line}, the vector of active power injection measurements as P_{bus}, the vector of the reactive power flow measurements as Q_{line}, the vector of the reactive power injection measurements as Q_{bus}, and the vector of voltage magnitude measurements as V_{bus}. For convenience of notation, the measurement vectors are denoted as $\mathbf{z} = [\mathbf{P}_{\text{line}}^{\top}, \mathbf{P}_{\text{bus}}^{\top}, \mathbf{Q}_{\text{line}}^{\top}, \mathbf{Q}_{\text{bus}}^{\top}, \mathbf{V}_{\text{bus}}^{\top}]^{\top}$. Rearrange the state variables as a vector $\mathbf{x} = [\delta^{\top}, \mathbf{V}^{\top}]^{\top}$, then the overall measurement model can be represented by (7.74) and the objective for multi-area SE can be denoted by (7.76). Based on (7.51), the distributed subgradient algorithm is utilized for solving (7.76). Each agent is initialized with the so-called flat-start condition, which is given as follows:

$$\mathbf{x}^i(0) = \begin{bmatrix} \mathbf{0} \\ \mathbf{1} \end{bmatrix} \tag{7.82}$$

where $\mathbf{0}$ and $\mathbf{1}$ are the initial values for the bus-phase angles and voltage magnitudes, respectively.

The subgradient in (7.78) is updated based on the up-to-date objective function, its corresponding Jacobian is given as:

$$\frac{\partial \mathbf{h}^i(\mathbf{x}^i)}{\partial \mathbf{x}^i} = \begin{bmatrix} \frac{\partial \mathbf{P}^i_{\text{line}}}{\partial \delta} & \frac{\partial \mathbf{P}^i_{\text{line}}}{\partial \mathbf{V}} \\ \frac{\partial \mathbf{P}^i_{\text{bus}}}{\partial \delta} & \frac{\partial \mathbf{P}^i_{\text{bus}}}{\partial \mathbf{V}} \\ \frac{\partial \mathbf{Q}^i_{\text{line}}}{\partial \delta} & \frac{\partial \mathbf{Q}^i_{\text{line}}}{\partial \mathbf{V}} \\ \frac{\partial \mathbf{Q}^i_{\text{bus}}}{\partial \delta} & \frac{\partial \mathbf{Q}^i_{\text{bus}}}{\partial \mathbf{V}} \\ \frac{\partial \mathbf{V}^i_{\text{bus}}}{\partial \delta} & \frac{\partial \mathbf{V}^i_{\text{bus}}}{\partial \mathbf{V}} \end{bmatrix} \tag{7.83}$$

It should be noted that the implementation of SE is similar to that of TI and can be conducted in a distributed manner. Moreover, the distributed subgradient algorithm only requires the convexity of the local objective functions and has no requirement for the observability of each local area [29, 30]. Therefore, this splitting of areas is flexible. Moreover, we only consider nonoverlapping subsystems in this chapter. The interconnection branch can belong to either of the connected areas and treated the same as the other branches.

7.2.5 Implementation of the Integrated MAS-Based Solution for TI and SE

For the proposed integrated solution, each area is associated with an intelligent agent. The agent is in charge of data acquisition within its own subsystem and communicating with its neighbors. The algorithm is implemented in the way that it has the properties of an MAS [34], e.g. the local control function and the communication style. The MAS can be implemented on Java Agent Development (JADE) Framework [35].

Figure 7.19 portrays the information stream and function units of an agent. Notice that only the estimated states $\mathbf{x}^i[k]$s are interchanged between the subsystems. Moreover, the volume of exchanged data is not affected by the configuration of the network or the placement of the measurements.

The MAS-based algorithm can be implemented by using different frameworks, e.g. JADE [35] or Presage2 [36]. Generally, JADE is a software platform that can provide basic middle-ware layer functionalities regardless of the application. The JADE-based applications are able to run various operating systems, such as Android, Windows, or IOS. We had implemented the JADE framework on windows-based NanoPCs and Android-based Tablets [37], and we also tested various distributed algorithms by using JADE. There are other hardware platforms

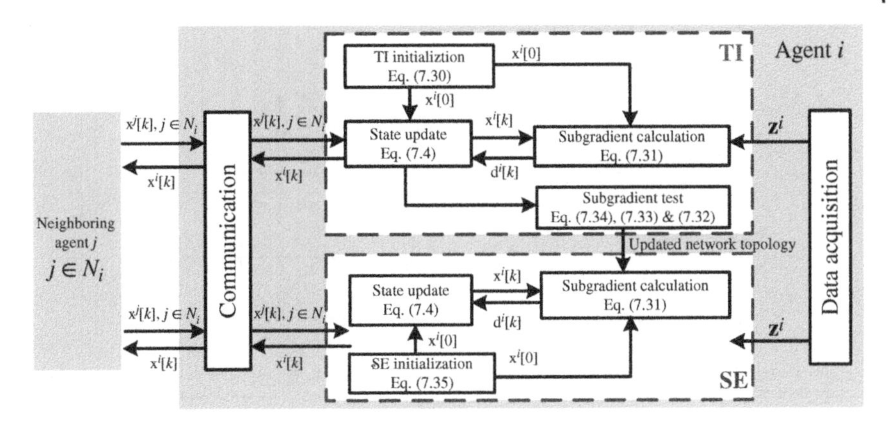

Figure 7.19 Operation of an agent for TI and SE.

such as the ARM-based and DSP-based control boards, which can also be adopted for implementation and algorithm evaluation.

In fact, the TI and SE are two function modules. Specifically, the outcome of TI is used in SE, but SE will not affect TI. Our test results demonstrate that the obtained phase angles deviate significantly from the real values if the changes in the topology are not recognized correctly. Therefore, TI, as a preprocessing step for SE, should be performed before SE because the communicational and computational mechanisms are the same for both TI and SE. It is not necessary to separate these two modules. In fact, TI can be regarded as a tool to improve the accuracy of results obtained from the SE. When the algorithm converges, only the estimated states are of interest, yet, the identified topology can be treated as a by-product that can be used for other advanced applications such as the contingency or security analysis.

7.2.6 Simulation Studies

In this section, we use four test cases to evaluate the performance of the proposed integrated TI and SE method. These four cases are an IEEE-14 bus system, a 120-bus system, a 590-bus system, and a 1062-bus system.

7.2.6.1 IEEE 14-bus System

The IEEE 14-bus system is divided into four areas. Figure 7.20 portrays the network topology and measurement locations. Power flow and injection sensors measure both active and reactive flows and injections. There are a total of 52 measurements, including 40 measurements of power flows on lines, 8 measurements of nodal power injections, and 4 measurements of voltage magnitudes. Assume that all measurements are corrupted by additive Gaussian noises with the same

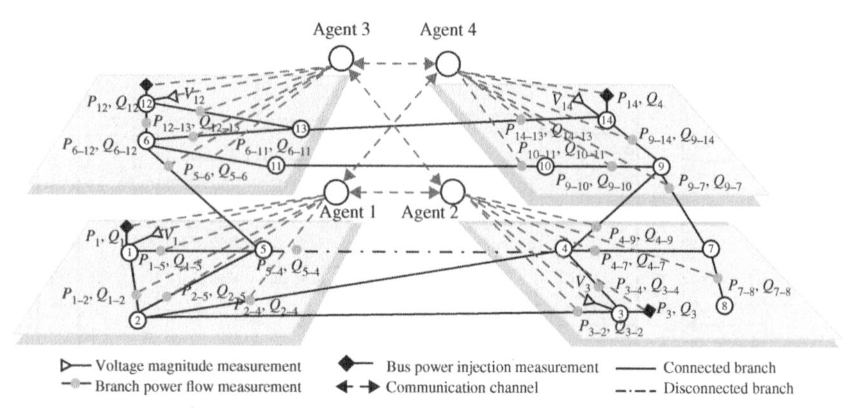

Figure 7.20 Algorithm implementation with IEEE 14-bus system.

variances of $\sigma^2 = 0.0001$. To test the proposed algorithm, line 4–5 is intentionally set as disconnected. Because there are four areas, the four-agent system is developed to implement the proposed algorithm, and the communication network for agent communication is provided in Figure 7.20.

Notice that the topology of the communication network can be designed to be different as long as the graph corresponding to the communication network is complete. In other words, the agents should be directly or indirectly linked [29]. In reality, the network topology should take both of the performances of the algorithm, i.e. speed and reliability, and the implementation cost, into account. Moreover, the topology of the communication network here satisfies the $N-1$ rule [37], which indicates that the distributed algorithm can still work properly if anyone of the communication links is lost.

Both the MAS-based integrated TI and SE methods are examined by numerical studies. The benchmark is chosen as the centralized WLS state estimator [12].

1) Test of Topology Identification

For the TI test, only active and reactive power measurements (48) are used. The actual state of the line 4–5 is "disconnected" and the goal of the TI to identify this change of topology.

The dimension of the state vectors for TI using the active or reactive power measurements is 20 because there are a total of 20 branches in the IEEE-14 bus system. Thus, the exchanged data between the two agents for each step of the update is a 20-dimensional vector. The volume of this exchanged data is relatively small, considering that there are a total of 48 measurements being used here.

By applying the distributed TI algorithm, four estimation active power flows P_{4-5}, P_{7-8}, P_{6-13}, and P_{9-14} during the iteration process are shown in Figure 7.21. For each area, the states associated with this area are initialized with measured

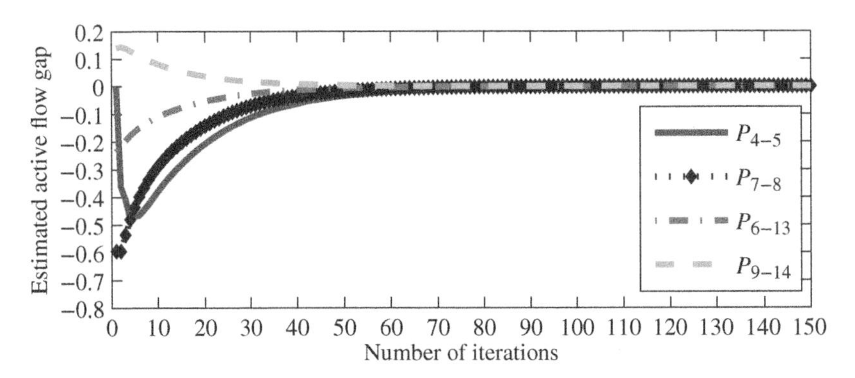

Figure 7.21 Estimated flow differences between distributed and centralized algorithms.

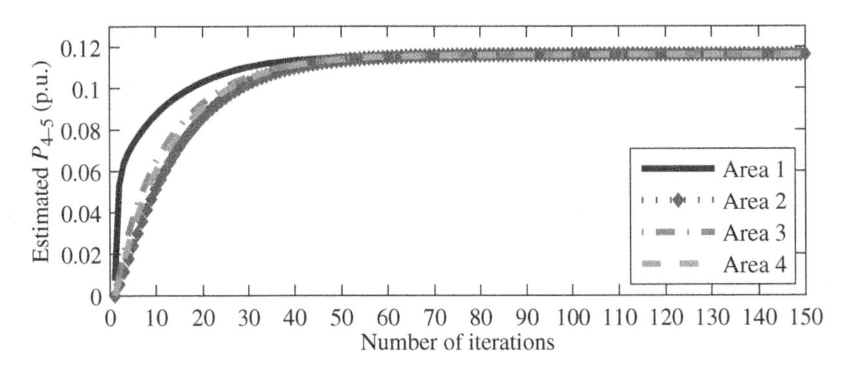

Figure 7.22 Estimated power flows (P_{4-5}) in all four areas.

flows while the other states relate to other areas are initiated with zeros. The figure shows that the algorithm takes around 100 iterations to convergence, and the converged values are comparable to that obtained with the centralized methods. The active flows for line 4–5 during the iteration process for all the four subsystems are shown in Figure 7.22. It can be observed that the flow (P_{4-5}) in all the areas achieves consensus at 100th iteration, which is one of the conditions for convergence for the distributed subgradient algorithm. Moreover, the values of subgradients for the local objective functions of all the areas converge to zeros, which indicates that the values of objective functions of these four areas stop decreasing, as shown in Figure 7.23. This is another convergence criterion for the distributed subgradient algorithm, as introduced previously.

As shown in Figure 7.22, the estimation of the active power flows of the line 4–5 are 0.12 p.u. Recall that the actual state of the line 4–5 is "disconnected"; thus, the active power flows on this line should be zero. This inaccuracy is due to occur because we have carried out the state estimation by applying the outdated topology

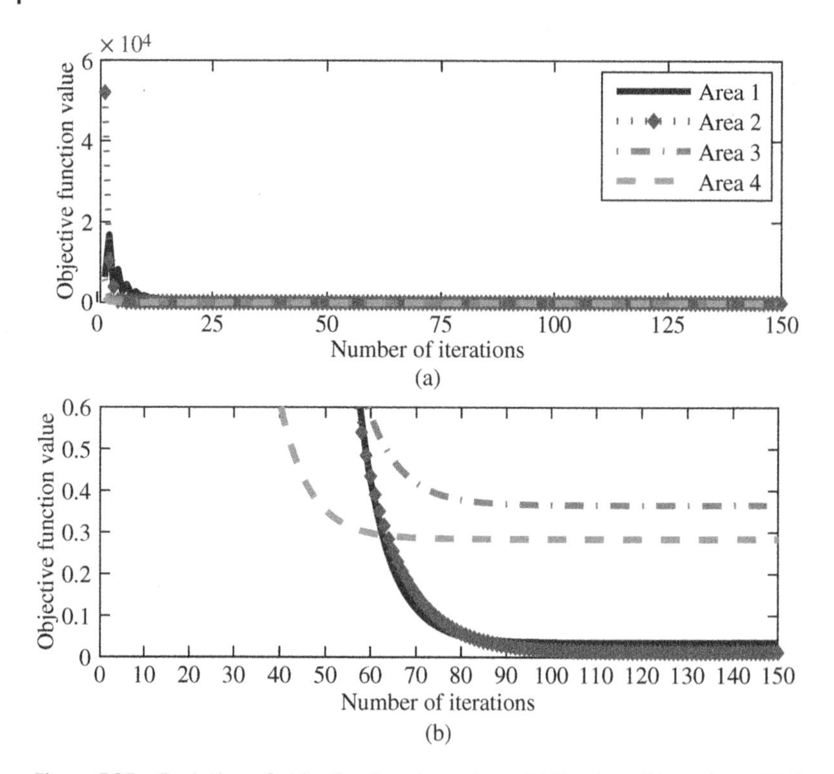

Figure 7.23 Evolution of objective function values. (a) Number of iterations. (b) Number of iterations (zoomed-in).

information, wherein the line 4–5 is assumed to be "connected." By applying the statistical test introduced previously, we can identify this topology error correctly. Tables 7.1 and 7.2 show the results of the statistical tests by using the measurements from P and Q, respectively. Here, the cutoff value for the standardized flow is set to 2.0, as discussed previously.

Because a branch (line) can transmit active or reactive power, so it is identified as "disconnected" only if statistical tests using both active and reactive measurements identify it as "disconnected." Tables 7.1 and 7.2 demonstrate that the cases for line 4–5 meet this condition. Therefore, branch 4–5 is identified as "disconnected," which reveals the real network topology. The test results shown in the tables also demonstrate that the distributed algorithm obtains the same network topology results as the centralized methods.

2) Test of State Estimation

The SE are carried out based on the identified topology from TI, wherein the branch 4–5 is disconnected. All the 52 measurements are used to improve the

Table 7.1 Test results of TI using active power flow measurements.

		Distributed			Centralized		
Area	Branch	$\hat{P}_{kl}$	$\hat{P}_{kl}/\hat{s}_{z,j}$	Status	$\hat{P}_{kl}$	$\hat{P}_{kl}/\hat{s}_{z,j}$	Status
1	1–2	1.77	198.27	Connected	1.77	198.27	Connected
	1–5	0.57	64.13	Connected	0.57	64.13	Connected
	2–4	0.90	82.27	Connected	0.90	82.27	Connected
	2–5	0.90	10.61	Connected	0.90	10.61	Connected
	4–5	0.12	0.66	**Disconnected**	0.12	0.66	**Disconnected**
2	2–3	−0.08	98.32	Connected	−0.08	98.32	Connected
	3–4	0.01	8.98	Connected	0.01	8.98	Connected
	4–7	0.19	17.32	Connected	0.19	17.32	Connected
	4–9	0.09	8.16	Connected	0.09	8.16	Connected
	7–8	0.60	0.49	**Disconnected**	0.60	0.49	**Disconnected**
3	5–6	0.19	54.34	Connected	0.19	54.34	Connected
	6–11	0.10	17.03	Connected	0.10	17.03	Connected
	6–13	0.23	10.67	Connected	0.23	10.67	Connected
	6–12	0.01	21.44	Connected	0.01	21.44	Connected
	12–13	0.18	4.52	Connected	0.18	4.52	Connected
4	7–9	−0.04	16.37	Connected	−0.04	16.37	Connected
	9–10	0.03	3.62	Connected	0.03	3.62	Connected
	9–14	−0.14	3.86	Connected	−0.14	3.86	Connected
	10–11	0.04	12.75	Connected	0.04	12.75	Connected
	13–14	0.11	12.53	Connected	0.11	12.53	Connected

accuracy of the estimation. We set the state variables as follows:

$$\begin{aligned} \mathbf{x}^{\top} = [&\delta_2, \delta_5, V_2, V_5, && \text{Area 1} \\ &\delta_3, \delta_4, \delta_7, \delta_8, V_3, V_4, V_7, V_8, && \text{Area 2} \\ &\delta_6, \delta_{11}, \delta_{12}, \delta_{13}, V_6, V_{11}, V_{12}, V_{13}, && \text{Area 3} \\ &\delta_9, \delta_{10}, \delta_{14}, V_9, V_{10}, V_{14}] && \text{Area 4} \end{aligned} \tag{7.84}$$

The voltage phase angle of node #5, δ_5 is depicted in Figure 7.24. The figure shows that the proposed solution for SE takes approximately 800 iterations till convergence, which is more than the iterations of TI. This is because that AC-PF model is used in this case with more measurements being also adopted for estimation.

Table 7.2 Test results of TI using reactive power flow measurements.

		Distributed			Centralized		
Area	Branch	$\hat{Q}_{kl}$	$\hat{Q}_{kl}/\hat{s}_{z,j}$	Status	$\hat{Q}_{kl}$	$\hat{Q}_{kl}/\hat{s}_{z,j}$	Status
1	1–2	−0.27	29.27	Connected	−0.27	29.27	Connected
	1–5	0.06	6.17	Connected	0.06	6.17	Connected
	2–4	0.03	6.16	Connected	0.03	6.16	Connected
	2–5	−0.07	6.96	Connected	−0.07	6.96	Connected
	4–5	0.08	0.71	**Disconnected**	0.08	0.71	**Disconnected**
2	2–3	0.03	3.11	Connected	0.03	3.11	Connected
	3–4	−0.01	3.83	Connected	−0.01	3.83	Connected
	4–7	−0.11	9.95	Connected	−0.11	9.95	Connected
	4–9	0.01	0.66	**Disconnected**	0.09	0.66	**Disconnected**
	7–8	0.15	19.40	Connected	0.15	19.40	Connected
3	5–6	0.02	13.69	Connected	0.02	13.69	Connected
	6–11	0.10	17.03	Connected	0.10	17.03	Connected
	6–13	0.08	0.10	**Disconnected**	0.08	0.10	**Disconnected**
	6–12	−0.21	7.68	Connected	−0.21	7.68	Connected
	12–13	0.09	1.39	**Disconnected**	0.09	1.39	**Disconnected**
4	7–9	0.07	7.86	Connected	0.07	7.86	Connected
	9–10	0.05	6.74	Connected	0.05	6.74	Connected
	9–14	0.01	5.65	Connected	0.01	5.65	Connected
	10–11	−0.01	0.49	**Disconnected**	−0.01	0.49	**Disconnected**
	13–14	0.00	0.36	**Disconnected**	0.00	0.36	**Disconnected**

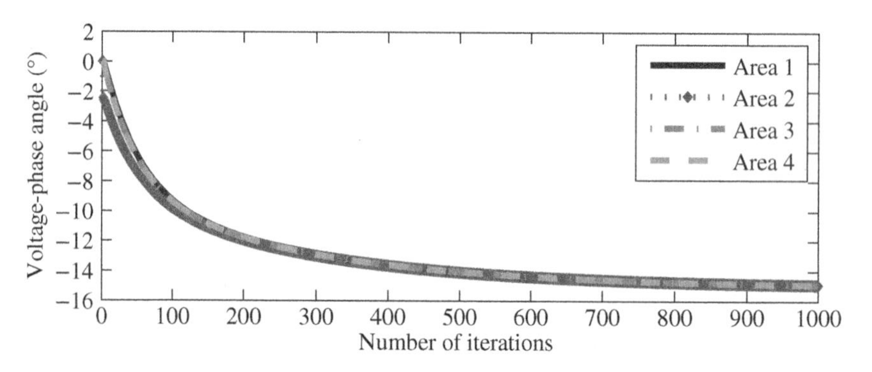

Figure 7.24 Evolution of the bus voltage phase angle (δ_5) and magnitude (V_5).

Table 7.3 Comparison of state estimate solutions.

Bus	Voltage phase angle $\hat{\delta}_i$ (°)			Voltage magnitude $\hat{V}_i$ (p.u.)		
	W/O TI	With TI	True value	W/O TI	With TI	True value
1	0.00	0.00	0.00	1.060	1.060	1.060
2	−5.60	−5.72	−5.68	1.045	1.045	1.045
3	−11.34	−14.47	−15.18	1.010	1.010	1.010
4	−8.53	−13.24	−14.33	1.026	1.020	1.014
5	−7.07	−6.23	−6.52	1.028	1.025	1.024
6	−8.09	−13.65	−13.98	1.070	1.070	1.070
7	−10.34	−16.04	−16.31	1.053	1.055	1.057
8	−10.24	−15.92	−16.31	1.090	1.090	1.090
9	−11.41	−17.03	−17.35	1.042	1.047	1.048
10	−11.10	−16.69	−17.04	1.037	1.045	1.044
11	−9.58	−15.15	−15.65	1.047	1.048	1.052
12	−9.15	−14.69	−15.02	1.056	1.054	1.052
13	−9.28	−14.83	−15.29	1.050	1.049	1.049
14	−11.59	−17.16	−17.46	1.026	1.026	1.030

The estimated states using the integrated TI and SE method are provided in Table 7.3. State estimation results without integrating TI are provided here for comparison.

Table 7.3 shows that the estimated states obtained from the integrated TI & SE solution are very close to true values. However, the estimated phase angles have great deviations without TI being carried out first because the disconnection of line 4–5 leads to the change of active power flows on the branches. Nevertheless, the voltage magnitudes are still close to the true values even without TI. This is because the reactive power flow of line 4–5 is so small that the connection of line 4–5 has an insignificant influence on the reactive power flows through this branch.

3) Observability Analysis

The observability of a subsystem is determined by the rank of the Jacobian matrix corresponding to the local measurements, which is denoted by $\mathbf{H}^i$ [11]. An area A_i is observable if and only if

$$\text{rank}(\mathbf{H}^i) = n_s^i - 1 \tag{7.85}$$

where n_s^i is the number of the states to be estimated in A_i.

Based on the placements of the measurements shown in Figure 7.20, the ranks of the Jacobian matrix corresponding to the local measurements, $\mathbf{H}^i$, are as follows:

$$\begin{cases} \text{rank}(\mathbf{H}^1) = 3, & \text{rank}(\mathbf{H}^2) = 5 \\ \text{rank}(\mathbf{H}^3) = 6, & \text{rank}(\mathbf{H}^4) = 5 \end{cases} \tag{7.86}$$

According to the state variables defined in (7.84), we can verify that all of these four areas are observable. Moreover, the rank of the overall Jacobian of the entire system is 22 without the voltage magnitudes of the PV buses being excluded. Consequently, the system is observable.

Moreover, if we exclude the active and reactive power flow measurements associated with the line 6–11 the ranks of the local measurements Jacobian matrix under these circumstances are then as follows:

$$\begin{cases} \text{rank}(\mathbf{H}^1) = 3, & \text{rank}(\mathbf{H}^2) = 5 \\ \text{rank}(\mathbf{H}^3) = 5, & \text{rank}(\mathbf{H}^4) = 5 \end{cases} \tag{7.87}$$

Here, we can observe that area A_3 becomes locally unobservable. In fact, when the measurements associated with line 6–11 are excluded, the states' associate with bus #11 (δ_{11}, V_{11}) are no longer observable in Area #3. Nevertheless, the observability of the overall system still holds as the rank of the system-wide measurement Jacobian matrix is still 22. Figure 7.25 demonstrates that for the case when A_3 is not observable, the estimated state (δ_{11}) is the same as that of the case without the removal of measurements associated with line 6–11. Similar phenomena can also be observed for other estimated states. In fact, for our proposed distributed solution, the observability for the subsystems is not required, which allows for the more adaptable decomposition of subsystems for the distributed solution of the SE problems.

For our implementation, an agent shares its estimation with its neighboring agents only, thereby it can greatly reduce the amount of measured data

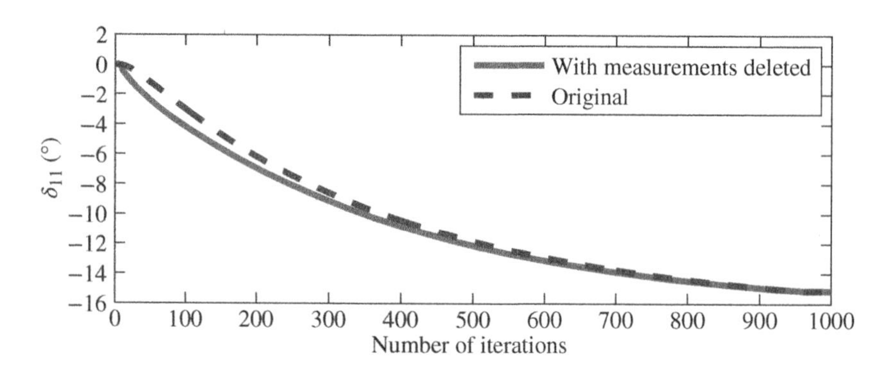

Figure 7.25 Convergence of δ_{11} with or W/O deletion of measurements.

Table 7.4 Configuration of the test systems.

Test system	n_b	n_A	n_{bp}	n_{tp}
14-Bus	20	4	5	1
120-Bus	172	4	43	2
590-Bus	908	5	182	4
1062-Bus	1635	9	182	6

transmission. Moreover, the communication time can be greatly shortened without gathering measurement data from multiple distant areas. Furthermore, the algorithm does not need to calculate the inverse of the gain matrix (the majority of the centralized methods do). This is helpful for reducing the computation time and dealing with the situation with ill-conditioned gain matrices. Therefore, the proposed distributed solution is robust and computation-efficient.

7.2.6.2 Large Test Systems

Here, we provide more test results of TI from three other systems to evaluate our proposed distributed approach.

Table 7.4 summarizes the configurations of these three test systems, where n_b is the number of lines, n_A is the number of areas, n_{bp} is the average number of branches (lines) per area, and n_{tp} is the number of topology errors we aim to identify.

The graph of the communication network for the agents' communication is shown in Figure 7.26.

We choose one of the state variables for each of these three system for demonstration, and the evolution of the corresponding variable during the iteration process is shown in Figure 7.27. As can be seen in the figure, the algorithm takes 50 iterations to converge for the 120-bus system, 120 iterations for the 590-bus system, and 120 iterations for the 1062-bus system. The test results of these systems are summarized in Table 7.5. It can be observed that the time consumed for a single iteration is the same for the 590-bus and 1062-bus system. Moreover, the computation time is greatly shortened compared with the centralized algorithms because the computational task is distributed to multiple agents instead of being taken only by a central processor.

For the implementation of the distributed algorithm, we had tested the configuration with the agents being implemented on a single computer as well as multiple computers [37]. Current tests for the TI and SE are carried out with a single computer. The message exchanged between two agents should contain two parts, i.e. the header and the actual state vector. The header includes two

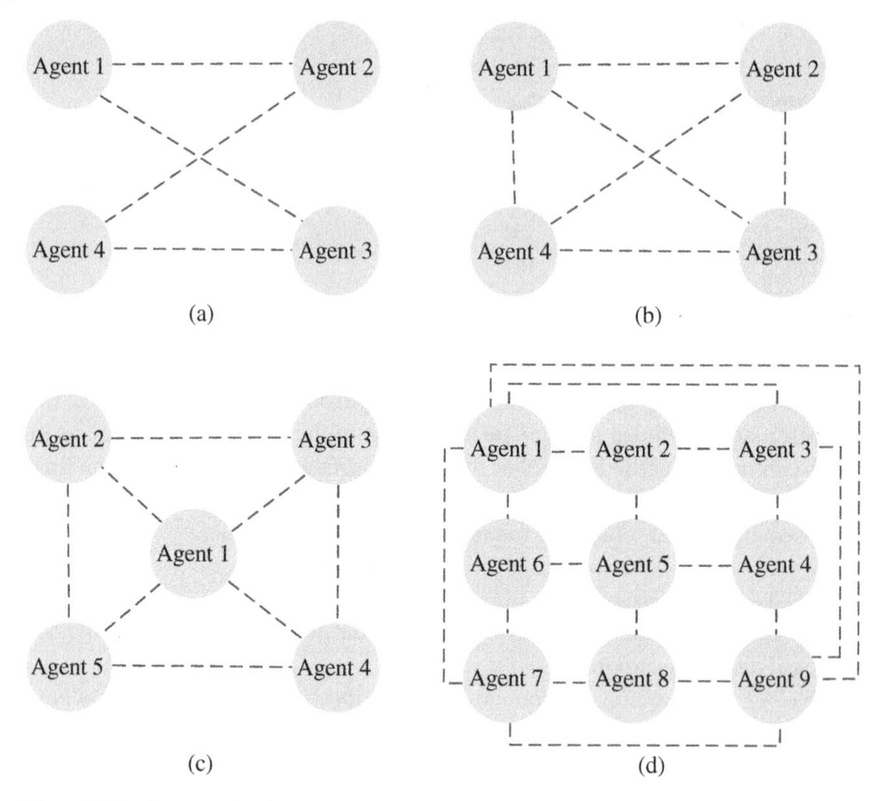

Figure 7.26 Communication network configurations of the test systems. (a) 4-Area 14-bus system. (b) 4-Area 120-bus system. (c) 5-Area 590-bus system. (d) 9-Area 1062-bus system.

parts, namely the agent ID and the iteration number. The state vector includes the voltage magnitudes and phase angles of all buses. For the tested IEEE 14-bus system, the state vector has $14 \times 2 = 28$ elements. Each element is stored using 16-bit data. Therefore, the size of the message data is $(2 + 28) \times 16 = 512$ bits.

The simulation results show that it takes 1000 iterations to converge for the 14-bus system because the mathematical operation of one iteration is so simple that the time consumed by the computation can be neglected. If we expected the algorithm to converge in three minutes, the minimum communication bandwidth is only $512 \times 1000/180 = 2.84$ kbit/s. Similarly, we can estimate bandwidth requirement of other systems by considering the scale of the system (the size of message data), convergence rate (the number of iterations), and the expected outcome (time in seconds). It turns out that the minimum communication bandwidth requirement is 21 kbit/s for the 120-bus system, 105.7 kbit/s for the 590-bus system, and 188 kbit/s for the 1062-bus system. We can see that the

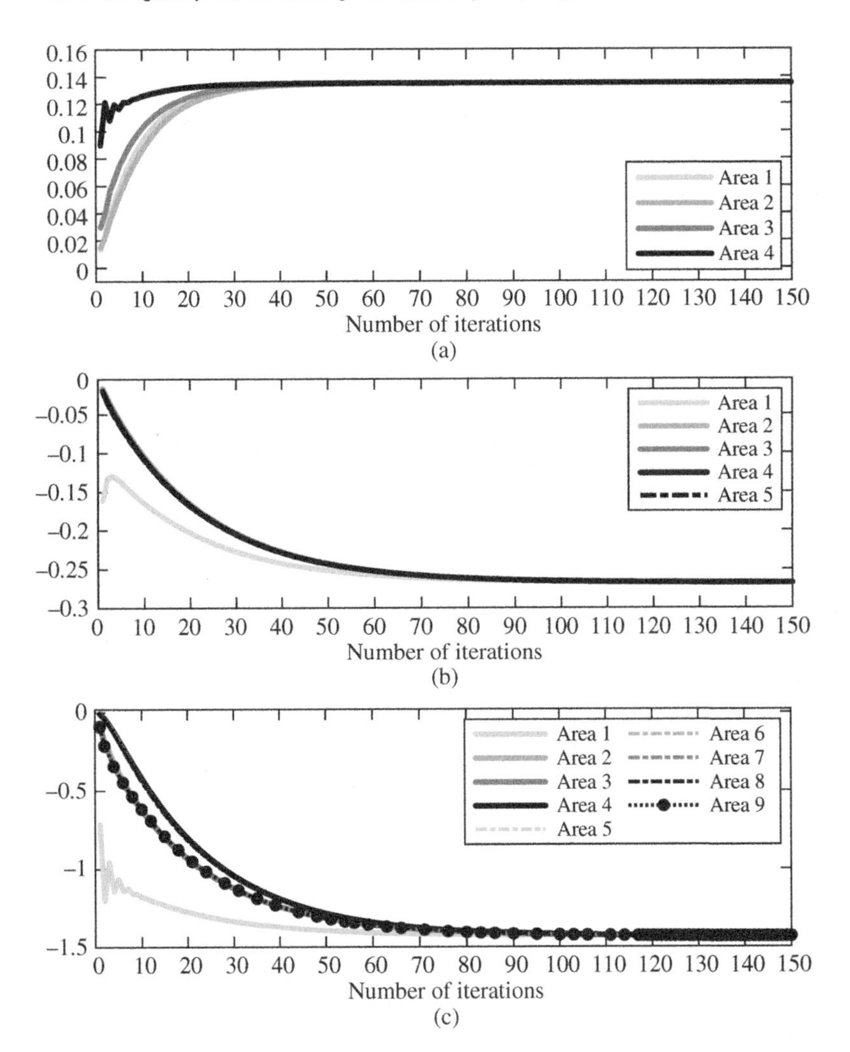

Figure 7.27 Convergence of a state variable for large test systems. (a) 120-Bus system. (b) 590-Bus system. (c) 1062-Bus system.

proposed distributed control scheme here does not impose strict requirements on communication.

In practice, we can integrate the proposed TI and SE solution in the energy management system (EMS) as one of its function modules. The EMS in each area coordinates with the Supervisory Control and Data Acquisition (SCADA) system and communicate with its neighboring areas for information exchange via the communication network.

Table 7.5 Configuration of the test systems.

Test system	Distributed (Local)		Centralized (Global)	
	Iterations	Time per iteration (ms)	Iterations	Time per iteration (ms)
14-Bus	20	0.1	80	1.51
120-Bus	50	4	80	86
590-Bus	120	69	100	3465
1062-Bus	120	72	85	10 990

Notice that the actual processing time of our designed state estimator is affected by many factors including the scale of the power system, the structure and the latency of the communication network, and the capability of the data processor of the control center. Our experience indicates that the processing time of the state estimator is not likely to exceed two minutes. For example, in the largest 1062-bus test system, the simulation conducted in Matlab takes about 1000 iterations till convergence, which takes 72 seconds in a Laptop with an Intel(R) Core(TM) i7-4770 CPU and 8.0 GB RAM. If a communication network with a speed of 10 MBs/s is adopted, the time for communication only costs three seconds. Therefore, the state estimation can be completed within 75 seconds under ideal conditions. If the algorithm is programmed in C, the time will be further shortened. Notice that the state estimator is designed to work with the SCADA system coordinately. Because the update interval of SCADA ranges from 3 to 10 minutes, our designed algorithm is fast enough to obtain the SE solution within one update interval of the SCADA system.

The integration of the intermittent distributed generations, such as solar and wind energies, greatly increases the uncertainty of the power grid. To handle uncertainty, more precise and rapid SE methods that do not depend largely upon computational burden are highly desirable. We can further integrate other techniques, e.g. event-triggered mechanism [38], to improve the performance of the proposed integrated TI and SE solution.

7.3 Conclusion and Discussion

In this chapter, we investigated two distributed algorithms for state estimation. The first one the the consensus based algorithm, which discovers the gain matrix information via the consensus fusion techniques. It still need the calculate the inverse of the gain matrix, yet it avoids of collecting global information. Accordingly, we can see this method as distributed information gather with centralized-like optimization.

The second one is the distributed subgradient based algorithm, which actually decomposes the original optimization problem into multiple optimization problem. This algorithm can be implemented by following the rules of distributed information gathering as well as distributed computation. In practice, one needs to systematically consider the hardware, software as well as communication network configuration to decide which one is better for deployment.

References

1 Zhu, H. and Giannakis, G.B. (2012). Multi-area state estimation using distributed SDP for nonlinear power systems. *2012 IEEE 3rd International Conference on Smart Grid Communications (SmartGridComm)*. IEEE, pp. 623–628.

2 Korres, G.N., Tzavellas, A., and Galinas, E. (2013). A distributed implementation of multi-area power system state estimation on a cluster of computers. *Electric Power Systems Research* 102: 20–32.

3 Kekatos, V. and Giannakis, G.B. (2012). Distributed robust power system state estimation. *IEEE Transactions on Power Systems* 28 (2): 1617–1626.

4 Korres, G.N. (2010). A distributed multiarea state estimation. *IEEE Transactions on Power Systems* 26 (1): 73–84.

5 Gomez-Exposito, A. and de la Villa Jaen, A. (2009). Two-level state estimation with local measurement pre-processing. *IEEE Transactions on Power Systems* 24 (2): 676–684.

6 Caro, E., Conejo, A.J., and Minguez, R. (2011). Decentralized state estimation and bad measurement identification: an efficient Lagrangian relaxation approach. *IEEE Transactions on Power Systems* 26 (4): 2500–2508.

7 Xie, L., Choi, D.H., Kar, S. and Poor, H.V. (2012). Fully distributed state estimation for wide-area monitoring systems. *IEEE Transactions on Smart Grid* 3 (3): 1154–1169.

8 Kar, S., Hug, G., Mohammadi, J. and Moura, J.M. (2014). Distributed state estimation and energy management in smart grids: a consensus + innovations approach. *IEEE Journal of Selected Topics in Signal Processing* 8 (6): 1022–1038.

9 Lobel, I. and Ozdaglar, A. (2010). Distributed subgradient methods for convex optimization over random networks. *IEEE Transactions on Automatic Control* 56 (6): 1291–1306.

10 Chavali, P. and Nehorai, A. (2015). Distributed power system state estimation using factor graphs. *IEEE Transactions on Signal Processing* 63 (11): 2864–2876.

11 Korres, G.N. (2011). A distributed multiarea state estimation. *IEEE Transactions on Power Systems* 26 (1): 73–84.

12 Abur, A. and Exposito, A.G. (2004). *Power System State Estimation: Theory and Implementation*. CRC Press.

13 Olfati-Saber, R., Fax, J.A., and Murray, R.M. (2007). Consensus and cooperation in networked multi-agent systems. *Proceedings of the IEEE* 95 (1): 215–233.

14 Schweppe, F.C. and Rom, D.B. (1970). Power system static-state estimation, Part II: Approximate model. *IEEE Transactions on Power Apparatus and Systems* 1: 125–130.

15 Choi, D.-H. and Xie, L. (2011). Fully distributed bad data processing for wide area state estimation. *2011 IEEE International Conference on Smart Grid Communications (SmartGridComm)*. IEEE, pp. 546–551.

16 University of Washington (1999). *ower System Test Cases Archive*. http://www.ee.washington.edu/research/pstca/ (visited on 10/02/2019).

17 Schweppe, F.C. and Wildes, J. (1970). Power system static-state estimation, Part I: Exact model. *IEEE Transactions on Power Apparatus and systems* 1: 120–125.

18 Goluh, G.H. and Van Loan, C.F. (1996). *Matrix Computations*. Johns Hopkins University Press.

19 Schweppe, F.C. and Wildes, J. (1970). Power system static-state estimation, Part I: Exact model. *IEEE Transactions on Power Apparatus and Systems* PAS-89 (1): 120–125. https://doi.org/10.1109/TPAS.1970.292678.

20 Zhang, L. and Abur, A. (2013). Strategic placement of phasor measurements for parameter error identification. *IEEE Transactions on Power Systems* 28 (1): 393–400. https://doi.org/10.1109/TPWRS.2012.2199139.

21 Monticelli, A. (1999). Fast decoupled state estimator. *State Estimation in Electric Power Systems*, 313–342. Springer.

22 Wu, F.F., Liu, W.H., Holten, L., Gjelsvik, L. and Aam, S. (1988). Observability analysis and bad data processing for state estimation using Hachtel's augmented matrix method. *IEEE Transactions on Power Systems* 3 (2): 604–611.

23 Kekatos, V. and Giannakis, G.B. (2013). Distributed robust power system state estimation. *IEEE Transactions on Power Systems* 28 (2): 1617–1626.

24 Falcao, D.M., Wu, F.F., and Murphy, L. (1995). Parallel and distributed state estimation. *IEEE Transactions on Power Systems* 10 (2): 724–730.

25 Dasgupta, K. and Swarup, K.S. (2011). Tie-line constrained distributed state estimation. *International Journal of Electrical Power & Energy Systems* 33 (3): 569–576.

26 Lakshminarasimhan, S. and Girgis, A.A. (2007). Hierarchical state estimation applied to wide-area power systems. *Power Engineering Society General Meeting, 2007. IEEE*. IEEE, pp. 1–6.

27 Yang, T., Sun, H., and Bose, A. (2011). Transition to a two-level linear state estimator–Part I: Architecture. *IEEE Transactions on Power Systems* 26 (1): 46–53.

28 Xie, L., Choi, D.-H., and Kar, S. (2011). Cooperative distributed state estimation: local observability relaxed. *Power and Energy Society General Meeting, 2011 IEEE*. IEEE, pp. 1–11.

29 Nedic, A. and Ozdaglar, A. (2009). Distributed subgradient methods for multi-agent optimization. *IEEE Transactions on Automation and Control* 54 (1): 48–61.

30 Lobel, I. and Ozdaglar, A. (2011). Distributed subgradient methods for convex optimization over random networks. *IEEE Transactions on Automation and Control* 56 (6): 1291–1306.

31 Nedic, A. and Ozdaglar, A.E. (2007). Approximate primal solutions and rate analysis for dual subgradient methods. *SIAM Journal on Optimization* 19 (4): 1757–1780.

32 Mili, L., Steeno, G., Dobraca, F. and French, D. (1999). A robust estimation method for topology error identification. *IEEE Transactions on Power Systems* 14 (4): 1469–1476.

33 Huber, P.J. (2011). *Robust Statistics*. Berlin, Heidelberg: Springer-Verlag.

34 Wooldridge, M. (2009). *An Introduction to Multiagent Systems*. Wiley.

35 Bellifemine, F.L., Caire, G., and Greenwood, D. (2007). *Developing Multi-Agent Systems with JADE*. Wiley.

36 Macbeth, S., Pitt, J., Schaumeier, J. and Busquets, D. (2012). Animation of self-organising resource allocation using Presage2. *2012 IEEE 6th International Conference on Self-Adaptive and Self-Organizing Systems (SASO)*, pp. 225–226.

37 Zhang, W. (2013). Mutiagent system based algorithm and their applicaitons in power systems. PhD dissertation. New Mexico Sate University: Las Cruces, NM.

38 Francy, R.C., Farid, A.M., and Youcef-Toumi, K. (2015). Event triggered state estimation techniques for power systems with integrated variable energy resources. *ISA Transactions* 56: 165–172.

8

Hardware-Based Algorithms Evaluation

8.1 Steps of Algorithm Evaluation

For engineering research, one should always bring real-world application requirements in mind. Once an algorithm has been designed, it is always desirable to perform hardware experimentation to estimate its real-world performance. For some highly application-oriented research areas, such as power electronics, experimentation studies have already become a requirement for journal publications. The requirement for hardware experimentation is being extended to other research areas, including microgrid control over the past years. Under such a condition, we have been working hard to gain hardware experimentation capability. With over one million USD funding from various sources, especially two Defense University Research Instrumentation Program (DURIP) grants from the US Department of Defense, our lab has developed an advance test-bed for experimental study of algorithms on operation and control of power systems. We have published a paper in IEEE Transactions on Industrial Informatics to introduce our hardware development.

Developing a good test-bed for power system study is technically challenging because of the big gaps among related research areas, including but not limited to the theoretical control, analytical power systems, and experimental power electrics. In addition to the technical challenges, hardware development also requires significant investment for purchasing components and subsystems and hiring persons with the right skills. With the fast development of power electronics and communication techniques, it is possible to develop a suitable test-bed based on the need of research and availability of resources. In this chapter, we will introduce our effort with algorithm evaluation, from hardware-in-the-loop (HIL) simulation to hardware experimentation, with an increase of technology readiness level (TRL). We believe that the introduction can benefit other researchers.

Distributed Energy Management of Electrical Power Systems, First Edition.
Yinliang Xu, Wei Zhang, Wenxin Liu, and Wen Yu.

Published 2021 by John Wiley & Sons, Inc.

During algorithm evaluation, one needs to evaluate an algorithm first through model-based simulation. For complicated dynamic systems whose simulations are unacceptably slow, real-time simulation has its advantages. The powerful hardware and specialized software of the real-time simulator (RTS) can simulate a complex model in real time if the RTS is powerful enough and the simulation is well implemented. However, RTS does not have any advantage in terms of simulation speed when a model can be simulated faster than real time. With the increasing speed of computers, more and more models can be simulated faster than real time. Under such conditions, a reasonable question one may ask is "Why should we use the expensive RTS?" As discussed below, we believe the best use of RTS is with fast prototyping of algorithms and HIL simulation.

HIL simulations can be classified into two types, controller-HIL (C-HIL) and power-HIL (P-HIL) simulations. There are many different ways of performing HIL simulations. Here, we will introduce our ways of HIL simulations. During C-HIL simulation, the RTS simulates a power system model in real time and interacts with external controllers (PC or DSP control board) that implement a control algorithm. During P-HIL simulation, a controller (RTS or DSP control board) simulates the dynamic model of equipment in real time and emulates the response of the equipment through interaction with external power circuit. In our lab, we use C-HIL simulation to test system- and equipment-level algorithms and P-HIL simulation to emulate equipment that we do not have, such as synchronous generator (SG), large energy storage system (ESS), etc.

Even though HIL simulation is very useful, it can never replace hardware experimentation. This is mainly because the model-based HIL simulation is unable to evaluate the performance of an algorithm under model inaccuracy. Thus, we have developed many power electronics-based equipment, including multiple microgrid test-beds, modular multilevel converters, etc. After C-HIL simulation, both equipment- and system-level control algorithms have been implemented in a way similar to that with hardware experimentation, i.e. equipment-level algorithm being implemented using DSP control board and system-level algorithm being implemented using PC, respectively. After P-HIL simulation, even the hard-to-obtain equipment becomes available. Finally, we can integrate the developed subsystems as a multiple-bus power system test-bed for hardware experimentation.

In this chapter, we will introduce our ways of algorithm evaluations that are enabled through years of hardware and software development. Figure 8.1 introduces the steps of algorithm evaluation, ranging from C-HIL simulation, P-HIL simulation, until hardware experimentation. Details are given in the following sections.

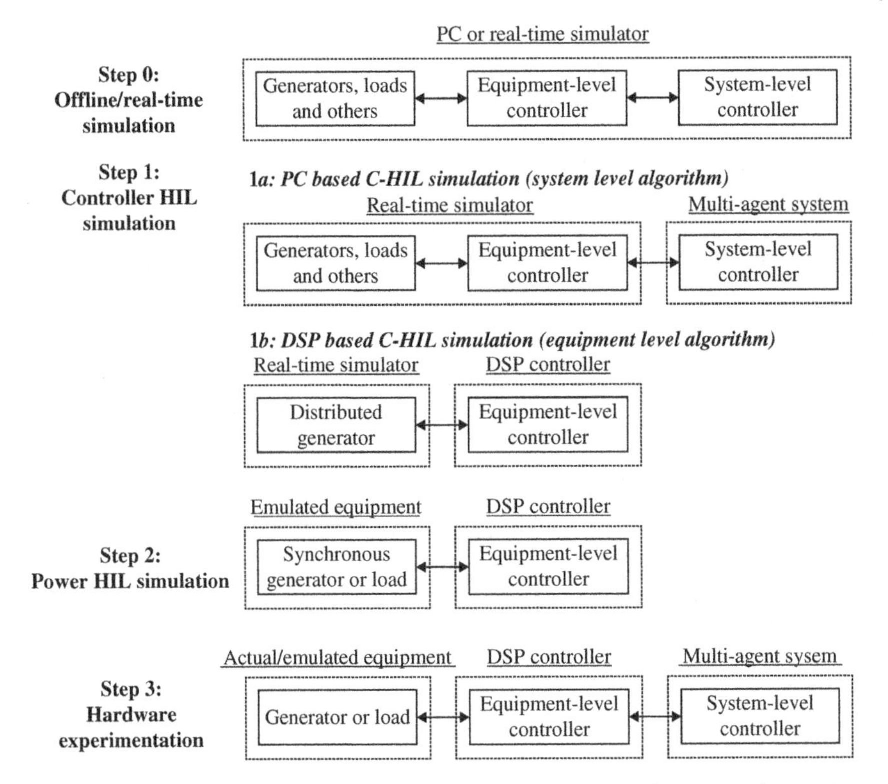

Figure 8.1 Steps of algorithm evaluation, from simulation to hardware experimentation.

8.2 Controller Hardware-In-the-Loop Simulation

During C-HIL simulation, hardware controller can interact with power system models simulated in real time through wireless or wired communications. Through such study, both capability of controller hardware and performance of control algorithms can be tested. We have performed two types of C-HIL simulations with the RTS, which are PC-based and DSP control board-based, respectively. It should be noted that the C-HIL simulations might require significant effort with hardware development and algorithm implementation. Before we introduce the details of our C-HIL simulations, we will introduce the RTS that we use.

Our eMEGAsim series RTS was made by OPAL-RT. It can provide high-fidelity modeling, simulation, and HIL simulations. Our RTS has 5 out of 12 CPU cores (2 hexa-core CPUs) unlocked. This enables the system being able to simulate 100+ nodes power systems in sufficient details. The RTS has a number of advantages. First, it uses the popular Matlab/Simulink software for modeling and algorithm implementations, which is a very attractive feature for control researcher. Second, the RTS has sufficient analog and digital (A/D) input and output (I/O) ports. In addition to the I/O cards provided by the manufacturer, our lab has designed customized electrical-to-optical (E2O) conversion boards for reliable high-frequency pulse width modulation (PWM) signal communications between RTS and power converter.

8.2.1 PC-Based C-HIL Simulation

To implement the MAS-based algorithms, we choose to use the JADE (Java Agent Development) framework. JADE is a software framework fully implemented in Java. It simplifies the implementation of MAS through a middle-ware that complies with the Foundation for Intelligent Physical Agents (FIPA) standard specifications and through a set of graphical tools that support the debugging and deployment phases. A JADE-based system can be distributed across machines, which do not even need to run the same operating system. JADE is completely implemented in Java language and the minimal system requirement is the version 5 of Java (the run time environment or the JDK). Over the past years, we have successfully implemented the MAS-based algorithms with devices running Windows, Android, and Linux.

To meet the requirement, the C API provided by the manufacturer is used. The two solutions that can realize such interactions are illustrated in Figure 8.2. As illustrated in Figure 8.2a, the MAS is interacting with the host computer that monitors and controls the power system model simulated with the RTS (target computer). Because the interactions indirectly go through the host computer, the solution has a one critical disadvantage. Because the host computer processes all communicated data as a central hub, the communication speed is slow, and the autonomous and asynchronous properties of the MAS are violated. In order to

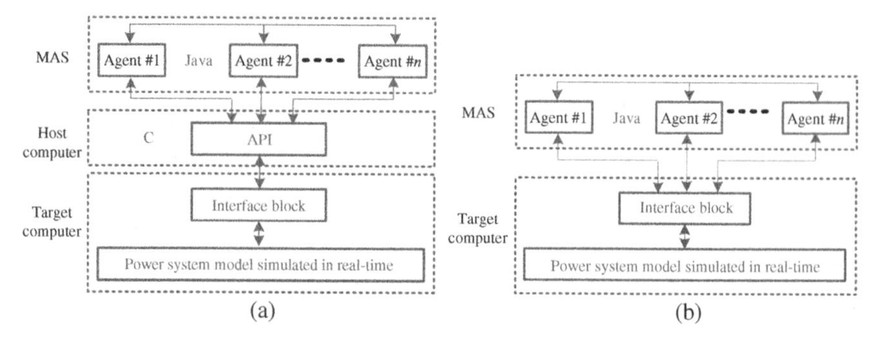

Figure 8.2 Modes of interaction between the MAS and the RTS. (a) Indirect communication scheme. (b) Direct communication scheme.

overcome this problem, we have realize direct interaction between the MAS and the target computer through advanced programming, as illustrated in Figure 8.2b.

Thus, there is a hidden requirement for a common triggering signal for synchronization, which could be from either the same internet time server or the GPS. Because the microgrid is not grid-tied, we select one agent to generate the reference signals for triggering system-wide control updates. This way of implementation might violate the distributed or decentralized property of an algorithm. However, this is necessary to maintain reliability operation of a distributed control system.

Figure 8.3 shows the block diagram for reliable implementation of a distributed control algorithm. Because communication and calculation both take time to complete, we intentionally add small delays between operations. The delays are much smaller than the step size (updating interval) of the distributed algorithm,

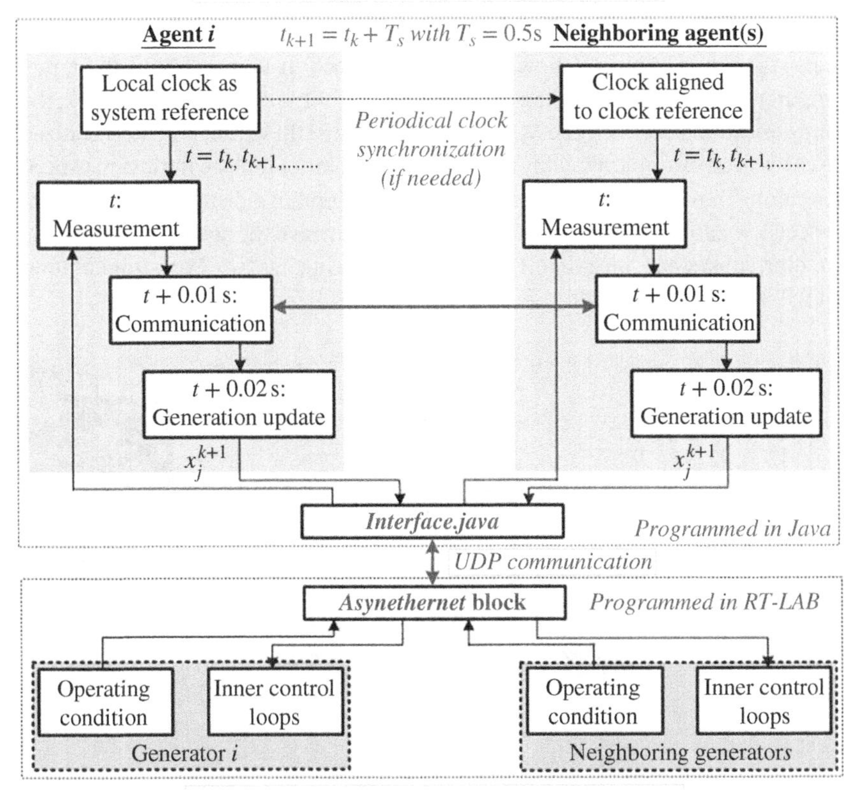

Figure 8.3 Reliably realizing a discrete-distributed algorithm.

for example, 0.01 seconds vs. 0.5 seconds, to wait for the completion of previous operations. This can make sure that the correct information can be received and processed. The time step of 0.5 seconds is just an example, which should work for traditional large-scale power system but might be too large for power electronic-based microgrids because of their small inertia. One needs to carefully decide the time step size based on application requirements.

In one of our previous projects, we implemented a proportional load sharing algorithm for both medium-voltage AC (MVAC) shipboard power system (SPS) and medium-voltage DC (MVDC) SPS. The control objective is to share the total demand among multiple distributed generators (DGs) according to their generation capacities. Because of the similarities MVAC and MVDC SPS, we just need to feed the same control algorithm $x_i(k+1) = x_i(k) - \sum_{j=1}^{n_i} [k_{ij}(x_i(k) - x_j(k))]$ with different inputs, i.e. active power generation for MVAC SPS and output current for MVDC SPS, respectively. The algorithm implementation and interaction between MAS and RTS are illustrated in Figure 8.4.

It should be noted that the way of implementation is still bottlenecked by the single *Asynethernet* block (shown as interface block in Figure 8.2) in the target computer. Because one real-time simulated model only allows one such block, the communications between the MAS and the RTS are still handled in a centralized way. This is an example showing that simulation cannot replace hardware experimentation. To realize the ideal way of implementation of a distributed algorithm, we need to establish direct communication links between the agents and the corresponding subsystem controller. This is realized during hardware experimentation and will be introduced later.

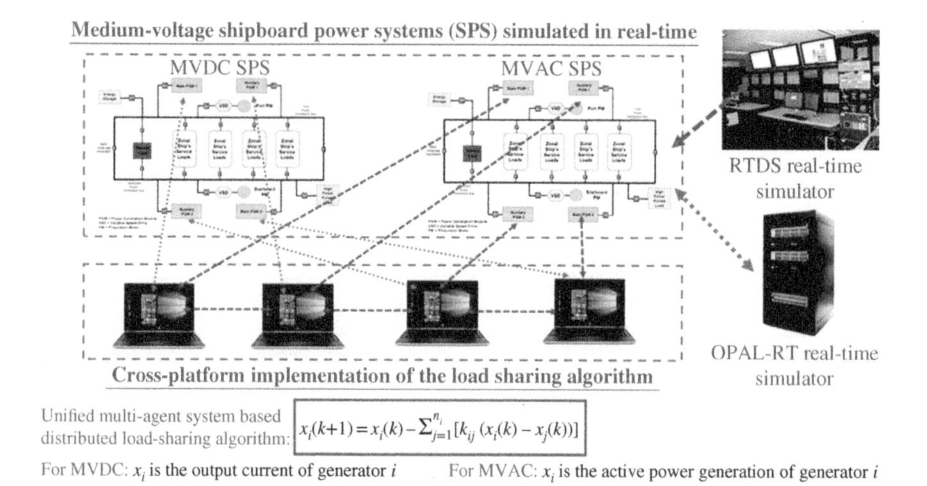

Figure 8.4 An example interaction between the MAS and RTS.

8.2.2 DSP-Based C-HIL Simulation

In addition to performing PC-based C-HIL simulation for evaluation of system-level algorithms, we also performed DSP-based C-HIL simulation for evaluation of equipment-level algorithms. This is because most modern power electronic equipment are controlled by DSP control boards. Thus, implementing the equipment-level control algorithms with DSP control boards is a necessary step toward hardware experimentation.

The DSP control board shown in Figure 8.5 was developed based on a dual-core (DSP+ARM) System on Chip (SoC), TI OAMP L138. The motivation was to implement both component-level and system-level algorithms on one platform. In addition to many A/D channels for sensor inputs and D/A channels for signal visualization using oscilloscope, it also features a FPGA of CYCLONE IV for PWM signal generation and communication channel extension. However, the SoC (TI OMAP L138) that we selected has some limitations. The two cores share a lot of resources together, including the 456 MHz clock speed and the 128 kB shared SRAM. When both cores are in use, the two core's processing speed will be significantly lowered. In our experimentation, it is used for equipment-level algorithm implementation only, for which it has been proven very reliable.

The DSP-based C-HIL simulation is illustrated in Figure 8.6. The DSP control board implements converter control algorithm and generates the PWM signals through the onboard FPGA. The gating signals are connected to the analog input card (OP5340K1) of our RTS. The RTS simulates switch-level power electronic systems with the real-time computer (OP5644) and the Virtex 7 FPGA processor (OP5607). During implementation, OP5644 simulates power electronic system and produces voltage and current measurements through its analog output card (OP5330K1). OP5607 is used for mux and de-mux of multiple signals

Figure 8.5 DSP control board used for test-bed development.

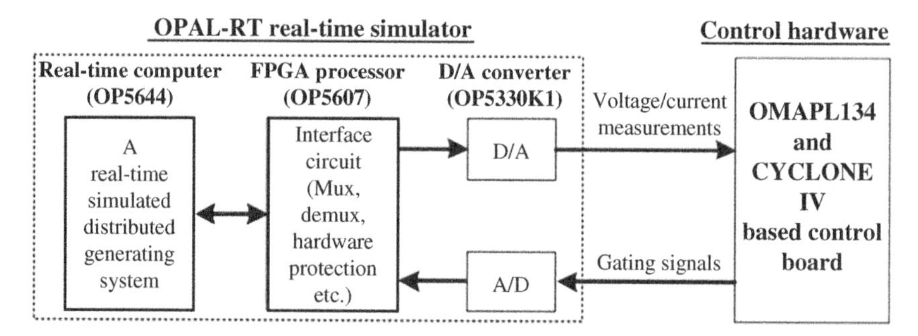

Figure 8.6 Controller HIL simulation of power electronic system.

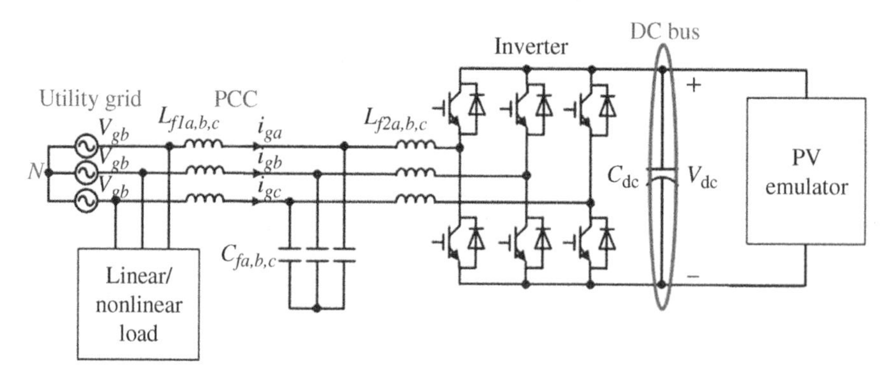

Figure 8.7 Schematic of the distributed generator plant implemented in the CPU.

and block the gating signals once overcurrent or over/low-voltage faults is detected.

During experimentation, a simple grid-connected PV system is simulated with the RTS, and its control is implemented with a DSP control board. The topology of the simulated system is shown in Figure 8.7, and the actual experimental setup is shown in Figure 8.8. The time step for the real-time simulation is 12 microseconds. Off-line simulation result is shown in Figure 8.9a, and the C-HIL simulation results captured with an oscilloscope are shown in Figure 8.9b. Considering that the two plots match each other reasonably well, we can say that the C-HIL simulation is effective. The subsequent hardware experimentation will benefit from the C-HIL simulations. The C and Verilog codes will not need significant modifications. If the real-time simulated model is accurate, the hardware experimentation results will be very similar to the simulation results.

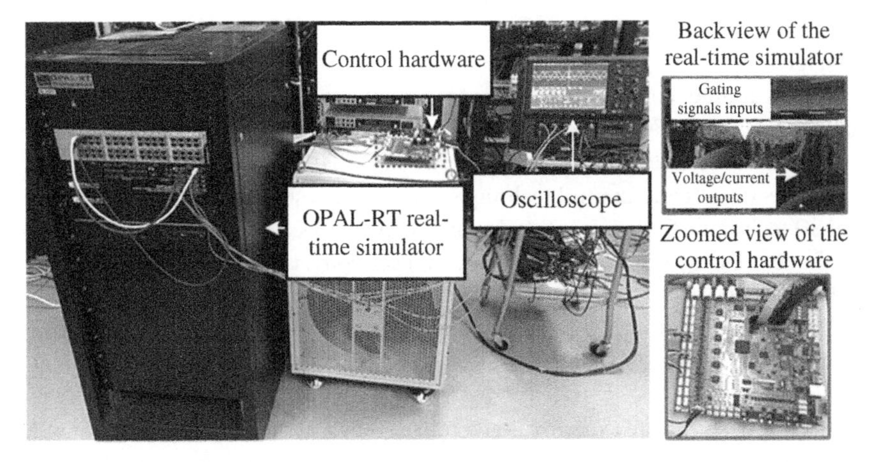

Figure 8.8 Prototype of the controller HIL simulation.

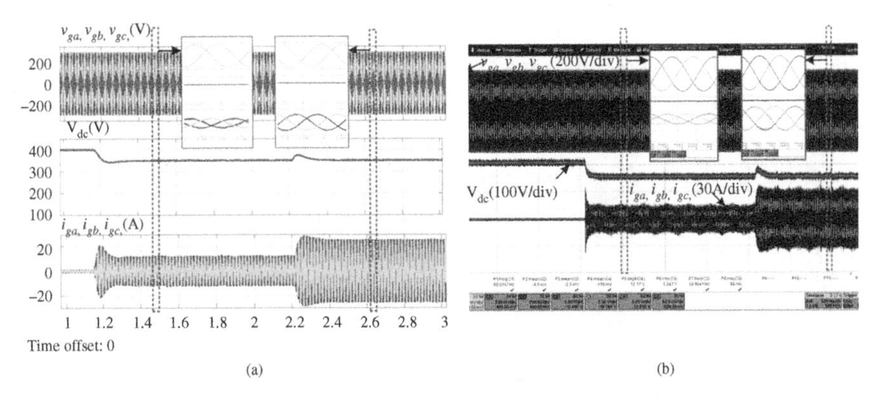

Figure 8.9 Comparison of experimental and simulation results. (a) Off-line simulation result. (b) Oscilloscope screenshot of C-HIL simulation.

8.3 Power Hardware-In-the-Loop Simulation

In power HIL simulation, a DSP controller interacts with an emulated power circuit. In our lab, we use the technique to emulate equipment that is expensive to obtain and operate, such as traditional SG and special ESS (flywheel and ultra-capacitor). In the past, we had thought to purchase a 20 kW motor-generator set to emulate a traditional SG. After knowing that it costs $30k USD, USD, weights 2000 lbs, and measures 6 ft long, we have to give up the idea. Because multiple SGs are usually needed for experiment with a large-scale power system, the expense will be beyond the capability of most research groups. It should be noted that using a motor-generator set for SG emulation is still model based and different from a real SG.

There is another consideration for not buying the motor-generator set. Such specialized equipment lacks flexibility and can only be used for certain studies. Thus, we decide to emulate such equipment through P-HIL simulation with modern power electronics and control techniques. The idea is somewhat close to the virtual SG technique for low-inertia microgrid control. Because we have a lot of programmable DC power supplies, modular power converters, DSP control boards, and other necessary accessories, we can emulate many different types of equipment that we do not have. As long as simulation of the equipment is not beyond the capability of the DSP control board, we can emulate the equipment with reasonable accuracy.

Figure 8.10 shows the experimental setup to study a control algorithm for pulsed power load accommodation on a SPS. The circuit diagram of the experimental setup shown in Figure 8.10 is illustrated by Figure 8.11. The P-HIL simulation platform consisted of an emulated SG and a real supercapacitor. To emulate an SG, two function modules, one for signal calculation and one for signal realization, need to be implemented on a DSP control board. During signal calculation,

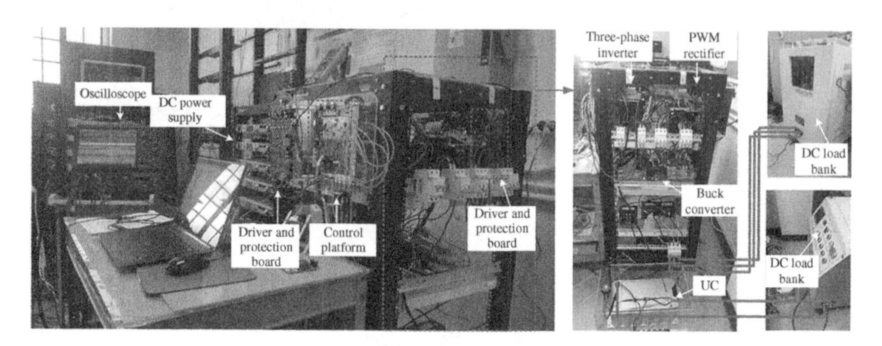

Figure 8.10 A photo of the experimental setup for P-HIL simulation.

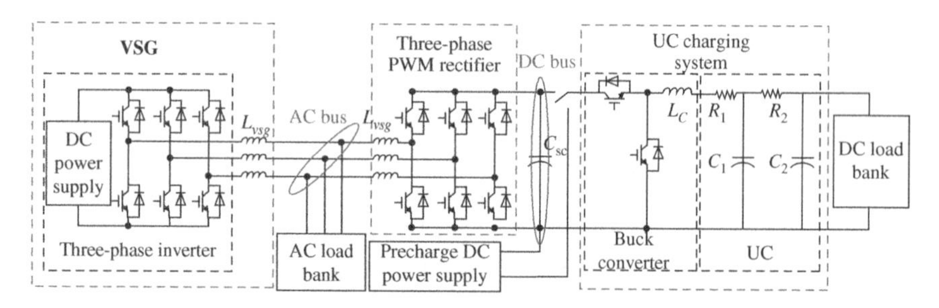

Figure 8.11 Circuit diagram of the experimental setup for P-HIL simulation.

a fifth-order SG model is simulated in real time using DSP. The calculated signals of terminal voltages are amplified using a three-phase inverter together with a programmable DC power supply. The measured currents are then fed back to the simulated models for next step calculation. In this way, the developed algorithms can be tested in a way close to reality. In our lab, the P-HIL simulation can be implemented either with the DSP control board or with the much more powerful RTS. We can choose which one to be used based on the complexity of the model to simulate.

8.4 Hardware Experimentation

Over the past decade, we have developed various equipment for different projects. An introduction of our hardware development is presented in a journal paper.[1] Most of the function modules in those equipment, such as power converter, control boards, driver, and protection boards, power sand sensor boards, and so on are all developed by ourselves. In this way, we not only can reduce the development and maintenance costs but also can maximize the usability and flexibility. For experimental studies of multiple-bus microgrids, we decided to develop two new test-beds. One reason is that the existing test-bed is basically a one-bus microgrid that has limited usage. The other reason is that the existing equipment has good looking but is very difficult to modify. Instead of modifying existing test-beds and recovering them in the future, we decided to develop two new microgrid test-beds by practicing our idea of "LEGO-like development of power electronic systems." One test-bed is for three-phase AC microgrid study and the other one for single-phase AC and DC microgrid study.

8.4.1 Test-bed Development

As shown in Figure 8.12, we have developed two modular DGs for a three-phase three-line microgrid study. Each DG consisted of a 2.4 kW programmable DC power supply (TDK Lambda GEN 150-16-LAN-3P208), an Intelligent Power Module (Mitsubishi PM100RLA120A IPM), an LC-type output filter, a TI OMAP L138SoC-based DSP control board, a driver and protection board, and many other boards (voltage sensor, current sensor, low-voltage power supply, etc.). We have successfully implemented all control loops of a voltage source converter and realized P-HIL simulation for SG emulation. Thus, the modular DG can be used

1 W. Liu, J. Kim, C. Wang, W. Im, L. Liu, and H. Xu, "Power converters based advanced experimental platform for integrated study of power and controls," IEEE Transactions on Industrial Informatics, vol. 14, no. 11, pp. 4940–4952, November 2018.

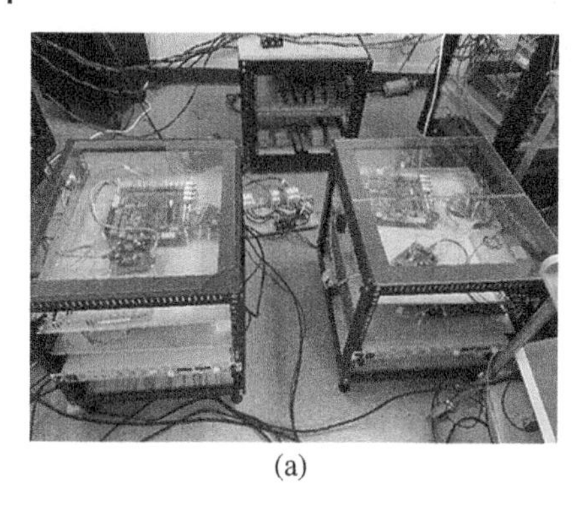

(a)

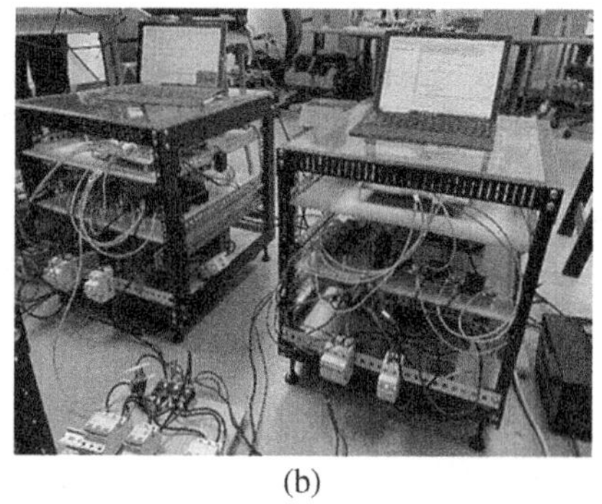

(b)

Figure 8.12 Two photos of the modular three-phase AC microgrid test-beds. (a) Front view. (b) Back view. Source: Yinliang Xu.

to study both power converter-based three-phase AC microgrids and traditional large-scale power systems. In the future, we can develop more of the DG modules. Then, we can connect the DGs through emulated power lines. To emulate a linear load, we can use large resistor as a constant load and use a programmable RLC load bank as a variable load. To emulate a nonlinear load, we can use power diode linked with a linear load or use our previously developed equipment.

Even though the IPM module (Mitsubishi PM100RLA120A) used in the above three-phase AC microgrid can work in single-phase mode and we have verified it through experimentation, controlling the three-phase test-bed as a single phase requires a lot of time for hardware reconfiguration, DSP and FPGA programming, and debugging. Thus, we decide to develop a separate microgrid test-bed that can be used to test the control algorithms for both single-phase AC microgrid and DC microgrid.

A photo of the new microgrid test-bed under development is shown in Figure 8.13. The test-bed consisted of four DGs that are connected through emulated distribution lines. Three DGs are connected as a ring and a fifth DG is connected to the ring. Because the topology is a combination of radial and ring structure, it can represent a general class of microgrid topologies. If needed, we can develop more of the modular DGs. Because of the modular hardware design, the effort for larger test-bed development is made easy. The modular DGs in this test-bed share a lot of components with the modular DG in the three-phase AC microgrid test-bed, including DSP control board, power supply board, and voltage and current sensors boards.

The modular converter boards used in the test-bed was initially developed for our modular multilevel converter. During design, we have already brought reconfiguration capability into consideration. A photo of the modular converter board is shown in Figure 8.14. The converter takes optical PWM signal from the DSP

Figure 8.13 A photo of the one-phase AC/DC microgrid test-bed under development. Source: Yinliang Xu.

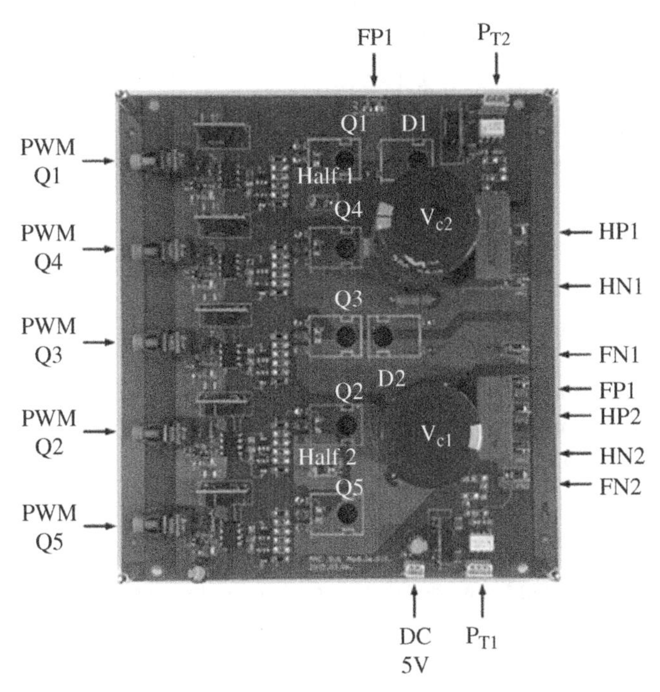

Figure 8.14 Top view of the modular power converter. Source: Yinliang Xu.

control board. The converter board integrates driver circuits for the five IGBTs (INFINEON IKW50N60T), DC link capacitors, and two voltage sensors. By simply connecting different terminals on the converter boards, i.e. HP2, HP1, FP2, FP1, HN2, HN2, FN1, and FN2 in Figure 8.12, different topologies can be realized,

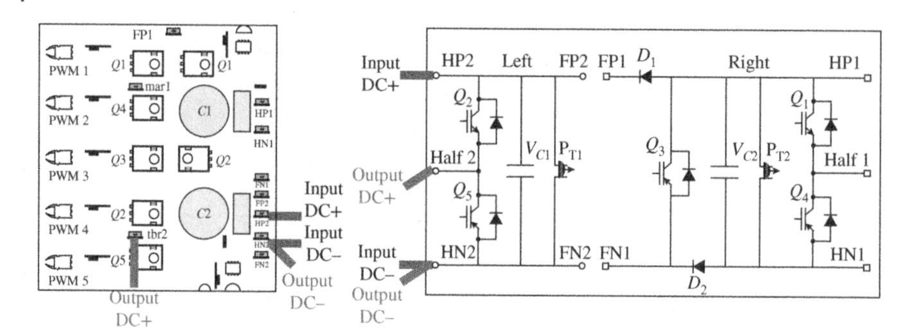

Figure 8.15 Configuration of the converter module as a DC–DC converter.

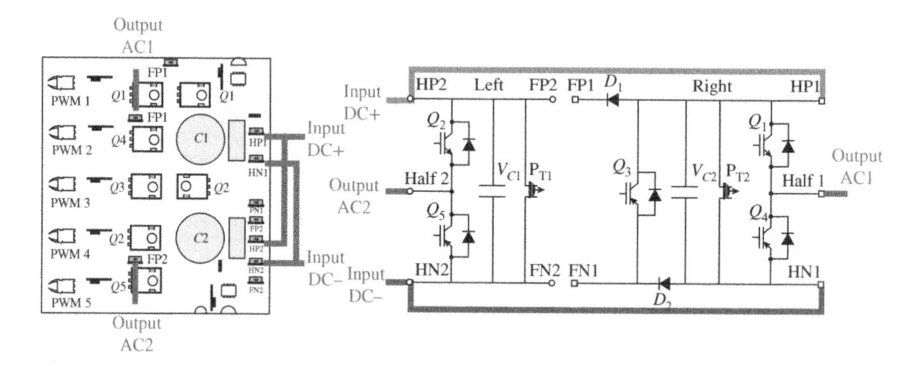

Figure 8.16 Configuration of the converter module as a DC–AC inverter.

including half-bridge, full-bridge, and clamp double half-bridge. The converter module can be configured as either a DC–DC converter or a DC–AC inverter as shown in Figures 8.15 and 8.16, respectively.

8.4.2 Algorithm Implementation

Based on our previous study of MAS, we decide to use PC to implement distributed system-level algorithms. This is because we need to use PC to program and debug the DSP control board anyway. It will be convenient and cost efficient to use the same PC for system-level algorithm implementation. Thus, we need to establish communications between the system-level controller (PC) and equipment-level controller (DSP control board). We decided to use the serial ports for information exchange between the PC and DSP control board. Certainly, the speed of serial communication is not very fast. However, we think that a good system-level algorithm should not require communication speed higher than serial. To test a distributed energy management algorithm through hardware experimentation,

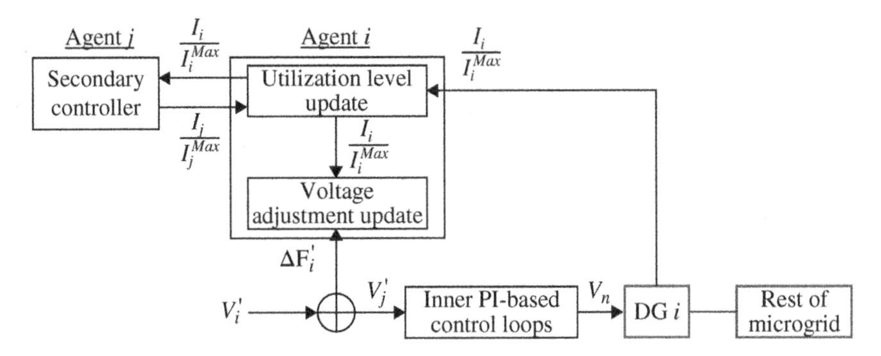

Figure 8.17 Block diagram of the distributed control solution for DC microgrid.

we need to implement a complete control solution with both system- and equipment-level algorithms. For equipment-level control, we implemented the well-established PI-based primary control algorithm. For system-level control, we implemented a distributed consensus-based secondary control algorithm represented in Eq. (8.1).The block diagram of the overall control solution is illustrated in Figure 8.17.

$$\Delta \dot{V}_i(t) = -k_i \sum_{j=1, j \neq i}^{N} a_{ij} \left(\frac{I_i(t)}{I_i^{Max}} - \frac{I_j(t)}{I_j^{Max}} \right) \tag{8.1}$$

After deciding to use serial communication between the PCs and DSP control boards, we need to write Java and C code on the two sides, respectively, to realize communications between the two control layers. Because the functions provided by the DSP manufacturer can only realize sending and receiving of one byte ASCII code, we need to code and decode the voltage and current information that is float type to and from two ASCII bytes during sending and receiving, respectively. Reliable communication is critical for closed-loop control. Thus, we also implemented some functions that can check the validness of the received data. The other issue is synchronization of distributed controllers. The need comes from the unavoidable unmatched clocks of subsystem controllers. To synchronize distributed control activities, we choose one agent as system reference. The agent will decide when to perform the next step of operation, which could be time based or event triggered. After reliable communication and effective synchronization have been established, we can start testing in steps toward hardware experimentation as shown in Figure 8.18. The details of the four steps and some experimentation results are given as follows.

Step1: Test the code for MAS only: In preparation for the following tests, we first implemented the consensus-based secondary control algorithm using PCs. Because we have realized C-HIL simulation in the past, this step is relatively

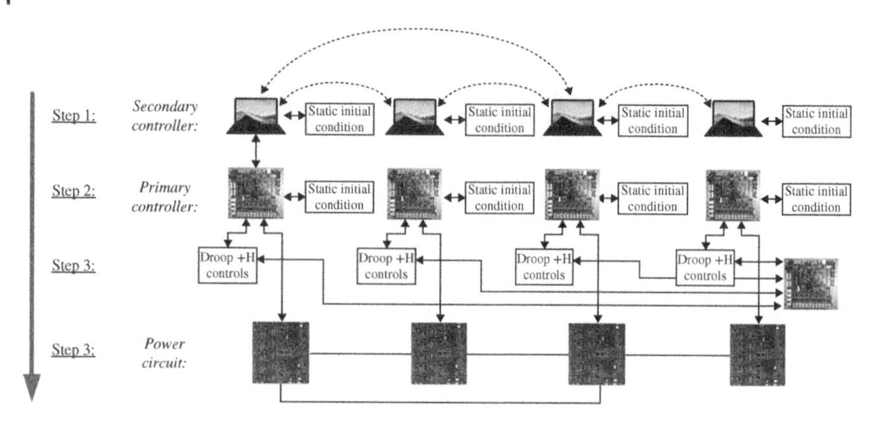

Figure 8.18 Steps for algorithm testing. Source: Yinliang Xu.

easier. We used a top of the line wireless router (NETGEAR – Nighthawk AX12 12-Stream AX6000 WiFi 6 Router) that can allow 12-Stream WiFi with up to 1200 + 4800 Mbps for ultrafast wireless speeds. To avoid uncertainties due to sudden internet traffic and hacker attack, the router is not connected to the Internet. At this step, we only tested the reliability of the upgraded MAS code. Each PC locally maintains a virtual operating data, which changes from an initial value through interactions with neighboring agents. After the MAS can reliably reach a consensus, we can move to the next step of the test. During experimentation, we found that the Ethernet/IP-based communication introduced significant delay and randomness. Communications between two agents take an average of 10 ms. This is because Ethernet/IP was not designed for real-time distributed control applications.

Step 2: Test the code for MAS and PC/DSP communication: In this test, we want to test if the communications between the system- and equipment-level controllers can reliably support system-level control. In this step, the DSP control boards hold the virtual operating data that is updated by the MAS. For every step, the DSP control board sends its local operating data through serial communication to the agent. Afterward, the MAS will send the updated data to the DSP control board. Then, the operating data locally maintained by DSP control board will be updated immediately for next step of operation. With the D/A interface built in to our DSP control board, we can check the convergence with oscilloscope. During experimentation, we found that the series communication, even though is slower in data rate, can provide much faster response compared to Ethernet/IP, for example, 2 ms vs. 10 ms. This is because the serial communication is fully under our control instead of going through all seven open system interconnection (OSI) layers of Ethernet/IP. After optimizing the communication strategy and the C/Java code, we can reliably realize the control update interval of 40 ms.

Step 3: Test of complete control hardware and software (without power circuit): During this test, the complete loops of primary control are included in test. However, the input and output of the primary controller are not connected to external power circuit. To close the control loop, the control input to the inverter is immediately converted to bus voltage without considering the dynamics of output filters and power lines by the four DGs. Because there is no actual power network to link the DGs, we have to use a fifth DSP control board to model the power network. During implementation, the bus voltages are sent to the fifth DSP, which will calculate the output current of the four DGs and send the information back. This test can check the robustness of the control algorithm against the unavoidable delays with communications. If necessary, we need to adjust control frequency and control gains of secondary control. This step is critical and necessary before performing hardware experimentation with power circuit. Some experimentation results are shown in Figure 8.19.

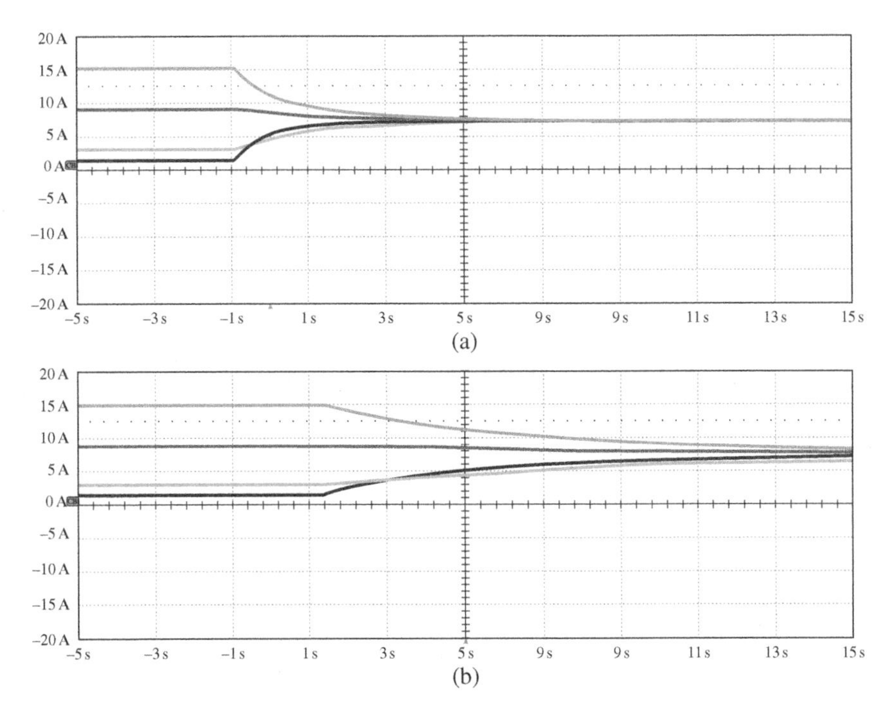

Figure 8.19 Experimentation results with the distributed control solution (TS = 40 ms). (a) Responses of utilization levels. (b) Responses of utilization levels (zoomed).

Step 4: Test of the complete control system (with power circuit): Above tests neither involve real power circuit nor consider their dynamics. In reality, a power converter-based DG, even though has very fast response speed, still has dynamics associated with output filter and line dynamics. In addition, there are many unavoidable imperfections in the real world. Because measurements go through several stages of analog-to-analog and analog-to-digital conversion, a tiny deviation of the component parameter will result in unmatched observation of the same signal. In power electronic system, the voltage and current signals always have certain degree of harmonics. The filtering impact of sensor will introduce another degree of inaccuracy to measurements. In addition, the test-bed has thousands of components, even just for the control board. It is impossible to guarantee everything work perfectly before experiment. Any hardware problem during experimentation could be very hard to locate and fix. Thus, experimental study is very technically challenging and time-consuming. After months of hard work, we finally completed the hardware experimentation.

8.5 Future Work

Hardware experimentation can tell us whether an algorithm will work in reality or not and how to make it work and work better. It can also tell us the implementation requirements and inspire us of future research. Below we will discuss our plan on future hardware development.

As introduced earlier, the Ethernet/IP communication is designed for point-to-point communication of large amount of data. It is not specialized for real-time distributed control. The large and random delays caused by Ethernet/IP communication have very negative impact on the robustness of the system under control. To overcome the problem, we plan to use EtherCAT (Ethernet for Control Automation Technology) for deterministic, real-time communication between the distributed controllers. EtherCAT can realize cycle times of 100 μs even 10 μs. It will be capable to implement any advanced control algorithms in the future.

In addition to communication, we also want to improve the control board design. Currently, the system-level algorithm was implemented using PC, which is not very practical. It is desirable to implement both the distributed system-level control algorithm and equipment-level control algorithm with the same controller, i.e. DSP control board. Thus, we plan to develop a much more powerful dual-core DSP-based control board with EtherCAT interfaces. Because we probably will not use JADE for system-wide algorithm implementation, we will write our own code for everything in C. This will require significant effort on hardware and software developments.

Impressed by our hardware development, we were encouraged by the funding agency to develop a medium-voltage (5000 V DC and 4160 V AC) SPS test-bed with the promising wide band gap devices, gallium nitride (GaN), and silicon carbide (SiC). Because most of our previous experiences are with lower switching frequency silicon-based devices, developing high-performance power converters using the new devices will be very challenging for us. Once all problems have been overcome, we will publish another paper to share the experiences and lessons learned during the new test-bed development.

9

Discussion and Future Work

Different control problems require different control solutions. By and large, a control problem can be formulated as an optimization problem with a control objective function and corresponding constraints. Distributed control solutions seek for distributed optimization algorithms to solve this optimization problem. As for algorithms for distributed optimization, it mainly concerns two factors, one is the optimality and feasibility and the other is convergences. A distributed solution of an optimization problem always involves decomposing the original problem, which leads the problem to the loss of optimality. Not all distributed optimization algorithms can yield the same optimal optima (if exists) as the original problem. For example, the generalized distributed subgradient algorithm not only requires the original optimization problem to be convex but also requires each decomposed optimization problem (sub-problem) to be convex [1]. For example, for the following optimization problem:

$$\begin{cases} \min f(x_1, x_2, x_3) = x_1^2 + x_1 x_2 + x_2^2 + x_2 x_3 + x_3^2 \\ \mathbf{x} = [x_1, x_2, x_3]^T \end{cases} \tag{9.1}$$

If we desire to solve it using a three-agent system, we can decompose the original objective function into three parts, corresponding three objective functions of three agents, as follows:

$$\begin{cases} f^1(\mathbf{x}) = x_1^2 + x_1 x_2 + \frac{1}{2} x_2^2 \\ f^2(\mathbf{x}) = \frac{1}{4} x_2^2 \\ f^3(\mathbf{x}) = \frac{1}{4} x_2^2 + x_2 x_3 + x_3^2 \end{cases} \tag{9.2}$$

By applying distributed subgradient algorithm, one can yield the optimal solution of $\mathbf{x}^* = [000]^T$, which is exactly the optimal solution of the original problem. However, for decomposition given in Eq. (9.3), we cannot find the optimal

Distributed Energy Management of Electrical Power Systems, First Edition.
Yinliang Xu, Wei Zhang, Wenxin Liu, and Wen Yu.

Published 2021 by John Wiley & Sons, Inc.

solution for $f^1(\mathbf{x})$ because it is not a convex function.

$$\begin{cases} f^1(\mathbf{x}) = x_1^2 + x_1x_2 \\ f^2(\mathbf{x}) = \frac{3}{4}x_2^2 \\ f^3(\mathbf{x}) = \frac{1}{4}x_2^2 + x_2x_3 + x_3^2 \end{cases} \tag{9.3}$$

Thus, designing a proper decomposition technique is crucial for the distributed optimization algorithm. Yet, in the real world, the "true" optima is not what we really desire because the operating conditions of the system change from time to time, notwithstanding the modeling error. In this case, the sub-optima is usually acceptable as long as it improves the performance of the control solution to a certain extent. As for the feasibility, the distributed optimization algorithm should yield the solution that can be implemented in the practical control systems, which means the physical constraints of the power system must not be violated. To design a distributed optimization algorithm or control strategy, we dedicate most of the time in handling the constraints to ensure the feasibility of the obtained solutions. For example, to design the distributed algorithm to solve the *optimal active power dispatch problem* (Section 3.3), we carefully set the weight matrix $\mathbf{W}$ to ensure that the supply and demand of the power system are balanced out. Furthermore, to avoid the violation of the inequality constraints of generators, we even design a mechanism for virtual communication network configuration. To obtain the congestion prices in solving the *social welfare optimization problem* (Section 6.1), we resort to the consensus algorithm for discovering information of other agents. Currently, most of our designed distributed algorithms exploit some simplifications and assumptions. Yet, more constraints should be incorporated in these algorithms to signify the real power system world. As for the power system applications, the common state variables we constantly come across are voltage and power. In the DC power flow model, we assume that the active power of a component is a linear function of the voltage phase angle, and reactive power is a linear function with respect to the voltage magnitude (Section 7.1). By considering these assumptions, we do get good approximations of the states. However, the error is inevitable. For the AC power network, if we consider the node power injection equations given in Eq. (9.4):

$$\begin{cases} P_i = \sum_{j \in i} (G_{ij}V_iV_j \cos\theta_{ij} + B_{ij}V_iV_j \sin\theta_{ij}) \\ Q_i = \sum_{j \in i} (G_{ij}V_iV_j \sin\theta_{ij} - B_{ij}V_iV_j \cos\theta_{ij}) \end{cases} \tag{9.4}$$

where P_i and Q_i are active and reactive power injections, V_i and V_j are voltage magnitudes, and θ_{ij} is the voltage phase angle difference between node i and node j. $G_{ij} + jB_{ij} = Y_{ij}$ is the element of the admittance matrix. Obviously, these two functions are far more complex than linear. Similarly, the line flow of a branch in AC power networks is written as:

$$\begin{cases} P_{ij} = G_{ij}V_i^2 - G_{ij}V_iV_j\cos\theta_{ij} - B_{ij}V_iV_jc\sin\theta_{ij} \\ Q_{ij} = B_{ij}V_i^2 - B_{ij}V_iV_j\cos\theta_{ij} - G_{ij}V_iV_jc\sin\theta_{ij} \end{cases} \tag{9.5}$$

which is also a complex, nonlinear, and non-convex function. The good side of the applying assumptions and simplifications is obvious. We can theoretically know if the solution we seek exists or not; if it does, we can at least get one. Nevertheless, the good side is that we have to recognize that *the loss of optimality* is inevitable, even worse than the loss of feasibility. Loss of optimality is tolerable for most of the cases, but the *the loss of feasibility* is unacceptable under most of the circumstances. Hence, for a distributed algorithm design, it is necessary to design a fail-safe mechanism to ensure that the final control actions we enforce on the physical systems are physically feasible. Here, we provide an example to handle the non-convexity of Eqs. (9.4) and (9.5), which is the well-known second-order cone program (SOCP) relaxation [2]. Define u_i, v_{ij}, w_{ij} as:

$$\begin{cases} u_i = V_i^2 \\ v_{ij} = V_iV_j\cos\theta_{ij} \\ w_{ij} = V_iV_j\sin\theta_{ij} \end{cases} \tag{9.6}$$

v_{ij}, w_{ij} are not independent, they are correlated because of the triangle identity:

$$(\cos\theta_{ij})^2 + (\sin\theta_{ij})^2 = 1 \tag{9.7}$$

Accordingly, we have

$$v_{ij}^2 + w_{ij}^2 = u_iu_j \tag{9.8}$$

By further relaxing Eq. (9.8), we have:

$$v_{ij}^2 + w_{ij}^2 \leq u_iu_j \tag{9.9}$$

Accordingly, the original non-convex function transforms into the following convex function plus an additional inequality constraint.

$$\begin{cases} P_i = G_{ii}u_i + \sum_{j\in i, j\neq i}(G_{ij}v_{ij} + B_{ij}w_{ij}) \\ Q_i = -B_{ii}u_i + \sum_{j\in i, j\neq i}(G_{ij}w_{ij} - B_{ij}v_{ij}) \\ P_{ij} = G_{ij}u_i - G_{ij}v_{ij} - B_{ij}w_{ij} \\ Q_{ij} = B_{ij}u_i - B_{ij}v_{ij} - G_{ij}w_{ij} \\ v_{ij}^2 + w_{ij}^2 \leq u_iu_j \end{cases} \tag{9.10}$$

Obviously, the above constraints are convex. By applying this transformation, one can design a proper distributed algorithm based on convex decomposition technique (such as the alternating direction method of multipliers, ADMM), without loss of feasibility. This is beyond the scope of this book, details about this method can be found in Refs [3–5]. Regarding the convergence, we have demonstrated that the convergence of the distributed algorithm is greatly related to the connectivity of the communication network. Generally, the problem we aim to solve involve multiple variables, and the control objective need to consider quite a lot of components as well. Because these components are an inseparable part of the entire system, fully decoupling them is not possible. Hence, a fully distributed control solution without information exchange among distributed controllers is generally impractical. Information exchange among controllers is crucial for coordination and optimization. Higher connectivity of the communication network graph means not only the better convergence performance but also more reliable of the overall control system. In Chapter 2, we discussed that to improve the robustness. It is recommended to comply with the $N-1$ rule during the communication network design. However, in some applications, dynamic setting the parameters of a distributed algorithm can still ensure the optimality as well as the feasibility of the obtained solution. By designing a distributed algorithm that adapts to the communication network, we can provide an even better control solution that is less dependent on the communication network. Here, we provide an example of the consensus algorithm with switching topology of the communication network graph. The initial values of the agents are set as $\mathbf{x}^0 = [10\ 20\ 30\ 40]^T$. The communication network of four-agent system is shown in Figure 9.1. Note that the topology is no longer fixed here. We assume that there are three phases, i.e. phase 1: steps [1, 20]; phase 2: steps [21, 40]; and phase 3: steps [41, 100]. The topology in phase 1 follows Figure 9.1a, and it is changed to Figure 9.1b in phase 2 and finally restores to Figure 9.1a in phase 3. The Laplacians of these two graphs are given in Eqs. (9.11) and (9.12), respectively. We set $\Delta t = 0.15$ and calculate the weight matrix $\mathbf{W}$ of these two graphs according to Eq. (2.7), which are given in Eqs. (9.13) and (9.14), respectively.

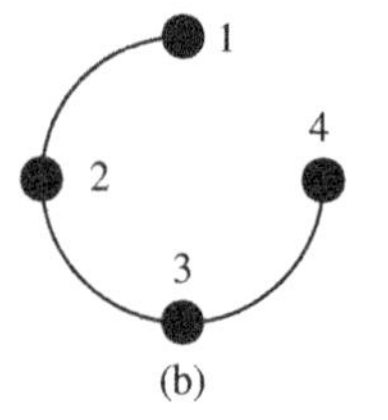

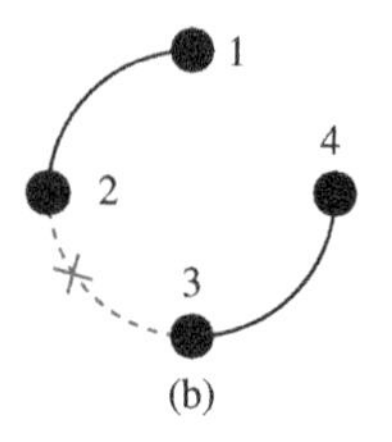

Figure 9.1 Communication network graphs of two scenarios: (a) Scenario 1. (b) Scenario 2.

$$\mathbf{L}_a = \begin{bmatrix} 1 & -1 & 0 & 0 \\ -1 & 2 & -1 & 0 \\ 0 & -1 & 2 & -1 \\ 0 & 0 & -1 & 1 \end{bmatrix} \quad (9.11)$$

$$\mathbf{L}_b = \begin{bmatrix} 1 & -1 & 0 & 0 \\ -1 & 1 & 0 & 0 \\ 0 & 0 & 1 & -1 \\ 0 & 0 & -1 & 1 \end{bmatrix} \quad (9.12)$$

$$\mathbf{W}_a = \begin{bmatrix} 0.85 & 0.15 & 0 & 0 \\ 0.15 & 0.70 & 0.15 & 0 \\ 0 & 0.15 & 7.0 & 0.15 \\ 0 & 0 & 0.15 & 0.85 \end{bmatrix} \quad (9.13)$$

$$\mathbf{W}_b = \begin{bmatrix} 0.85 & 0.15 & 0 & 0 \\ 0.15 & 0.85 & 0 & 0 \\ 0 & 0 & 0.85 & 0.15 \\ 0 & 0 & 0.15 & 0.85 \end{bmatrix} \quad (9.14)$$

As can be seen in Figure 9.1b, there are two communication islands in the graph. During phase 2 with this communication setting, it seems that the agents of one of the islands (e.g. agent 2 and agent 3) are independent of the agents of the other island. Nevertheless, the way to set the weight matrix **W** ensures that the convergence features of the overall algorithm are not hindered. When the communication is restored to the original settings with a connected graph as Figure 9.1a, the algorithm quickly converges. The iteration process is shown in Figure 9.2, indicating that the final converged values are $\mathbf{x}^{100} = [24.99\ 25\ 0.0025\ 0.0025\ 0.01]^T$, which are the expected values. It should be noted that the loss of connectivity of the graph does slowdown the convergence. Yet, it still obtains the solution without loss of optimality and feasibility here. This kind of algorithm design with adaption actually is the ideal design we desire for the control implementation in the real world. In practice, convergence performance is merely one of the factors we consider for control algorithm implementation. Aside from that, the investment, reliability, optimality, and feasibility are all the factors we need to take into account. A tradeoff must be made during the design because we cannot find a good-for-all solution. As for the distributed control solution implementation, many facets and technology details need to be investigated, and research efforts and contributions from all around the world are more than welcome.

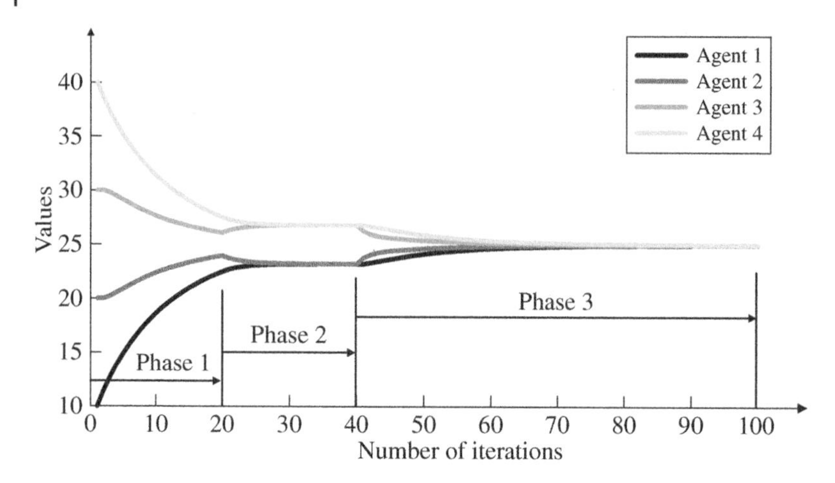

Figure 9.2 Iteration process of the consensus algorithm with switching topology.

References

1 Zhang, W. (2013). *Multiagent System Based Algorithms and their Applications in Power Systems*. State University: New Mexico

2 Jabr, R.A. (2006). Radial distribution load flow using conic programming. *IEEE transactions on Power Systems* 21 (3): 1458–1459.

3 Mhanna, S., Verbič, G. and Chapman, A.C. (2018). Accelerated methods for the SOCP-relaxed component-based distributed optimal power flow. *In 2018 Power Systems Computation Conference (PSCC)* (pp. 1–7). IEEE.

4 Zhang, W. and Xu, Y. (2018). Distributed optimal control for multiple microgrids in a distribution network. *IEEE Transactions on Smart Grid* 10 (4): 3765–3779.

5 Tian, Z. and Wu, W. (2019). Recover feasible solutions for SOCP relaxation of optimal power flow problems in mesh networks. *IET Generation, Transmission & Distribution* 13 (7): 1078–1087.

Index

Distributed Energy Management of Electrical Power Systems, First Edition.
Yinliang Xu, Wei Zhang, Wenxin Liu, and Wen Yu.

Published 2021 by John Wiley & Sons, Inc.

IEEE Press Series on Power Engineering

Series Editor: M. E. El-Hawary, Dalhousie University, Halifax, Nova Scotia, Canada

The mission of IEEE Press Series on Power Engineering is to publish leading-edge books that cover the broad spectrum of current and forward-looking technologies in this fast-moving area. The series attracts highly acclaimed authors from industry/academia to provide accessible coverage of current and emerging topics in power engineering and allied fields. Our target audience includes the power engineering professional who is interested in enhancing their knowledge and perspective in their areas of interest.

1. *Electric Power Systems: Design and Analysis, Revised Printing*
Mohamed E. El-Hawary

2. *Power System Stability*
Edward W. Kimbark

3. *Analysis of Faulted Power Systems*
Paul M. Anderson

4. *Inspection of Large Synchronous Machines: Checklists, Failure Identification, and Troubleshooting*
Isidor Kerszenbaum

5. *Electric Power Applications of Fuzzy Systems*
Mohamed E. El-Hawary

6. *Power System Protection*
Paul M. Anderson

7. *Subsynchronous Resonance in Power Systems*
Paul M. Anderson, B.L. Agrawal, J.E. Van Ness

8. *Understanding Power Quality Problems: Voltage Sags and Interruptions*
Math H. Bollen

9. *Analysis of Electric Machinery*
Paul C. Krause, Oleg Wasynczuk, and S.D. Sudhoff

10. *Power System Control and Stability, Revised Printing*

Paul M. Anderson, A.A. Fouad

11. *Principles of Electric Machines with Power Electronic Applications, Second Edition*

Mohamed E. El-Hawary

12. *Pulse Width Modulation for Power Converters: Principles and Practice*

D. Grahame Holmes and Thomas Lipo

13. *Analysis of Electric Machinery and Drive Systems, Second Edition*

Paul C. Krause, Oleg Wasynczuk, and S.D. Sudhoff

14. *Risk Assessment for Power Systems: Models, Methods, and Applications*

Wenyuan Li

15. *Optimization Principles: Practical Applications to the Operations of Markets of the Electric Power Industry*

Narayan S. Rau

16. *Electric Economics: Regulation and Deregulation*

Geoffrey Rothwell and Tomas Gomez

17. *Electric Power Systems: Analysis and Control*

Fabio Saccomanno

18. *Electrical Insulation for Rotating Machines: Design, Evaluation, Aging, Testing, and Repair*

Greg C. Stone, Edward A. Boulter, Ian Culbert, and Hussein Dhirani

19. *Signal Processing of Power Quality Disturbances*

Math H. J. Bollen and Irene Y. H. Gu

20. *Instantaneous Power Theory and Applications to Power Conditioning*

Hirofumi Akagi, Edson H. Watanabe and Mauricio Aredes

21. *Maintaining Mission Critical Systems in a 24/7 Environment*
Peter M. Curtis

22. *Elements of Tidal-Electric Engineering*
Robert H. Clark

23. *Handbook of Large Turbo-Generator Operation and Maintenance, Second Edition*
Geoff Klempner and Isidor Kerszenbaum

24. *Introduction to Electrical Power Systems*
Mohamed E. El-Hawary

25. *Modeling and Control of Fuel Cells: Distributed Generation Applications*
M. Hashem Nehrir and Caisheng Wang

26. *Power Distribution System Reliability: Practical Methods and Applications*
Ali A. Chowdhury and Don O. Koval

27. *Introduction to FACTS Controllers: Theory, Modeling, and Applications*
Kalyan K. Sen, Mey Ling Sen

28. *Economic Market Design and Planning for Electric Power Systems*
James Momoh and Lamine Mili

29. *Operation and Control of Electric Energy Processing Systems*
James Momoh and Lamine Mili

30. *Restructured Electric Power Systems: Analysis of Electricity Markets with Equilibrium Models*
Xiao-Ping Zhang

31. *An Introduction to Wavelet Modulated Inverters*
S.A. Saleh and M.A. Rahman

32. *Control of Electric Machine Drive Systems*
Seung-Ki Sul

33. *Probabilistic Transmission System Planning*
Wenyuan Li

34. *Electricity Power Generation: The Changing Dimensions*
Digambar M. Tagare

35. *Electric Distribution Systems*
Abdelhay A. Sallam and Om P. Malik

36. Practical Lighting Design with LEDs
Ron Lenk, Carol Lenk

37. *High Voltage and Electrical Insulation Engineering*
Ravindra Arora and Wolfgang Mosch

38. *Maintaining Mission Critical Systems in a 24/7 Environment, Second Edition*
Peter Curtis

39. *Power Conversion and Control of Wind Energy Systems*
Bin Wu, Yongqiang Lang, Navid Zargari, Samir Kouro

40. *Integration of Distributed Generation in the Power System*
Math H. Bollen, Fainan Hassan

41. *Doubly Fed Induction Machine: Modeling and Control for Wind Energy Generation Applications*
Gonzalo Abad, Jesús López, Miguel Rodrigues, Luis Marroyo, and Grzegorz Iwanski

42. *High Voltage Protection for Telecommunications*
Steven W. Blume

43. *Smart Grid: Fundamentals of Design and Analysis*
James Momoh

44. *Electromechanical Motion Devices, Second Edition*
Paul Krause, Oleg Wasynczuk, Steven Pekarek

45. *Electrical Energy Conversion and Transport: An Interactive Computer-Based Approach, Second Edition*
George G. Karady, Keith E. Holbert

46. *ARC Flash Hazard and Analysis and Mitigation*
J.C. Das

47. *Handbook of Electrical Power System Dynamics: Modeling, Stability, and Control*
Mircea Eremia, Mohammad Shahidehpour

48. *Analysis of Electric Machinery and Drive Systems, Third Edition*
Paul C. Krause, Oleg Wasynczuk, S.D. Sudhoff, Steven D. Pekarek

49. *Extruded Cables for High-Voltage Direct-Current Transmission: Advances in Research and Development*
Giovanni Mazzanti, Massimo Marzinotto

50. *Power Magnetic Devices: A Multi-Objective Design Approach*
S.D. Sudhoff

51. *Risk Assessment of Power Systems: Models, Methods, and Applications, Second Edition*
Wenyuan Li

52. *Practical Power System Operation*
Ebrahim Vaahedi

53. *The Selection Process of Biomass Materials for the Production of Bio-Fuels and Co-Firing*
Najib Altawell

54. *Electrical Insulation for Rotating Machines: Design, Evaluation, Aging, Testing, and Repair, Second Edition*
Greg C. Stone, Ian Culbert, Edward A. Boulter, and Hussein Dhirani

55. *Principles of Electrical Safety*
Peter E. Sutherland

56. *Advanced Power Electronics Converters: PWM Converters Processing AC Voltages*
Euzeli Cipriano dos Santos Jr., Edison Roberto Cabral da Silva

57. *Optimization of Power System Operation, Second Edition*
Jizhong Zhu

58. *Power System Harmonics and Passive Filter Designs*
J.C. Das

59. *Digital Control of High-Frequency Switched-Mode Power Converters*
Luca Corradini, Dragan Maksimoviæ, Paolo Mattavelli, Regan Zane

60. *Industrial Power Distribution, Second Edition*
Ralph E. Fehr, III

61. *HVDC Grids: For Offshore and Supergrid of the Future*
Dirk Van Hertem, Oriol Gomis-Bellmunt, Jun Liang

62. *Advanced Solutions in Power Systems: HVDC, FACTS, and Artificial Intelligence*
Mircea Eremia, Chen-Ching Liu, Abdel-Aty Edris

63. *Operation and Maintenance of Large Turbo-Generators*
Geoff Klempner, Isidor Kerszenbaum

64. *Electrical Energy Conversion and Transport: An Interactive Computer-Based Approach*
George G. Karady, Keith E. Holbert

65. *Modeling and High-Performance Control of Electric Machines*
John Chiasson

66. *Rating of Electric Power Cables in Unfavorable Thermal Environment*
George J. Anders

67. *Electric Power System Basics for the Nonelectrical Professional*
Steven W. Blume

68. *Modern Heuristic Optimization Techniques: Theory and Applications to Power Systems*
Kwang Y. Lee, Mohamed A. El-Sharkawi

69. *Real-Time Stability Assessment in Modern Power System Control Centers*
Savu C. Savulescu

70. *Optimization of Power System Operation*
Jizhong Zhu

71. *Insulators for Icing and Polluted Environments*
Masoud Farzaneh, William A. Chisholm

72. *PID and Predictive Control of Electric Devices and Power Converters Using MATLAB®/Simulink®*
Liuping Wang, Shan Chai, Dae Yoo, Lu Gan, Ki Ng

73. *Power Grid Operation in a Market Environment: Economic Efficiency and Risk Mitigation*
Hong Chen

74. *Electric Power System Basics for Nonelectrical Professional, Second Edition*
Steven W. Blume

75. *Energy Production Systems Engineering*
Thomas Howard Blair

76. *Model Predictive Control of Wind Energy Conversion Systems*
Venkata Yaramasu, Bin Wu

77. *Understanding Symmetrical Components for Power System Modeling*
J.C. Das

78. *High-Power Converters and AC Drives, Second Edition*
Bin Wu, Mehdi Narimani

79. *Current Signature Analysis for Condition Monitoring of Cage Induction Motors: Industrial Application and Case Histories*
William T. Thomson, Ian Culbert

80. *Introduction to Electric Power and Drive Systems*
Paul Krause, Oleg Wasynczuk, Timothy O'Connell, Maher Hasan

81. *Instantaneous Power Theory and Applications to Power Conditioning, Second Edition*
Hirofumi, Edson Hirokazu Watanabe, Mauricio Aredes

82. *Practical Lighting Design with LEDs, Second Edition*
Ron Lenk, Carol Lenk

83. *Introduction to AC Machine Design*
Thomas A. Lipo

84. *Advances in Electric Power and Energy Systems: Load and Price Forecasting*
Mohamed E. El-Hawary

85. *Electricity Markets: Theories and Applications*
Jeremy Lin, Jernando H. Magnago

86. *Multiphysics Simulation by Design for Electrical Machines, Power Electronics and Drives*
Marius Rosu, Ping Zhou, Dingsheng Lin, Dan M. Ionel, Mircea Popescu, Frede Blaabjerg, Vandana Rallabandi, David Staton

87. *Modular Multilevel Converters: Analysis, Control, and Applications*
Sixing Du, Apparao Dekka, Bin Wu, Navid Zargari

88. *Electrical Railway Transportation Systems*
Morris Brenna, Federica Foiadelli, Dario Zaninelli

89. *Energy Processing and Smart Grid*
James A. Momoh

90. *Handbook of Large Turbo-Generator Operation and Maintenance, 3rd Edition*
Geoff Klempner, Isidor Kerszenbaum

91. *Advanced Control of Doubly Fed Induction Generator for Wind Power Systems*
Dehong Xu, Dr. Frede Blaabjerg, Wenjie Chen, Nan Zhu

92. *Electric Distribution Systems, 2nd Edition*
Abdelhay A. Sallam, Om P. Malik

93. *Power Electronics in Renewable Energy Systems and Smart Grid: Technology and Applications*
Bimal K. Bose

94. *Distributed Fiber Sensing and Dynamic Rating of Power Cables*
Sudhakar Cherukupalli, George J. Anders

95. *Power System Control and Stability, Third Edition*
Vijay Vittal, James D. McCalley, Paul M. Anderson, A. A. Fouad

96. *Electromechanical Motion Devices: Rotating Magnetic Field-Based Analysis with Online Animations, Third Edition*
Paul Krause, Oleg Wasynczuk, Steven Pekarek, Timothy O'Connell

97. *Application of Modern Heuristic Optimization Methods in Power and Energy Systems*
Kwang Y. Lee, Zita Vale

98. *Handbook of Large Hydro Generators: Operation and Maintenance*
Glenn Mottershead, Stefano Bomben, Isidor Kerszenbaum, Geoff Klempner

99. *Advances in Electric Power and Energy: Static State Estimation*
Mohamed E. El-Hawary

100. *Arc Flash Hazard Analysis and Mitigation, Second Edition*
J.C. Das

101. *Distributed Energy Management of Electrical Power Systems*
Yinliang Xu, Wei Zhang, Wenxin Liu, Wen Yu

Printed and bound by CPI Group (UK) Ltd, Croydon, CR0 4YY